Student Solutions Manual
for Kotz and Treichel's

Chemistry
and
Chemical Reactivity

Fifth Edition

Alton J. Banks,
Northern Carolina State University

THOMSON

BROOKS/COLE

Australia • Canada • Mexico • Singapore • Spain • United Kingdom • United States

**For more information about our products,
contact us at:**
Thomson Learning Academic Resource Center
1-800-423-0563

**For permission to use material from this text,
contact us by:**
Phone: 1-800-730-2214
Fax: 1-800-731-2215
Web: www.thomsonrights.com

Asia
Thomson Learning
5 Shenton Way #01-01
UIC Building
Singapore 068808

Australia
Nelson Thomson Learning
102 Dodds Street
South Street
South Melbourne, Victoria 3205
Australia

Canada
Nelson Thomson Learning
1120 Birchmount Road
Toronto, Ontario M1K 5G4
Canada

Europe/Middle East/South Africa
Thomson Learning
High Holborn House
50-51 Bedford Row
London WC1R 4LR
United Kingdom

Latin America
Thomson Learning
Seneca, 53
Colonia Polanco
11560 Mexico D.F.
Mexico

Spain
Paraninfo Thomson Learning
Calle/Magallanes, 25
28015 Madrid, Spain

To the student:

The skills involved in solving chemistry problems are acquired only by discovering the conceptual paths which connect the available data to the desired piece(s) of information. These paths are discovered in different ways by different people. What is true is that those discoveries frequently require repetition. Working multiple problems is one very good approach to clarifying and solidifying the fundamental concepts. I would suggest that this **Solutions Manual** will provide maximum benefit if you consult it *after* you have attempted to solve a problem.

The selected Study Questions have been chosen by the authors of your text to allow you to discover the range and depth of your understanding of chemical concepts. The importance of mastering the "basics" cannot be overemphasized. You will find that the text, **Chemistry & Chemical Reactivity, 5th Edition**, has a wealth of study questions to assist you in your study of the science we call Chemistry.

Many of the questions contained in your book—and this solutions manual—have multiple parts. In many cases, comments have been added to aid you in the process of gathering available data and applicable conversion factors, and connecting them via fundamental concepts. In working these multiple-step questions, you may find an answer which differs slightly from those given here. This may be a result of "rounding" intermediate answers. The procedure followed in this manual was to report intermediate answers to the appropriate number of significant figures, and to calculate the "final" answer without any intermediate rounding. In cases involving atomic and molecular masses, those quantities were expressed with at least one digit more than the number of digits needed for the data provided.

A word of appreciation is due to several people. Thanks go to the authors, especially Dr. John C. Kotz, for conversations held during the development of this manual. The many fine folks at the publishers have been helpful as well. Special thanks go to Peter McGahey and Alyssa White. I also want to thank my wife, Dr. Catherine Hamrick Banks, for her invaluable assistance in typing and proofreading this manuscript. One of my many blessings is to have such a person as a patient wife and a chemical colleague. With this edition of the manual, I have also had the able assistance of Jennifer C. Banks and Jonathan B. Banks as proofreaders, and I hereby thank them for their abilities and patience with me.

While we have worked diligently to remove all errors from this text, I am certain that some have escaped the many inspections. I accept responsibility for all those errors.

Alton J. Banks
Department of Chemistry
North Carolina State University
Raleigh, North Carolina 27695

Table of Contents

Chapter 1
Matter and Measurement

Reviewing Important Concepts

1. Symbols of Calcium: **Ca**   and of Fluorine: **F**

   The shape of the fluorite crystals can be described as interwoven cubes.

   The overall shape of the crystals indicates that the ions in the solid matrix arrange themselves with alternating calcium and fluoride ions to produce the crystal appearance—sometimes called the crystal habit. Another example of a cubic solid is common table salt—sodium chloride.

3. The non-uniform appearance of the mixture indicates that samples taken from different regions of that mixture would be different—a characteristic of a **heterogeneous mixture**. The components of a mixture retain their physical properties, Recalling that iron is attracted to a magnetic field, while sand is generally not attracted in this way suggests that passing a magnet through the mixture would separate the sand and iron, as the iron adhered to the magnet and left the sand behind.

5. Determine if the property is physical or chemical property for the following:

(a)	color	a physical property
(b)	transformed into rust	a chemical property
(c)	explode	a chemical property
(d)	density	a physical property
(e)	melts	a physical property
(f)	green	a physical property

   Physical properties are those that can be observed or measured without changing the composition of the substance. Exploding or transforming into rust results in substances which are **different** from the original substances— and represent chemical properties.

7. Liquids: **mercury** and **water**   Solid: **copper**

   Note that the liquid and mercury both conform to the shape of the test tube. This is one characteristic of the liquid phase. Of the substances shown, **mercury** is most dense and **water** is least dense.

8. Significant figures in the following:

   (a) 6.2348--5 sf – all digits to right of decimal point should also be counted.

   (b) 20,600 – 3sf—trailing zeros are only placekeepers (unless decimal point follows.

   (c) 0.00823 —3sf—zeros to right of decimal serve only as placekeepers.

   (d) 1.670 x 10^{-6} –4 sf—in exponential notation, all digits are counted (even zeros)

9. For the gemstone aquamarine:

   (a) Elemental symbols: Aluminum: **Al**   ; Silicon: **Si**   ;Oxygen: **O**

   (b) Physical properties of elements and mineral: Oxygen is a gas, while aluminum, silicon, and aquamarine are solids at room temperature. Regarding color, oxygen is—in the concentration shown here, colorless, while aluminum and silicon are gray. The gemstone is a bluish color.

11. The large colorless block of salt represents the **macroscopic** view-the one that we can see without additional instrumentation (e.g. a microscope). The spheres represent the **microscopic or particulate** view—the view that would require additional aids for an individual to "see". If one can imagine producing multiple "copies" of the particulate view, the macroscopic view will result.

Practicing Skills

Elements and Atoms, Compounds and Molecules

12. Names for the following elements:

(a) C - carbon	(c) Cl - chlorine	(e) Mg - magnesium
(b) K - potassium	(d) P - phosphorus	(f) Ni - nickel

14. Symbols for the following elements:

(a) barium - Ba	(c) chromium - Cr	(e) arsenic - As
(b) titanium - Ti	(d) lead - Pb	(f) zinc - Zn

16. In each of the pairs, which is an element and which is a compound?

   Compounds contain **more than one element**, so for the pairs given:

   (a) NaCl is a compound while sodium (Na) is an element of that compound.

   (b) Sugar is a compound (with C,H,O) while carbon (C) is an element.

   (c) Gold chloride is a compound (with Au and Cl) while gold (Au) is an element.

Physical and Chemical Properties

18. Descriptors of physical versus chemical properties:

    Table 1.1 lists some physical properties.

    (a) Color and physical state are physical properties. (colorless, liquid) while **burning** reflects a chemical property.

    (b) Shiny, metal, orange, and liquid are physical properties while **reacts readily** describes a chemical property.

Using Density

20. What mass of ethylene glycol (in grams) possesses a volume of 500. mL of the liquid?

$$\frac{500.\ mL}{1} \cdot \frac{1\ cm^3}{1\ mL} \cdot \frac{1.11\ g}{cm^3} = 555\ g$$

22. What volume of a liquid compound with a density of 0.718 g/cm^3 has a mass of 2.00 g?

$$\frac{2.00\ g}{1} \cdot \frac{1\ cm^3}{0.718\ g} \cdot \frac{1\ mL}{1\ cm^3} = 2.79\,mL$$

24. The metal will displace a volume of water that is equal to the volume of the metal.

    Hence the difference in volumes of water (20.2-6.9) corresponds to the volume of metal.

    Since 1 mL = 1 cm^3, the density of the metal is then:

$$\frac{Mass}{Volume} = \frac{37.5\ g}{13.3\ cm^3}\ or\ 2.82\ \frac{g}{cm^3}$$

    From the list of metals provided, the metal with a density closest to this is **Aluminum**.

Temperature

26. Express 25 °C in kelvins:

    K = (25 °C + 273) or 298 kelvins

28. Make the following temperature conversions:

°C	K
(a) 16	16 + 273.15 = 289
(b) 370 - 273 or 97	370
(c) 40	40 + 273.15 = 310

    Note no decimal point after 40

Units and Unit Conversions

30. Express 40.0 km in meters; in miles:

$$\frac{40.0 \text{ km}}{1} \cdot \frac{1000 \text{ m}}{1 \text{ km}} = 40,000 \text{ m (or } 4.00 \times 10^4 \text{ to 3 sf)}$$

$$\frac{40.0 \text{ km}}{1} \cdot \frac{0.62137 \text{ miles}}{1 \text{ km}} = 24.85 \text{ (or 24.9 miles to 3 sf)}$$

The factor (0.62137 mi/km is found inside the back cover of the text)

32. Express the area of a 2.5 cm x 2.1 cm stamp in cm^2 ; in m^2 :

$$2.5 \text{ cm} \cdot 2.1 \text{ cm} = 5.3 \text{ cm}^2$$

$$5.3 \text{ cm}^2 \cdot \left(\frac{1 \text{ m}}{100 \text{ cm}}\right)^2 = 5.3 \times 10^{-4} \text{ m}^2$$

34. Express 250 mL in cm^3; in liters (L); in m^3 ; in dm^3:

$$\frac{250 \text{ mL}}{1 \text{ beaker}} \cdot \frac{1 \text{ cm}^3}{1 \text{ mL}} = \frac{250 \text{ cm}^3}{1 \text{ beaker}}$$

$$\frac{250 \text{ cm}^3}{1 \text{ beaker}} \cdot \frac{1 \text{ L}}{1000 \text{ cm}^3} = \frac{0.25 \text{ L}}{1 \text{ beaker}}$$

$$\frac{250 \text{ cm}^3}{1 \text{ beaker}} \cdot \frac{1 \text{ m}^3}{1 \times 10^6 \text{ cm}^3} = \frac{2.5 \times 10^{-4} \text{ m}^3}{1 \text{ beaker}}$$

$$\frac{250 \text{ cm}^3}{1 \text{ beaker}} \cdot \frac{1 \text{ L}}{1000 \text{ cm}^3} \cdot \frac{1 \text{ dm}^3}{1 \text{ L}} = \frac{0.25 \text{ dm}^3}{1 \text{ beaker}}$$

36. Convert book's mass of 2.52 kg into grams:

$$\frac{2.52 \text{ kg}}{1 \text{ book}} \cdot \frac{1 \times 10^3 \text{ g}}{1 \text{ kg}} = \frac{2.52 \times 10^3 \text{ g}}{\text{book}}$$

Accuracy, Precision, and Error

38. Using the data provided, the averages and their deviations are as follows:

Data point	Method A	deviation	Method B	deviation
1	2.2	0.2	2.703	0.777
2	2.3	0.1	2.701	0.779
3	2.7	0.3	2.705	0.775
4	2.4	0.0	5.811	2.331
Averages:	2.4	0.2	3.480	1.166

Note that the deviations for both methods are calculated by first determining the average of the four data points, and then subtracting the individual data points from the average (without regard to sign).

(a) The average density for method A is 2.4 ± 0.2 grams while the average density for method B is 3.480 ± 1.166 grams—if one includes all the data points. *Data point 4 in Method B* has a large deviation, and *should probably be excluded* from the calculation. If one omits data point 4, Method B gives a density of 2.703±0.001 g

(b) The error for each method :

Error = experimental value - accepted value

From Method A   error = (2.4 - 2.702) = 0.3

From Method B   error = (2.703 - 2.702) = 0.001 (omitting data point 4)

error = (3.480 - 2.702) = 0.778 (including all data points)

(c) Precision and Accuracy of each method:

If one counts all data points, the deviations **for all data points** of Method A are less than those for **the data points of** Method B, Method A offers *better precision* . On the other hand, omitting data point 4, Method B offers both *better accuracy* (average closer to the accepted value) and *better precision* (since the value is known to a greater number of significant figures).

Significant Figures

40. The number of significant figures in each of the following numbers:

The Rule numbers referred to are those found in the *Guidelines for Determining Significant Figures* in Chapter 1 of the text.

(a) 0.0123   3sf; Rule 1 – 0 to the left of the 1 locate the decimal point

(b) 3.40 x 10^3 3sf; All digits to the left of the "x" are considered significant

(c) 1.6402   5 sf; "trapped" zeros are considered significant

(d) 1.020   4 sf; Rule 1- when a number is greater than 1, all zeros to the right of the decimal point are significant.

42. Express the product of three numbers to the proper number of significant figures:

$$(0.0546)(16.0000)(\frac{7.779}{55.85}) = 0.122 \quad \text{(3 sf are allowed—owing to 0.0546)}$$

Problem Solving

44. Volume of a 1.50 carat diamond:

This problem is in essence (a) conversion of the unit "carat" to units of "grams", then using the density of diamond to convert from mass to volume.

$$\frac{1.50 \text{ carat}}{1} \cdot \frac{0.200 \text{ g}}{1 \text{ carat}} \cdot \frac{1 \text{ cm}^3}{3.513 \text{ g}} = 0.0854 \text{ cm}^3$$

46. Mass of a gold coin 2.2 cm in diameter and 3.0 mm thick:

To calculate the mass of the coin we'll need to determine the volume of the coin, and then use the density of gold.

Volume = $\pi r^2 \cdot$ thickness = $3.14159 \cdot (\frac{2.2 \text{ cm}}{2})^2 \cdot (\frac{3.0 \text{ mm}}{1} \cdot \frac{1 \text{ cm}}{10 \text{ mm}}) = 1.1 \text{ cm}^3$

Note that dividing the diameter (2.2cm) by 2 gives the radius (r). The last two terms are used to convert the thickness of the coin in millimeters to centimeters.

The product of $r^2 \cdot$ thickness gives the volume in cubic centimeters.

The mass of the coin is then: V • D = $1.1 \text{ cm}^3 \cdot \frac{19.3 \text{ g}}{\text{cm}^3}$ = 22 g (to 2 sf)

48. Fuel required for the trip: 22,300 kg. Calculating the volume occupied by the mass gives:

$$\frac{22,300 \text{ kg}}{1} \cdot \frac{1 \text{ lb}}{0.4536 \text{ kg}} \cdot \frac{1 \text{ L}}{1.77 \text{ lb}} = 27775 \text{ L required for the flight.}$$

Amount of fuel needed: (27775 L – 7682 L) = 20,093 L or 20,100 L to 3sf

General Questions

50. The normally accepted value for a human temperature is 98.6° F. On the Celsius scale, this corresponds to: °C = 5/9 (98.6 –32) = 37 °C. Since Gallium's melting point is 29.8 °C, the solid should melt in your hand.

52. Express the length 1.97 Angstroms in nanometers? In picometers?

$$\frac{1.97 \text{ Angstrom}}{1} \cdot \frac{1 \times 10^{-10} \text{m}}{1 \text{ Angstrom}} \cdot \frac{1 \times 10^9 \text{ nm}}{1 \text{ m}} = 0.197 \text{ nm}$$

$$\frac{1.97 \text{ Angstrom}}{1} \cdot \frac{1 \times 10^{-10} \text{m}}{1 \text{ Angstrom}} \cdot \frac{1 \times 10^{12} \text{pm}}{1 \text{m}} = 197 \text{ pm}$$

54. We begin by calculating the mass of the water in the 250. mL can.

(a) Mass of liquid water: $\dfrac{250. \text{ mL}}{1} \cdot \dfrac{1 \text{ cm}^3}{1 \text{ mL}} \cdot \dfrac{0.997 \text{ g}}{1 \text{ cm}^3} = 249.25 \text{ g}$

Assuming the water remains in the can as it begins to freeze, we can now use the density of solid water to calculate the volume that the solid water would assume.

Volume of solid water: $\dfrac{249.25 \text{ g water}}{1} \cdot \dfrac{1 \text{ cm}^3}{0.917 \text{ g}} = 271.8 \text{ or } 272 \text{ cm}^3$

(b) The increase in volume (272 mL –250. mL) represents an increase of 22 mL or 8.8% (22/250). The can will probably burst. (Happens frequently when sodas are placed in freezers and "forgotten".

56. Which occupies a larger volume: 600 g of water or 600 g of lead ?

$600 \text{ g H}_2\text{O} \cdot \dfrac{1 \text{ cm}^3}{0.995 \text{ g}}$ = 603 cm³   (600 cm³ to 1 sf)

$600 \text{ g Pb} \cdot \dfrac{1 \text{ cm}^3}{11.34 \text{ g}}$ = 52.9 cm³   (50 cm³ to 1 sf)

58. Estimate the density of water at 20 °C :

Below are two graphs in which the Density of water has been plotted versus the temperature for the range of data given. The graph on the left has been fit to a straight line, while the graph on the right shows the fit to an exponential equation.

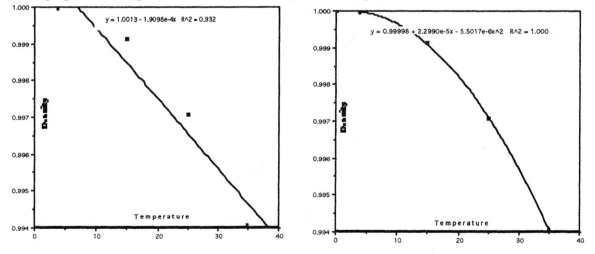

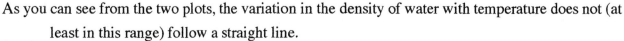

As you can see from the two plots, the variation in the density of water with temperature does not (at least in this range) follow a straight line.

The equation for this line is :

$$y = 2.2990 \times 10^{-5x} - 5.5017 \times 10^{-6x} + 0.99998$$

Your lab partner has assumed that the variation in the density of water with temperature is linear and would be the **average** of the densities between 15 and 25°C.
(0.99913 + 0.99707)/2. The graphs indicate that this *is not a reasonable assumption.*

60. Will aluminum (D = 2.70 g/cm³) and plastic (D = 1.37 g/cm³) float or sink in carbon tetrachloride (D = 1.58 g/cm³)?

Since substances in a fluid usually seek a density that is equal to their own, the plastic (with a much lower density than the carbon tetrachloride) will float. The aluminum (with a density that is greater) will sink. You probably most associate this phenomenon with helium-filled balloons "floating" in the atmosphere. The density of the helium is much less than the surrounding atmosphere, so the balloon "rises".

62. The mass of lead in a 250 g block of solder that is 67% lead:

$$\frac{250 \text{ g solder}}{1} \cdot \frac{67 \text{ g Pb}}{100 \text{ g solder}} = 167.5 \text{ g Pb } (170 \text{ g Pb to 2 sf})$$

64. Indicate the relative arrangements of the particles in each of the following:

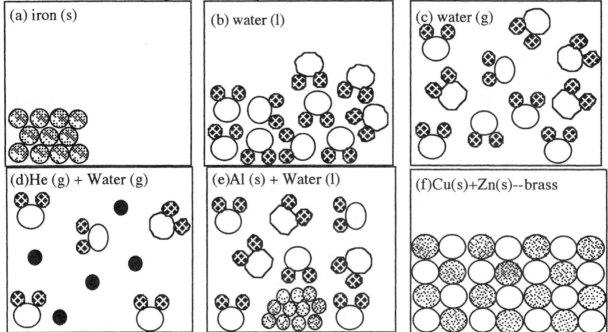

(a) iron (s)

(b) water (l)

(c) water (g)

(d) He (g) + Water (g)

(e) Al (s) + Water (l)

(f) Cu(s)+Zn(s)--brass

66. Experimental method to determine if a liquid is water:

There are several methods. One method would be to weigh an accurately known volume of the liquid. An empty dry weighed 10.0 mL graduated cylinder could be filled to the 10.0 mL mark with the liquid, and reweighed. The mass of liquid divided by the volume would provide the density of the liquid. That density could be compared with published values of the density of water *at that temperature.*

To determine if the water contains dissolved salts, test the electrical conductivity. Pure water is a very poor conductor. Water containing dissolved salts (and therefore the ions produced when that salt dissolves) would conduct an electric current.

To determine is there is salt dissolved in the water, one can boil the solution to dryness. Any dissolved salts would not boil (at water's normal boiling temperature) and would remain as a solid residue in the beaker.

68. Pouring three immiscible liquids into a test tube will result in three discrete layers in which the liquids arrange themselves from the most dense liquid (at bottom) to the least dense liquid (at top).

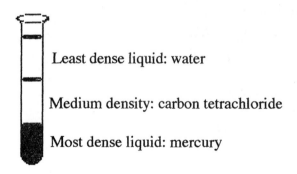

Least dense liquid: water

Medium density: carbon tetrachloride

Most dense liquid: mercury

70. To determine if a copper-colored metal is pure copper :

One can check some of the properties that identify copper. (1) Melting a sample of the wire and comparing the melting point of the sample to the melting point of pure copper. (2) Carefully determine the density of the wire, and compare that density with the literature value for the density of pure copper. (3) Test the electrical conductivity of the wire and compare that conductivity to copper's electrical conductivity.

72. For the reaction of elemental potassium reacting with water:
    (a) States of matter involved: **Solid** potassium reacts with **liquid** water to produce **gaseous** hydrogen and aqueous potassium hydroxide solution (a homogenous mixture).
    (b) The observed change is chemical. The products (hydrogen and potassium hydroxide) are quite different from elemental potassium and water. Litmus paper would also provide the information that while the original water was neither acidic nor basic, the solution produced would be basic. (It would change the color of red litmus paper to blue.)
    (c) The reactants: potassium and water
        The products: hydrogen and potassium hydroxide solution
    (d) Potassium reacts **vigorously** with water. Potassium is less dense than water, and floats atop the surface of the water. The reaction produces enough heat to ignite the hydrogen gas evolved. The flame observed is typically violet-purple in color. The potassium hydroxide formed is soluble in water (and therefore not visible). Refer to SQ 66 for information on isolating the KOH.

74. Calculate the mass of 12 ounces of aluminum in grams and from that mass, the volume:

$$\text{Volume} = \frac{\text{Mass}}{\text{Density}} \qquad = \frac{12 \text{ oz} \cdot \frac{454 \text{ g}}{16 \text{ oz}}}{2.70 \text{ g/cm}^3} \qquad = 130 \text{ cm}^3$$

$$\text{Volume} = \text{Area} \cdot \text{Thickness}$$

Express the area in units of cm^2, then calculate the thickness by dividing the volume by the

area:
$$\text{Area} = 75 \text{ ft}^2 \cdot \left(\frac{12 \text{ in}}{1 \text{ ft}}\right)2 \cdot \left(\frac{2.54 \text{ cm}}{1 \text{ in}}\right)2 = 7.0 \times 10^4 \text{ cm}^2$$

then
$$\text{Volume} = \text{Area} \cdot \text{Thickness}$$

$$130 \text{ cm}^3 = 7.0 \times 10^4 \text{ cm}^2 \cdot \text{Thickness}$$

and
$$\text{Thickness} = \frac{130 \text{ cm}^3}{7.0 \times 10^4 \text{ cm}^2} \qquad = 1.8 \times 10^{-3} \text{ cm} \quad \text{or} \quad 1.8 \times 10^{-2} \text{ mm}$$

76. To determine the thickness of the oil layer, we can think about the oil layer as having a certain volume, ($V = l \times w \times h$), and that our "task" is to determine the "thickness" -- or h in our formula. The volume of oil is 1 teaspoon (5 cm^3). The area covered ($l \times w$) is 0.5 acres. So if we divide the volume by the area, we should have the **thickness**.

$$\frac{5 \text{ cm}^3}{1 \text{ teaspoon}} \quad \cdot \quad \frac{1 \text{ teaspoon}}{0.5 \text{ acre}} \quad \cdot \quad \frac{2.47 \text{ acres}}{1.0 \times 10^4 \text{ m}^2} \quad \cdot \quad \left(\frac{1 \text{ m}}{100 \text{ cm}}\right)2 = 2 \times 10^{-7} \text{ cm}$$

Volume / Area		Acre converted to square meters	Conversion of sq. m to sq. cm

Chapter 2
Atoms and Elements

Reviewing Important Concepts

6. The diameter of an atom if the nucleus of the atom were approximately 6 cm:

    Your text (Section 2.1—specifically Exercise 2.1) provides the relative sizes of the nuclear and atomic diameters, with the **nuclear radius** on the order of 0.001pm and the **atomic radius** approximately 100pm.

Nuclear diameter	Atomic diameter	Ratio
0.002 pm	200 pm	1:100,000
6 cm (orange)	?	1:100,000

    If the nuclear diameter is 6 cm, then the atomic diameter is 600,000 cm.

    Translating that number into larger units: 600,000 cm = 6,000 m or 6 km.

8. Comparison of Titanium and Thallium:

Name	Symbol	Atomic #	Atomic Weight	Group #	Period #	Metal, Metalloid, or nonmetal
Titanium	Ti	22	47.87	4B (4)	4	Metal
Thallium	Tl	81	204.38	3A (3)	6	Metal

10. Which member of each pair represents more mass:

    (a) 0.5 mol of Na of 0.5 mol of Si: One mole of Na has a mass of approximately 23 g while a mole of Si has a mass of 28 g. So **0.5 mol of Si has a greater mass**.

    (b) 9.0 g of Na or 0.50 mol of Na: One-half mol of Na would have a mass of approximately 12.5 g, so **0.50 mol Na has a greater mass**.

    (c) 10 atoms of Fe or 10 atoms of K: The atomic weight of K is approximately 39 while that of Fe is approximately 56. Each atom of Fe has a greater mass than an atom of K., so **10 atoms of Fe have a greater mass**.

12. Difference between a group(family) and a period on the Periodic Table:

    Groups are the **columns** on the Periodic Table, while

    Periods are the horizontal **rows**.

14. Periodic Table designations:

(a) Three elements that are metals:

Name	Symbol	Group	Period
Lithium	Li	1A (1)	2
Silver	Ag	1B (11)	5
Lead	Pb	4A (14)	6

(b) Four elements that are nonmetals

Name	Symbol	Group	Period
Carbon	C	4A (14)	2
Phosphorus	P	5A (15)	3
Selenium	Se	6A (16)	4
Iodine	I	7A (17)	5

(c) Two elements that are metalloids.

Name	Symbol	Group	Period
Silicon	Si	4A (14)	3
Arsenic	As	5A (15)	4

16. Periodic Table designations:

Transition elements: Any element in Groups 3B (3) - 2B (12)

Halogens: Elements in Group 7A (17) – hydrogen excluded

Noble gases: Elements in Group 8A (18)

Alkali metals   Elements in Group 1A (1) –hydrogen excluded

18. An element discovered by Madame Curie:

Name	Symbol	Atomic Number	Origin of elemental name
Polonium	Po	84	Curie's homeland--Poland
Radium	Ra	88	Radius, ray(Latin)—with Pierre Curie

Practicing Skills

Composition of Atoms

20. Mass number for

(a) Mg (at. no. 12) with 15 neutrons :   27

(b) Ti (at. no. 22) with 26 neutrons :   48

(c) Zn (at. no. 30) with 32 neutrons :   62

The mass number represents the SUM of the protons + neutrons in the nucleus of an atom.

The atomic number represents the # of protons, so (atomic no. + # neutrons)=mass number

22. Mass number (A)   = no. of protons + no. of neutrons;

Atomic number (Z) = no. of protons

(a) $^{39}_{19}\text{K}$ (b) $^{84}_{36}\text{Kr}$ (c) $^{60}_{27}\text{Co}$

24.

substance	protons	neutrons	electrons
(a) magnesium-24	12	12	12
(b) tin-119	50	69	50
(c) thorium-232	90	142	90

Note that the number of protons and electrons are **equal** for any **neutral atom**. The number of protons is **always** equal to the atomic number. The mass number equals the sum of the numbers of protons and neutrons.

Isotopes

26. For technetium- 99 (at. no. 43) : # protons : 43

neutrons : (99 - 43) = 56

electrons : 43

28. Isotopes of cobalt (atomic number 27) with 30, 31, and 33 neutrons:

would have symbols of $^{57}_{27}\text{Co}, ^{58}_{27}\text{Co}$, and $^{60}_{27}\text{Co}$ respectively.

Isotope Abundance and Atomic Mass

30. Thallium has two stable isotopes ^{203}Tl and ^{205}Tl. The more abundant isotope is:_____

The atomic weight of thallium is 204.3833. The fact that this weight is closer to 205 than 203 indicates that the 205 isotope is the more abundant.

32. The atomic mass of lithium is:

$$(0.0750)(6.015121) + (0.9250)(7.016003) = 6.94 \text{ amu}$$

Recall that the atomic mass is a weighted average of all isotopes of an element,

and is obtained by **adding** the *product* of (relative abundance x mass) for all isotopes.

34. The average atomic weight of gallium is 69.723. If we let **x** represent the abundance of the lighter isotope, and (**1-x**) the abundance of the heavier isotope, the expression to calculate the atomic weight of gallium may be written:

$$(x)(68.9257) + (1 - x)(70.9249) = 69.723$$

[Note that the sum of all the isotopic abundances must add to 100% -- or 1 (in decimal notation).] Simplifying the equation gives:

$$68.9257 \text{ amu } x + 70.9249 \text{ amu} - 70.9249 \text{ amu } x = 69.723 \text{ amu}$$
$$-1.9992 \text{ amu } x = (69.723 \text{ amu} - 70.9249)$$
$$-1.9992 \text{ amu } x = -1.202 \text{ amu}$$
$$x = 0.6012$$

So the relative abundance of isotope 69 is 60.12 % and that of isotope 71 is 39.88 %.

Atoms and the Mole

36. The mass, in grams of:

(a) 2.5 mol Al:

$$\frac{2.5 \text{ mol Al}}{1} \cdot \frac{27.0 \text{ g Al}}{1 \text{ mol Al}} = 67.5 \text{ g Al (68 g Al to 2sf)}$$

(b) 1.25×10^{-3} mol Fe:

$$\frac{1.25 \times 10^{-3} \text{mol Fe}}{1} \cdot \frac{55.85 \text{ g Fe}}{1 \text{ mol Fe}} = 0.0698 \text{ g Fe (3 sf)}$$

(c) 0.015 mol Ca:

$$\frac{0.015 \text{ mol Ca}}{1} \cdot \frac{40.1 \text{ g Ca}}{1 \text{ mol Ca}} = 0.60 \text{ g Ca (2 sf)}$$

(d) 653 mol Ne:

$$\frac{653 \text{ mol Ne}}{1} \cdot \frac{20.18 \text{ g Ne}}{1 \text{ mol Ne}} = 1.32 \times 10^{4} \text{ g Ne (3 sf)}$$

Note that, whenever possible, one should use a molar mass of the substance that contains **one more** significant figure than the data, to reduce round-off error.

38. The amount (moles) of substance represented by:

(a) 127.08g Cu:

$$\frac{127.08 \text{ g Cu}}{1} \cdot \frac{1 \text{ mol Cu}}{63.546 \text{ g Cu}} = 1.9998 \text{ mol Cu (5 sf)}$$

(b) 0.012g Li:

$$\frac{0.012 \text{ g Li}}{1} \cdot \frac{1 \text{ mol Li}}{6.94 \text{ g Li}} = 1.7 \times 10^{-3} \text{ mol Li (2 sf)}$$

(c) 5.0mg Am:

$$\frac{5.0 \text{ mg Am}}{1} \cdot \frac{1 \text{ g Am}}{10^{3} \text{mg Am}} \cdot \frac{1 \text{ mol Am}}{243 \text{ g Am}} = 2.1 \times 10^{-5} \text{mol Am (2 sf)}$$

(d) 6.75g Al $\dfrac{6.75 \text{ g Al}}{1} \cdot \dfrac{1 \text{ mol Al}}{26.98 \text{ g Al}} = 0.250 \text{ mol Al}$

40. The average mass of one copper atom:

One mole of copper (with a mass of 63.546 g) contains 6.0221×10^{23} atoms. So the average mass of **one** copper atom is:

$\dfrac{63.546 \text{ g Cu}}{6.0221 \times 10^{23} \text{ atoms Cu}}$ $= 1.0552 \times 10^{-22}$ g/Cu atom

The Periodic Table

42. The elements in Group 5A are:

nitrogen (N - nonmetal),

phosphorus (P - nonmetal),

arsenic (As - metalloid),

antimony (Sb - metalloid),

bismuth (Bi - metal)

44. Periods with 8 elements: 2 Periods 2 (at.no. 3-10) and 3 (at.no. 11-18)

Periods with 18 elements: 2 Periods 4 (at.no 19-36) and 5 (at.no. 37-54)

Periods with 32 elements: 1 Period 6 (at.no. 55-86)

[and possibly Period 7 --although the full 32 are not presently known]

General Questions

46.

Symbol	^{58}Ni	^{33}S	^{20}Ne	^{55}Mn
Number of protons	28	16	10	25
Number of neutrons	30	17	10	30
Number of electrons in the neutral atom	28	16	10	25
Name of element	nickel	sulfur	neon	manganese

48. Given that the average atomic mass for potassium is 39.0983 amu, the lighter isotope, ^{39}K would be the more abundant of the **remaining** isotopes(ignoring K-40). Remember that atomic masses are **weighted averages**, that is the average atomic mass is closer to the mass of the most abundant isotope.

50. Regarding the elements:

(a) the most abundant metal: Fe (If you exclude the transition metals, it's Mg)

(b) the most abundant nonmetal: H

(c) the most abundant metalloid: Si

(d) most abundant transition element: Fe

(e) halogens included (and most abundant): F, Cl, Br (Cl is most abundant)

52. Of the following, the one that is impossible:

(a) Ag foil that is 1.2×10^{-4} m thick. This is equivalently 0.12 mm thick. This is greater than the diameter of a silver atom and therefore not impossible.

(b) A sample of K containing 1.784×10^{24} atoms. This is approximately 3 moles of potassium (113g) and therefore not impossible.

(c) A gold coin of mass 1.23×10^{-3} kg (alternatively stated: 1.23 g) — not impossible

(d) 3.43×10^{-27} mol S_8: This choice is impossible. If you multiply Avogadro's number (6.022×10^{23}) by the number of mol of sulfur given, the number of particles is: 0.00207 or LESS than one molecule of S_8.

54. Using the furnished plot of density as a function of atomic number:

(a) Three elements in the series with the greatest density: At. no. 27 (Cobalt), At. no. 28 (Nickel), and At. no. 29 (Copper). The density of all three **metals** is approximately 9 g/cm^3.

(b) The element in the second period with the largest density is Boron (atomic number 5) while the element in the third period with the largest density is Aluminum (atomic number 13). Both of these elements belong to group 3A.

(c) Elements from the first 36 elements with very **low densities** are **gases**. These include Hydrogen, Helium, Nitrogen, Oxygen, Fluorine, Neon, Chlorine, Argon and Krypton.

56. Reviewing the periodic table:

(a) An element in Group 2A : beryllium, magnesium, calcium, strontium, barium, radium

(b) An element in the third period: sodium, magnesium, aluminum, silicon, phosphorus, sulfur, chlorine, argon

(c) An element in the 2nd period in Group 4A: carbon

(d) An element in the third period in Group 6A: sulfur

(e) A halogen in the fifth period: iodine

(f) An alkaline earth element in the third period: magnesium

(g) A noble gas element in the fourth period: krypton

(h) A nonmetal in Group 6A and the third period: sulfur

(i) A metalloid in the fourth period: germanium or arsenic

58. Number of moles of Kr in 0.00789 g Kr:

$$0.00789 \text{ g Kr} \cdot \frac{1 \text{ mol Kr}}{83.80 \text{ g Kr}} = 9.42 \times 10^{-5} \text{ mol Kr (3 sf)}$$

Number of atoms: 6.02217×10^{23} atoms Kr /mol Kr $\cdot$ 9.42×10^{-5} mol Kr= 5.67×10^{19}

60. The volume of a cube of Na containing 0.125 mol Na:

First we need to know the **mass** of 0.125 mol Na:

$$0.125 \text{mol Na} \cdot \frac{22.99 \text{ g Na}}{1 \text{ mol Na}} = 2.87 \text{ g Na} \quad (3 \text{ sf})$$

Now we can calculate the volume that contains 2.87 g Na: (using the density given)

$$2.87 \text{g Na} \cdot \frac{1 \text{ cm}^3}{0.971 \text{ g Na}} = 2.96 \text{ m}^3$$

If the cube is a perfect cube (that is each side is equivalent in length to any other side), what is the length of one edge?

$$2.96 \text{ cm}^3 = l \times l \times l \quad \text{so} \quad 1.44 \text{ cm} = \text{length of one edge}$$

62. The number of Cr atoms in a coating that is 0.015 cm thick with a surface area of 15.3 cm^2

The volume of the coating will be: 0.2295 cm^3 (the surface area x thickness)

Given the density of Cr of 7.10 g/cm^3, we can calculate the **mass** of this volume:

$$\text{Mass Cr} = \frac{0.2295 \text{ cm}^3}{1} \cdot \frac{7.10 \text{ g Cr}}{1 \text{ cm}^3} = 1.63 \text{ g Cr (rounded to 3sf)}$$

Given that Avogadro's number of Cr atoms (6.02217×10^{23}) has a mass of 52.00g (at.wt), we can calculate the number of Cr atoms with this mass:

$$\frac{0.2295 \text{ cm}^3}{1} \cdot \frac{7.10 \text{ g Cr}}{1 \text{ cm}^3} \cdot \frac{1 \text{ mol Cr}}{52.00 \text{ g Cr}} \cdot \frac{6.02217 \times 10^{23} \text{atoms}}{1 \text{ mol Cr}} = 1.9 \times 10^{22} \text{ atoms Cr (2sf)}$$

64. The ratio of the atomic masses of P/O :

0.744 g of P combined with 0.960 g of O. (1.704 - 0.744)

Given that the compound formed is P$_4$O$_{10}$, the masses of phosphorus and oxygen correspond to the ratio of 4 phosphorus atoms to 10 oxygen atoms. Dividing the masses by the appropriate number of atoms

$$\frac{0.744 \text{ g P}}{4} = 0.186 \text{ g P} \quad \text{and} \quad \frac{0.960 \text{ g O}}{10} = 0.0960 \text{ g O}$$

So the mass of $\frac{P}{O}$ is $\frac{0.186 \text{ g P atoms}}{0.0960 \text{ g O atoms}} = 1.94$

If the atomic mass of oxygen is 16.000 amu, the atomic mass of phosphorus is

$$16.000 \text{ amu} \cdot \frac{1.94 \text{ g P atoms}}{1.0 \text{ g O atoms}} = 31.0 \text{ amu}$$

<u>66</u>. Number of atoms of C in 2.0000 g sample of C.

$$2.0000 \text{ g C} \cdot \frac{1 \text{ mol C}}{12.011 \text{ g C}} \cdot \frac{6.0221 \times 10^{23} \text{ atoms C}}{1 \text{ mol C}} = 1.0028 \times 10^{23} \text{ atoms C}$$

If the accuracy of the balance is +/- 0.0001g then the mass could be 2.0001g C, so we would do a similar calculation to obtain 1.00281×10^{23} atoms C—a difference of 5×10^{18} atoms C.

68. Begin by asking about the relationship between g-mol and ton-mol:

1 ton = 1000 kg = 1,000,000 (or 10^6)grams, so 1 ton-mol = 10^6 g-mol

If 1 g-mol contains Avogadro's number of Al atoms (6.02217×10^{23} atoms), then

$$\frac{6.02217 \times 10^{23} \text{ atoms Al}}{1 \text{ g-mol}} \cdot \frac{1 \times 10^6 \text{ g-mol}}{1 \text{ ton-mol}} = \frac{6.0 \times 10^{29} \text{ atoms Al}}{1 \text{ ton-mol}}$$

70. Simulation indicates Ca has five stable isotopes, with 20, 22, 23, 24 and 26 neutrons. As the number of protons increases, the ratio of n/p increases. The isotope Xenon-108 is not stable, having only 1 n/p.

Chapter 3
Molecules, Ions, and Their Compounds

Reviewing Important Concepts

1. The molecular formula for cis-Platin is : $Pt(NH_3)_2Cl_2$ contains

    N atoms in one molecule: 2 nitrogen atoms (1 in each of two ammonia groups)

    H atoms in one molecule: 6 hydrogen atoms (3 in each of two ammonia groups)

    H atoms in one mole of cis-platin:

        One mole of cis-platin contains Avogradro's # of molecules (6.022×10^{23} molecules)

        so the number of hydrogen atoms is $6 \times 6.022 \times 10^{23} = 3.6 \times 10^{24}$.

    The molar mass of $Pt(NH_3)_2Cl_2$ is 300.1 g/mol.

3. The **number of electrons** in a strontium atom is **38**, the same as the atomic number for the element. When an atom of strontium forms an ion, it **loses TWO ELECTRONS**, forming an ion having the same electron configuration as the noble gas **krypton**.

5. Regarding adenine:

    (a) The number of molecules cited (3.0×10^{23}) represents 1/2 mole of adenine. The molar mass of adenine is 135.14, so half of that mass is about 67.57 g. Forty (40) grams of adenine would have less mass.

    (b) Structural features of adenine common to nucleotides: Nucleotides are combinations of three types of molecules. A nitrogen base, a sugar (ribose or deoxyribose) and a phosphate group. Nitrogen bases form bonds to the sugars through the H atom on the N atom in the ring. Adenine has that hydrogen atom (on the 5-membered ring).

7. To determine whether 0.5 mol of $BaCl_2$ or 0.5 mol of $SiCl_4$ has the **larger mass**, one needs to determine the molar mass of each compound. The molar masses are as follows:

    $BaCl_2 = (137.3 + 2(35.45)) = 208.2$

    $SiCl_4 = (28.1 + 4(35.45)) = 169.9$

    One-half mole of $BaCl_2$ would have a mass of about 104 g while the same amount of $SiCl_4$ would have a mass of about 85 g—so 0.5 mol of $BaCl_2$ has the **larger mass**.

9. Larger weight percentage of oxygen:

    For H_2O, molar mass = approximately 18.0 [$\%O = (16/18) \cdot 100 = 88.7\ \%$]

    For CH_3OH, molar mass = approximately 32.0 [$\%O = (16/32) \cdot 100 = 49.9\ \%$]

    So water has the **larger weight percentage** of oxygen.

Practicing Skills

10. The molecular formula for the compounds:

    (a) 7 carbon atoms and 16 hydrogen atoms, and 1 oxygen atom per molecule: $C_7H_{16}O$

    (b) 6 carbon atoms, 8 hydrogen atoms, and 6 oxygen atoms per molecule: $C_6H_8O_6$

    (c) 14 carbon atoms, 18 hydrogen atoms, 2 nitrogen atoms, and 5 oxygen atoms per molecule: $C_{14}H_{18}N_2O_5$

12. Total atoms of each element in a formula unit:

	element	# atoms
(a) CaC_2O_4	Ca	1
	C	2
	O	4
(b) C_6H_5CHO	C	7
	H	6
	O	1
(c) $Co(NH_3)_5(NO_2)Cl_2$	Co	1
	N	6
	H	15
	O	2
	Cl	2
(d) $K_4Fe(CN)_6$	K	4
	Fe	1
	C	6
	N	6

Molecular Models

14. The formula for sulfuric acid is H_2SO_4. The molecule is **not flat**. The O atoms are arranged around the sulfur at the corners of a tetrahedron—that is the O-S-O angle would be about 109 degrees. The hydrogen atoms are connected to two of the oxygen atoms also with angles (H-O-S) of approximately 109 degrees.

Ions and Ion Charges

16. Most commonly observed ion for:

    (a) Magnesium: 2+ —like all the alkaline earth metals

    (b) Zinc : 2+

    (c) Nickel: 2+

    (d) Gallium: 3+ (an analog of Aluminum)

18. The symbol and charge for the following ions:
    (a) barium ion Ba^{2+}
    (b) titanium(IV) ion Ti^{4+}
    (c) phosphate ion PO_4^{3-}
    (d) hydrogen carbonate ion HCO_3^-
    (e) sulfide ion S^{2-}
    (f) perchlorate ion ClO_4^-
    (g) cobalt(II) ion Co^{2+}
    (h) sulfate ion SO_4^{2-}

20. When potassium becomes a monatomic ion, potassium—like all alkali metals—**loses 1 electron.** The noble gas atom with the same number of electrons as the potassium ion is **argon**.

Ionic Compounds

22. Barium is in Group 2A, and is expected to form a 2+ ion while bromine is in group 7A and expected to form a 1- ion. Since the compound would have to have an **equal amount** of negative and positive charges, the formula would be $BaBr_2$.

24. Formula, Charge, and Number of ions in:

	cation	# of	anion	# of
(a) K_2S	K^+	2	S^{2-}	1
(b) $CoSO_4$	Co^{2+}	1	SO_4^{2-}	1
(c) $KMnO_4$	K^+	1	MnO_4^-	1
(d) $(NH_4)_3PO_4$	NH_4^+	3	PO_4^{3-}	1
(e) $Ca(ClO)_2$	Ca^{2+}	1	ClO^-	2

26. Cobalt oxide

    Cobalt(II) oxide   CoO   cobalt ion : Co^{2+}
    Cobalt(III) oxide Co_2O_3 Co^{3+}

28. Provide correct formulas for compounds:
    (a) $AlCl_3$ The tripositive aluminum ion requires three chloride ions.
    (b) KF   Potassium is a monopositive cation. Fluoride is a mononegative anion.
    (c) Ga_2O_3 is correct; Ga is a 3+ ion and O forms a 2- ion
    (d) MgS   is correct; Mg forms a 2+ ion and S forms a 2- ion

30. Formulas for compounds from Mg^{2+} and Al^{3+} with O^{2-} and PO_4^{3-}

	Mg^{2+}	Al^{3+}
O^{2-}	MgO	Al_2O_3
PO_4^{3-}	$Mg_3(PO_4)_2$	$AlPO_4$

Naming Ionic Compounds

32. Names for the ionic compounds
   - (a) K_2S potassium sulfide
   - (b) $CoSO_4$ cobalt(II) sulfate
   - (c) $(NH_4)_3PO_4$ ammonium phosphate
   - (d) $Ca(ClO)_2$ calcium hypochlorite

34. Formulas for the ionic compounds
   - (a) ammonium carbonate $(NH_4)_2CO_3$
   - (b) calcium iodide CaI_2
   - (c) copper(II) bromide $CuBr_2$
   - (d) aluminum phosphate $AlPO_4$
   - (e) silver(I) acetate $AgCH_3CO_2$

36. Names and formulas for ionic compounds:

	anion	anion
cation	CO_3^{2-}	I^-
Na^+	Na_2CO_3 sodium carbonate	NaI sodium iodide
Ba^{2+}	$BaCO_3$ barium carbonate	BaI_2 barium iodide

Coulomb's Law

38. The fluoride ion has a smaller radius than the iodide ion. Hence the distance between the sodium and fluoride ions will be less than the comparable distance between sodium and iodide. Coulomb's Law indicates that the attractive force becomes greater as the distance between the charges grows smaller—hence NaF will have stronger forces of attraction.

Naming Binary, Nonmetal Compounds

40. Names of binary nonionic compounds
    (a) NF_3 nitrogen trifluoride

    (b) HI   hydrogen iodide

    (c) BI_3 boron triiodide

    (d) PF_5 phosphorus pentafluoride

42. Formulas for:

    (a) sulfur dichloride SCl_2

    (b) dinitrogen pentaoxide N_2O_5

    (c) silicon tetrachloride $SiCl_4$

    (d) diboron trioxide B_2O_3

Molecules, Compounds, and the Mole

44. Molar mass of the following: (with atomic weights expressed to 4 significant figures)
    (a) Fe_2O_3 $(2)(55.85) + (3)(16.00) = 159.7$

    (b) BCl_3 $(1)(10.81) + (3)(35.45) = 117.2$

    (c) $C_6H_8O_6$ $(6)(12.01) + (8)(1.008) + (6)(16.00) = 176.1$

46. Molar mass of the following: (with atomic weights expressed to 4 significant figures)
    (a) $Ni(NO_3)_2 \cdot 6H_2O$ $(1)(58.69) + (2)(14.01) + 6(16.00) + (12)(1.008) + (6)(16.00)$
    $$= 290.8$$

    (b) $CuSO_4 \cdot 5H_2O$ $(1)(63.55) + (1)(32.07) + 4(16.00) + (10)(1.008) + (5)(16.00)$
    $$= 249.7$$

48. Moles represented by 1.00 g of the compounds:

Compound	Molar mass	Moles in 1.00 g
(a) C_3H_7OH	60.10	0.0166
(b) $C_{11}H_{16}O_2$	180.2	0.00555
(c) $C_9H_8O_4$	180.2	0.00555

50. Moles of acetonitrile in 2.50 kg:
    1. Molar mass of CH_3CN:

    $$(2)(12.01) + (3)(1.008) + (1)(14.01) = 41.05 \text{ g/mol}$$

    2. Moles:

    $$2.50 \times 10^3 g \cdot \frac{1 \text{ mol } CH_3CN}{41.05 \text{ g}} = 60.9 \text{ mol } CH_3CN$$

23

52. Regarding sulfur trioxide:

1. Amount of SO_3 in 1.00 kg: $\dfrac{1.00 \times 10^3 \text{g } SO_3}{1} \cdot \dfrac{1 \text{ mol } SO_3}{80.07 \text{ g } SO_3} = 12.5 \text{ mol } SO_3$

2. Number of SO_3 molecules: $\dfrac{12.5 \text{ mol } SO_3}{1} \cdot \dfrac{6.022 \times 10^{23} \text{ molecules}}{1 \text{ mol } SO_3} = 7.52 \times 10^{24}$

3. Number of S atoms: With 1 S atom per SO_3 molecule-- 7.52×10^{24} S atoms

4. Number of O atoms: With 3 O atoms per SO_3 molecule—$3 \times 7.52 \times 10^{24}$ O atoms

or 2.26×10^{25} O atoms

Percent Composition

54. Mass percent for: [4 significant figures]

(a) PbS: $(1)(207.2) + (1)(32.06) = 239.3$ g/mol

$\%Pb = \dfrac{207.2 \text{ g Pb}}{239.3 \text{ g PbS}} \times 100 = 86.60 \%$

$\%S = 100.00 - 86.60 = 13.40 \%$

(b) C_3H_8: $(3)(12.01)+(8)(1.008) = 44.09$ g/mol

$\%C = \dfrac{36.03 \text{ g C}}{44.09 \text{ g } C_3H_8} \times 100 = 81.71 \%$

$\%H = 100.00 - 81.71 = 18.29 \%$

(c) $C_{10}H_{14}O$: $(10)(12.01) + (14)(1.008) + (1)(16.00) = 150.21$ g/mol

$\%C = \dfrac{120.1 \text{ g C}}{150.21 \text{ g } C_{10}H_{14}O} \times 100 = 79.96 \%$

$\%H = \dfrac{14.112 \text{ g H}}{150.21 \text{ g } C_{10}H_{14}O} \times 100 = 9.394 \%$

$\%O = 100.00 - (79.96 + 9.394) = 10.65 \%$

56. Mass of lead present in 10.0 g of PbS:

From SQ 54(a), the % of Pb in PbS is 86.59%, so in 10.0g of PbS there would be:

10.0 g PbS x 86.59% = 8.66 g of Pb

58. Mass of CuS to provide 10.0 g of Cu:

$\dfrac{10.0 \text{ g Cu}}{1} \cdot \dfrac{95.62 \text{ g CuS}}{63.55 \text{ g Cu}} = 15.0 \text{ g CuS}$

Note that the second term is the reciprocal of the term by which one would calculate the percent of Cu in CuS.

Empirical and Molecular Formulas

60. The empirical formula ($C_2H_3O_2$) would have a mass of 59.04 g.

Since the molar mass is 118.1 g/mol we can write

$$\frac{1 \text{ empirical formula}}{59.04 \text{ g succinic acid}} \cdot \frac{118.1 \text{ g succinic acid}}{1 \text{ mol succinic acid}} = \frac{2.0 \text{ empirical formulas}}{1 \text{ mol succinic acid}}$$

So the molecular formula contains 2 empirical formulas (2 x $C_2H_3O_2$) or $C_4H_6O_4$.

62. Provide the empirical or molecular formula for the following, as requested:

	Empirical Formula	Molar Mass (g/mol)	Molecular Formula
(a)	CH	26.0	C_2H_2
(b)	CHO	116.1	$C_4H_4O_4$
(c)	CH_2	112.2	C_8H_{16}

64. Calculate the empirical formula of acetylene by calculating the atomic ratios of carbon and hydrogen in 100 g of the compound.

$$92.26 \text{ g C} \cdot \frac{1 \text{ mol C}}{12.011 \text{ g C}} \qquad = 7.681 \text{ mol C}$$

$$7.74 \text{ g H} \cdot \frac{1 \text{ mol H}}{1.008 \text{ g H}} \qquad = 7.678 \text{ mol H}$$

Calculate the atomic ratio: $\quad \dfrac{7.68 \text{ mol C}}{7.68 \text{ mol H}} \quad = \dfrac{1 \text{ mol C}}{1 \text{ mol H}}$

The atomic ratio indicates that there is 1 C atom for 1 H atom (1:1). The **empirical formula is then CH**. The formula mass is 13.01. Given that the molar mass of the compound is 26.02 g/mol, there are two formula units per molecular unit, hence the **molecular formula for acetylene is C_2H_2**.

66. Determine the empirical and molecular formulas of cumene:

The percentage composition of cumene is 89.94% C and (100.00-89.94) or 10.06%H.

We can calculate the ratio of mol C: mol H as done in SQ64.

$$89.95 \text{ g C} \cdot \frac{1 \text{ mol C}}{12.011 \text{ g C}} \qquad = 7.489 \text{ mol C}$$

$$10.06 \text{ g H} \cdot \frac{1 \text{ mol H}}{1.008 \text{ g H}} \qquad = 9.981 \text{ mol H}$$

Calculating the atomic ratio:

$$\frac{9.981 \text{ mol H}}{7.489 \text{ mol C}} = \frac{1.33 \text{ mol H}}{1.00 \text{ mol C}} \text{ or a ratio of } 3C : 4H$$

So the empirical formula for cumene is C_3H_4 , with a formula mass of 40.06

If the molar mass of 120.2 g/mol, then dividing the "empirical formula mass" into the molar mass gives: 120.2/40.06 or 3 empirical formulas **per** molar mass. The **molecular formula** is then 3 x C_3H_4 or C_9H_{12}.

68. Empirical and Molecular formula for Mandelic Acid:

$$63.15 \text{ g C} \cdot \frac{1 \text{ mol C}}{12.0115 \text{ g C}} = 5.258 \text{ mol C}$$

$$5.30 \text{ g H} \cdot \frac{1 \text{ mol}}{1.0079 \text{ g H}} = 5.28 \text{ mol H}$$

$$31.55 \text{ g O} \cdot \frac{1 \text{ mol O}}{15.9994 \text{ g O}} = 1.972 \text{ mol O}$$

Using the smallest number of atoms, we calculate the ratio of atoms:

$$\frac{5.258 \text{ mol C}}{1.972 \text{ mol O}} = \frac{2.666 \text{ mol C}}{1 \text{ mol O}} \text{ or } \frac{22/3 \text{ mol C}}{1 \text{mol O}} \text{ or } \frac{8/3 \text{ mol C}}{1 \text{ mol O}}$$

So 3 mol O combine with 8 mol C and 8 mol H so the empirical formula is $C_8H_8O_3$. The formula mass of $C_8H_8O_3$ is 152.15. Given the data that the molar mass is 152.15 g/mL, the molecular formula for mandelic acid is $C_8H_8O_3$.

70. Molecules of water per formula unit of $MgSO_4$:

From 1.687 g of the hydrate, only 0.824 g of the magnesium sulfate remain.

The mass of water contained in the solid is: (1.687-0.824) or 0.863 grams

Use the molar masses of the solid and water to calculate the number of moles of each substance present:

$$0.824\text{g} \cdot \frac{1 \text{ mol MgSO}_4}{120.36 \text{ g MgSO}_4} = 6.85 \times 10^{-3} \text{ mol of magnesium sulfate}$$

$$0.863 \text{ g H}_2\text{O} \cdot \frac{1 \text{ mol H}_2\text{O}}{18.02 \text{ g H}_2\text{O}} = 4.79 \times 10^{-2} \text{ mol of water} \text{ ; the ratio of water to MgSO}_4 \text{ is:}$$

$$\frac{4.79 \times 10^{-2} \text{ mol water}}{6.85 \times 10^{-3} \text{ mol magnesium sulfate}} = 6.99$$

So we write the formula as $MgSO_4 \cdot 7 H_2O$.

72. Given the masses of xenon involved, we can calculate the number of moles of the element:

$$0.526 \text{ g S} \bullet \frac{1 \text{ mol Xe}}{131.29 \text{ g Xe}} = 0.00401 \text{ mol Xe}$$

The mass of fluorine present is: 0.678 g compound - 0.526 g Xe = 0.152 g F

$$0.152 \text{ g F} \bullet \frac{1 \text{ mol F}}{19.00 \text{ g F}} = 0.00800 \text{ mol F}$$

Calculating atomic ratios:

$$\frac{0.00800 \text{ mol F}}{0.00401 \text{ mol Xe}} = \frac{2 \text{ mol F}}{1 \text{ mol Xe}}$$

indicating that the empirical formula is XeF_2

74.   Formula of compound formed between zinc and iodine:

Calculate the amount of zinc and iodine present:

$$\frac{2.50 \text{ g Zn}}{1} \bullet \frac{1 \text{ mol Zn}}{65.39 \text{ g Zn}} = 3.82 \times 10^{-02} \text{mol Zn} \text{ and}$$

$$\frac{9.70 \text{ g I}_2}{1} \bullet \frac{1 \text{ mol I}_2}{253.8 \text{ g I}_2} = 3.82 \times 10^{-02} \text{ mol I}_2 \text{ (recall that the iodine is a diatomic}$$

specie, and would be the form of iodine reacting). Note that the amount of Zinc and I_2 combined are identical, making the formula for the compound ZnI_2 . An

alternative way of solving the problem would be to use the **atomic mass** of iodine (126.9g/ mol I) to represent 7.64×10^{-02} mol of I. The ratio of Zn:I would then be 1:2 with the formula being the same as that determined above.

General Questions on Molecules and Compounds

76. The number of water molecules in 0.05 mL of water:

$$0.05 \text{ mL water} \bullet \frac{1 \text{ cm}^3}{1 \text{ mL}} \bullet \frac{1.00 \text{ g water}}{1 \text{ cm}^3 \text{ water}} \bullet \frac{1 \text{ mol water}}{18.02 \text{ g water}} \bullet \frac{6.02 \times 10^{23} \text{ molecules}}{1 \text{ mol}}$$

$= 1.67 \times 10^{21}$ molecules of water (2×10^{21} — to 1 sf)

78. Molar mass and mass percent of the elements in $Cu(NH_3)_4SO_4 \bullet H_2O$:

Molar Mass: (1)(Cu) + (4)(N) + 12(H) + (1)(S) + (4)(O) + (2)(H) + (1)(O).

Combining the hydrogens and oxygen from water with the compound:

(1)(Cu) +   (4)(N) +   14(H) +   (1)(S) +   (5)(O) =

(1)(63.546) + (4)(14.0067) + 14(1.0079) + (1)(32.066) + (5)(15.9994) = 245.72 g/mol

The mass percents are:

Cu: (63.546/245.75) x 100 =   25.86% Cu

N: (56.027/245.75) x 100 =   22.80% N

H: (14.111/245.75) x 100 =   5.742% H

S: (32.066/245.75) x 100 =   13.05% S

O: (79.997/245.75) x 100 =   32.55% O

The mass of Copper and of Water in 10.5 g of the compound:

For Copper: $\dfrac{10.5 \text{ g compound}}{1} \bullet \dfrac{25.86 \text{ g Cu}}{100.00 \text{ g compound}} = 2.72 \text{ g Cu}$

For Water: $\dfrac{10.5 \text{ g compound}}{1} \bullet \dfrac{18.02 \text{ g H}_2\text{O}}{245.72 \text{ g compound}} = 0.770 \text{ g H}_2\text{O}$

80. Mass of Cr_2O_3 to produce 850 kg of Cr:

One can write an equation to describe this process: $Cr_2O_3 + C \rightarrow 2Cr + 3/2\ O_2$

However, one can also recognize that ALL of the Cr produced is coming from the oxide

One can answer the question by asking about the weight percent of chromium in the oxide:

The molar mass of the oxide is:151.99. Of that mass chromium represents (2x52.00) parts.

The percent of Cr in the oxide is then 104.00/151.99 or 68.43%.

The mass of the oxide needed to produce 850 kg is then:

$$\frac{850 \text{ kg Cr}}{1} \bullet \frac{151.99 \text{ g Cr}_2\text{O}_3}{104.00 \text{ g Cr}} = 1,242 \text{ kg Cr}_2\text{O}_3$$

Given that the answer is limited to 2sf (850 kg), we report 1200 kg Cr_2O_3.

82. The empirical formula of malic acid, if the ratio is: $C_1H_{1.50}O_{1.25}$

Since we prefer all our subscripts to be integers, we ask what "multiplier" we can use to convert each of these subscripts to integers **while** retaining the ratio of C:H:O that we're given.   Multiplying each subscript by 4 (we need to convert the 0.25 to an integer) gives a ratio of $C_4H_6O_5$.

84. The empirical formula for iron oxalate:

0.109 g of the compound contain 38.82% iron or (0.3882 x 0.109 = 0.0423) g iron

The law of mass action tells us that the rest is oxalate:

(0.109 g compound - 0.0423 g iron) = 0.0667 g oxalate ($C_2O_4{}^{2-}$)

Convert these masses into moles of iron and oxalate:

$$\frac{0.0423 \text{ g Fe}}{55.847 \text{ g/mol}} = 7.57 \times 10^{-4} \text{ moles of iron}$$

$$\frac{0.0667 \text{ g oxalate}}{88.02 \text{ g/mol}} = 7.57 \times 10^{-4} \text{ moles of } C_2O_4{}^{2-}$$

The ratio of oxalate to iron is: 1:1 and the empirical formula for the compound is FeC_2O_4

86. For ephedrine:

   (a) The molecular formula and molar mass:

   $C_{10}H_{15}NO$ with a corresponding molar mass of: 165.23

   (b) The weight percent of Carbon is: $[(10 \times 12.011)/165.23] \times 100$ or 72.69% C

   (c) The amount of ephedrine in 0.125 g sample:

   $$\frac{0.125 \text{ g ephedrine}}{1} \cdot \frac{1 \text{ mol ephedrine}}{165.2 \text{ g ephedrine}} = 7.57 \times 10^{-4} \text{ mol}$$

   (d) Molecules of ephedrine in 0.125 g sample:

   7.57×10^{-4} mol $\cdot$ 6.022×10^{23} molecules/mol $= 4.56 \times 10^{20}$ molecules

   How many Carbon atoms?

   Given that there are 10C atoms per molecule of ephedrine, the number of C atoms is then

   10 times the number of ephedrine molecules or 4.56×10^{21} C atoms

88. Ionic compounds; formulas and names

   (c) Li_2S lithium sulfide

   (d) In_2O_3 indium oxide

   (g) CaF_2 calcium fluoride

   Pair a & b & f: consist of two non-metals — a covalent compound is anticipated

   Pair e : Argon doesn't typically form ionic compounds

90. Formulas for compounds; identify the ionic compounds

(a)	sodium hypochlorite	NaClO	ionic
(b)	boron triiodide	BI_3	
(c)	aluminum perchlorate	$Al(ClO_4)_3$	ionic
(d)	calcium acetate	$Ca(CH_3CO_2)_2$	ionic
(e)	potassium permanganate	$KMnO_4$	ionic
(f)	ammonium sulfite	$(NH_4)_2SO_3$	ionic
(g)	potassium dihydrogen phosphate	KH_2PO_4	ionic
(h)	disulfur dichloride	S_2Cl_2	
(i)	chlorine trifluoride	ClF_3	
(j)	phosphorus trifluoride	PF_3	

92. Empirical and Molecular formula of Azulene:

Given the information that azulene is a hydrocarbon, if it is 93.71 % C, it is also (100.00 - 93.71) or 6.29 % H.

In a 100.00 g sample of azulene there are

$$93.71 \text{ g C} \cdot \frac{1 \text{ mol C}}{12.011 \text{ g C}} = 7.802 \text{ mol C} \quad \text{and}$$

$$6.29 \text{ g H} \cdot \frac{1 \text{ mol H}}{1.0079 \text{ g H}} = 6.241 \text{ mol H}$$

The ratio of C to H atoms is: 1.25 mol C : 1 mol H   or a ratio of 5 mol C: 4 mol H (C_5H_4).

The mass of such an empirical formula is ≈ 64. Given that the molar mass is ~128 g/mol, the molecular formula for azulene is $C_{10}H_8$.

94. Molecular formula of cadaverine:

Calculate the amount of each element in the compound (assuming that you have 100 g)

$$58.77 \text{ g C} \cdot \frac{1 \text{ mol C}}{12.0115 \text{ g C}} = 4.893 \text{ mol C}$$

$$13.81 \text{ g H} \cdot \frac{1 \text{ mol H}}{1.0079 \text{ g H}} = 13.70 \text{ mol H}$$

$$27.40 \text{ g N} \cdot \frac{1 \text{ mol N}}{14.0067 \text{ g N}} = 1.956 \text{ mol N}$$

The ratio of C:H:N can be found by dividing each by the smallest amount (1.956): to give $C_{2.50}H_7N_1$ and converting each subscript to an integer (multiplying by 2) $C_5H_{14}N_2$. The weight of this "empirical formula" would be approximately 102, hence the molecular formula is also $C_5H_{14}N_2$.

96. The empirical formula for MMT:

Moles of each atom present in 100. g of MMT:

$$49.5 \text{ g C} \cdot \frac{1 \text{ mol C}}{12.0115 \text{ g C}} = 4.13 \text{ mol C}$$

$$3.2 \text{ g H} \cdot \frac{1 \text{ mol H}}{1.0079 \text{ g H}} = 3.2 \text{ mol H}$$

$$22.0 \text{ g O} \cdot \frac{1 \text{ mol O}}{15.9994 \text{ g O}} = 1.38 \text{ mol O}$$

$$25.2 \text{ g Mn} \cdot \frac{1 \text{ mol Mn}}{54.938 \text{ g Mn}} = 0.459 \text{ mol Mn}$$

The ratio of C:H:O:Mn can be found by dividing each by the smallest amount (0.459): to give $MnC_9H_7O_3$.

98. I_2 + Cl_2 $\rightarrow$ I_xCl_y

    0.678 g   (1.246 - 0.678)   1.246 g

Calculate the ratio of I : Cl atoms

$$0.678 \text{ g I} \cdot \frac{1 \text{ mol I}}{126.9 \text{ g I}} = 5.34 \times 10^{-3} \text{ mol I atoms}$$

$$0.568 \text{ g Cl} \cdot \frac{1 \text{ mol Cl}}{35.45 \text{ g Cl}} = 1.6 \times 10^{-2} \text{ mol Cl atoms}$$

The ratio of Cl : I is : $\dfrac{1.6 \times 10^{-2} \text{ mol Cl atoms}}{5.34 \times 10^{-3} \text{ mol I atoms}} = 3.00 \dfrac{\text{Cl atoms}}{\text{I atoms}}$

The empirical formula is ICl_3 (FW = 233.3)

Given that the molar mass of I_xCl_y was 467 g/mol, we can calculate the number of empirical formulas per mole:

$$\frac{467 \text{ g/mol}}{233.3 \text{ g/empirical formula}} = 2 \frac{\text{empirical formula}}{\text{mol}}$$

for a molecular formula of I_2Cl_6.

100. Mass of Fe in 15.8 kg of FeS_2 :

$$\% \text{ Fe in } FeS_2 = \frac{55.85 \text{ g Fe}}{119.97 \text{ g FeS}_2} \times 100 = 46.55 \% \text{ Fe}$$

and in 15.8 kg FeS_2

$$15.8 \text{ kg FeS}_2 \cdot \frac{46.55 \text{ kg Fe}}{100.00 \text{ kg FeS}_2} = 7.35 \text{ kg Fe}$$

102. Statements about 57.1 g of octane(C_8H_{18}):

(a) **True** 57.1 g = 0.500 mol octane. The molar mass of C_8H_{18} is 114.2, so 57.1g of octane **does** correspond to 0.500 mol.

(b) **True** Compound is 84.1% C by weight. Eight C atoms correspond to 96.088 g. The percent C is then (96.088/114.2) x 100 or 84.1%.

(c) **True**. The empirical formula **is C_4H_9**. This ratio is the smallest **integral** ratio of C:H with the ratio of 4 carbon atoms to 9 hydrogen atoms.

(d) **False** 57.1g of octane **does not** contain 28.0 g of hydrogen atoms. Since the compound is 84.1%C, it is also (100-84) or 16%H. This percentage of 57.1 g would correspond to 9.08 g.

31

104. Mass of Bi in two tablets of Pepto-Bismol™ ($C_{21}H_{15}Bi_3O_{12}$):

$$\frac{2 \text{ tablets}}{1} \cdot \frac{300. \times 10^{-3} \text{ g } C_{21}H_{15}Bi_3O_{12}}{1 \text{ tablet}} \cdot \frac{1 \text{ mol } C_{21}H_{15}Bi_3O_{12}}{1086 \text{ g } C_2H_{15}Bi_3O_{12}} \cdot \frac{3 \text{ mol Bi}}{1 \text{ mol } C_2H_{15}Bi_3O_{12}} \cdot \frac{208.98 \text{ g Bi}}{1 \text{ mol Bi}}$$

$$= 0.346 \text{ g Bi}$$

106. What is the molar mass of ECl_4 and the identity of E?

2.50 mol of ECl_4 has a mass of 385 grams. The molar mass of ECl_4 would be:

$$\frac{385 \text{ g } ECl_4}{2.50 \text{ mol } ECl_4} = 154 \text{ g/mol } ECl_4.$$

Since the molar mass is 154, and we know that there are 4 chlorine atoms per mole of the compound, we can subtract the mass of 4 chlorine atoms to determine the mass of E. 154 - 4(35.5) = 12. The element with an atomic mass of 12 g/mol is **carbon**.

108. Of the ions Na^+, Mg^{2+}, Al^{3+}, which has the strongest attraction for water molecules in the vicinity? We can calculate the proportional charge density by dividing the ionic charges by the ionic radii:

$$Na^+ \qquad\qquad Mg^{2+} \qquad\qquad Al^{3+}$$

$$\frac{1}{116 \text{ pm}} \qquad\qquad \frac{2}{86 \text{ pm}} \qquad\qquad \frac{3}{68 \text{ pm}}$$

The charge density of Al^{3+} is greatest, and so this cation would have the greatest attraction to the dipole, H_2O.

110. Using the student data, let's calculate the number of moles of $CaCl_2$ and mole of H_2O:

$$0.739 \text{ g } CaCl_2 \cdot \frac{1 \text{ mol } CaCl_2}{111.0 \text{ g } CaCl_2} = 0.00666 \text{ mol } CaCl_2$$

$$(0.832 \text{ g} - 0.739 \text{ g}) \text{ or } 0.093 \text{ g } H_2O \cdot \frac{1 \text{ mol } H_2O}{18.02 \text{ g } H_2O} = 0.0052 \text{ mol } H_2O$$

The ratio of mol water/mol calcium chloride is then $\dfrac{0.0052 \text{ mol } H_2O}{0.00666 \text{ mol } CaCl_2} = 0.78$

This is a sure sign that they should **heat the crucible again, and then reweigh it**.

112. Calculate:

(a) moles of nickel—found by density **once** the volume of foil is calculated.

$$V = 1.25 \text{ cm} \times 1.25 \text{ cm} \times 0.0550 \text{ cm} = 8.59 \times 10^{-2} \text{ cm}^3$$

$$\text{Mass} = \frac{8.908 \text{ g}}{1 \text{ cm}^3} \cdot 8.59 \times 10^{-2} \text{ cm}^3 = 0.766 \text{ g Ni}$$

$$0.766 \text{ g Ni} \cdot \frac{1 \text{ mol Ni}}{58.69 \text{ g Ni}} = 1.30 \times 10^{-2} \text{ mol Ni}$$

(b) Formula for the fluoride salt:

Mass F = (1.261 g salt - 0.766 g Ni) = 0.495 g F

Moles F = 0.495 g F $\cdot$ $\dfrac{1 \text{ mol F}}{19.00 \text{ g F}}$ = 2.60 x 10^{-2} mol F,

so 1.30 x 10^{-2} mol Ni combines with 2.60 x 10^{-2} mol F , indicating a formula of NiF_2

(c) Name: Nickel(II) fluoride

Using Electronic Resources

114. The hydrated salt loses water upon heating. The anhydrous cobalt salt is a deep blue, and therefore visible.

116. The simulation shows that the forces between two ions are proportional to the charges of the ions and inversely proportional to the distance between them.

Chapter 4
Chemical Equations and Stoichiometry

Reviewing Important Concepts

2. A balanced equation for the production of ammonia:

    The formula for ammonia, NH_3, indicates **four** atoms per molecule—one atom of nitrogen and three atoms of hydrogen. Given that this substance is being formed from two diatomic substances—that is substances that have *two atoms per molecule* , we'll definitely need **odd** numbers of molecules of elemental nitrogen and elemental hydrogen to balance the equation. Answer (c) provides such a solution:

    $$N_2 (g) + 3 H_2 (g) \rightarrow 2 NH_3 (g)$$

    One molecule of nitrogen (contributes 2 nitrogen atoms), 3 molecules of hydrogen (contribute 6 hydrogen atoms), and two molecules of ammonia contain a total of 2 nitrogen atoms and 6 hydrogen atoms.

4. Stoichiometric factor needed for the relationship of **ammonia** to **nitrogen**:

    Since ammonia is our *desired substance* and *nitrogen* is our *given substance* , the ratio we need would involve those two—with ammonia in the numerator, viz. $\dfrac{2 \text{ moles } NH_3}{1 \text{ mol } N_2}$

    hence if we begin with **3 moles of nitrogen, we can calculate** :

    $$3 \text{ moles } N_2 \bullet \frac{2 \text{ moles } NH_3}{1 \text{ mol } N_2} = 6 \text{ moles } NH_3$$

6. Limiting Reactant and # Moles of Fe expected from 25 mol of Fe_2O_3? From 65 mol of CO?

    To identify the limiting reactant, determine the stochiometric ratio of reactants (from the balanced equation—the coefficients of the respective reactants): $\dfrac{3 \text{ mol CO}}{1 \text{ mol } Fe_2O_3}$

    Use the amounts provided to determine the "actually available" ratio:

    $$\frac{65 \text{ mol CO}}{25 \text{ mol } Fe_2O_3} = \frac{2.6 \text{ mol CO}}{1 \text{ mol } Fe_2O_3}$$

    As the ratio of CO/ Fe_2O_3 is **smaller** than that required by the balanced equation, **CO is the limiting reactant.**

    Amount of Fe expected from 25 mol of Fe_2O_3 :

    The amount of Fe in Fe_2O_3 and in Fe is in a 1:2 ratio—for each mole of Fe_2O_3 one can expect 2 moles of Fe. With 25 mol of Fe_2O_3 we would expect (2 • 25) or 50 mol Fe.

Amount of Fe expected from 65 mol of CO:

The ratio of CO to Fe (using the coefficients of the balanced equation) is a

3 mol CO:2 mol Fe. Said another way, For 1.5 mol of CO, we anticipate 1 mol of Fe.

From 65 mol of CO we anticipate: $65 \text{ mol CO} \cdot \dfrac{1 \text{ mol Fe}}{1.5 \text{ mol CO}} = 43 \text{ mol Fe}$

NOTE: Based on the past two calculations, we anticipate 50 mol of Fe (based upon our iron oxide calculation) and 43 mol of Fe (based upon the CO calculation)—further evidence that CO **limits** the amount of product.

Practicing Skills

Balancing Equations

Balancing equations can be a matter of "running in circles" if a reasonable methodology is not employed. While there isn't one "right place" to begin, generally you will suffer fewer complications if you begin the balancing process using a substance that contains the **greatest number** of elements **or the largest subscript** values. Noting that you must have at least that many atoms of each element involved, coefficients can be used to increase the "atomic inventory". In the next few questions, you will see one **emboldened** substance in each equation. This emboldened substance is the one that I judge to be a "good" starting place. One last hint--modify the coefficients of uncombined elements, i.e. those not in compounds, <u>after</u> you modify the coefficients for compounds containing those elements -- <u>not before</u>!

8.   (a) $4 \text{ Cr (s)} + 3 \text{ O}_2 \text{ (g)} \rightarrow 2 \textbf{ Cr}_2\textbf{O}_3 \text{ (s)}$
       1. Note the need for <u>at least</u> 2 Cr and 3 O atoms.
       2. Oxygen is diatomic -- we'll need an <u>even</u> number of oxygen atoms, so try : $2 \text{ Cr}_2\text{O}_3$.
       3. 3 O_2 would give 6 O atoms on both sides of the equation.
       4. 4 Cr would give 4 Cr atoms on both sides of the equation.
     (b) $\text{Cu}_2\text{S (s)} + \text{O}_2 \text{ (g)} \rightarrow 2 \text{ Cu(s)} + \textbf{SO}_2 \text{ (g)}$
       1. A minimum of 2 O in SO_2 is required, and is provided with one molecule of elemental oxygen.
       2. 2 Cu atoms (on the right) indicates 2 Cu (on the left).
     (c) $\textbf{C}_6\textbf{H}_5\textbf{CH}_3 \text{ (}\ell\text{)} + 9 \text{ O}_2 \text{ (g)} \rightarrow 4 \text{ H}_2\text{O (}\ell\text{)} + 7 \text{ CO}_2 \text{ (g)}$
       1. A minimum of 7 C and 8 H is required.
       2. 7 CO_2 furnishes 7 C and $4 \text{ H}_2\text{O}$ furnishes 8 H atoms.
       3. $4 \text{ H}_2\text{O}$ and 7 CO_2 furnish a total of 18 O atoms, making the coefficient of $\text{O}_2 = 9$.

10. Balance and name the reactants and products:

(a) **Fe$_2$O$_3$** (s) + 3 Mg(s) → 3 MgO (s) + 2 Fe (s)

    1. Note the need for <u>at least</u> 2 Fe and 3 O atoms.

    2. 2 Fe atoms would provide the proper iron atom inventory.

    3. 3 MgO would give 3 O atoms on both sides of the equation.

    4. 3 Mg would give 3 Mg atoms on both sides of the equation.

Reactants: iron(III) oxide and magnesium

Products: magnesium oxide and iron

(b) **AlCl$_3$** (s) + 3 H$_2$O (ℓ) → Al(OH)$_3$ (s) + 3 HCl (aq)

    1. Note the need for <u>at least</u> 1 Al and 3 Cl atoms.

    2. 3 HCl molecules would provide the proper Cl atom inventory.

    3. 3 H atoms (from HCl) and 3 H atoms (from Al(OH)$_3$) would give 6 H atoms

        needed on both sides of the equation—so a coefficient of 3 for water is

        needed to provide that balance.

Reactants: aluminum chloride and water

Products: aluminum hydroxide and hydrochloric acid.

    {Recall that *aqueous* HCl is a "hydro"-chlor-"ic" acid.

    Can you think of at least one other example ? Hint: Examine Group 7A !}

(c) 2 NaNO$_3$ (s) + H$_2$SO$_4$ (ℓ) → **Na$_2$SO$_4$** (s) + 2 HNO$_3$ (ℓ)

    1. Note the need for <u>at least</u> 2 Na and 1 S and 4 O atoms.

    2. 2 NaNO$_3$ will provide the proper Na atom inventory.

    3. The coefficient of 2 in front of NaNO$_3$ requires a coefficient of 2 for HNO$_3$

        —providing a balance for N atoms.

    4. The implied coefficient of 1 for Na$_2$SO$_4$ suggests a similar coefficient for

        H$_2$SO$_4$ —to balance the S atom inventory.

    5. O atom inventory is done "automatically" when we balanced N and S inventories.

Reactants: sodium nitrate and sulfuric acid

Products: sodium sulfate and nitric acid

    [....although nitric acid typically exists as an aqueous solution.]

(d) **NiCO$_3$** (s) + 2 HNO$_3$ (aq) → Ni(NO$_3$)$_2$ (aq) + CO$_2$ (g) + H$_2$O (ℓ)

    1. Note the need for <u>at least</u> 1 Ni atom on both sides. This inventory will mandate 2

        NO$_3$ groups on the right —and also on the left. Since these come from

        HNO$_3$ molecules, we'll need 2 HNO$_3$ on the left.

2. The 2 H from the acid and the CO_3 from nickel carbonate, provide 2H, 1 C and 3 O atoms. 1 H_2O takes care of the 2H, and **one** of the O atoms, 1 CO_2 consumes the 1 C and the remaining 2 O atoms.

Reactants: nickel(II) carbonate and nitric acid

Products: nickel(II) nitrate, carbon dioxide, and water

Mass Relationships in Chemical Reactions: Basic Stoichiometry

12. Moles of oxygen needed to reach with 6.0 mol of Al:

$$4\ Al\ (s) + 3\ O_2\ (g) \rightarrow 2\ Al_2O_3\ (s)$$

$$6.0\ mol\ Al \cdot \frac{3\ mol\ O_2}{4\ mol\ Al} = 4.5\ mol\ O_2$$

What mass of Al_2O_3 should be produced ?

$$6.0\ mol\ Al \cdot \frac{2\ mol\ Al_2O_3}{4\ mol\ Al} \cdot \frac{102\ g\ Al_2O_3}{1\ mol\ Al_2O_3} = 310\ g\ Al_2O_3\ (\ to\ 2\ sf)$$

14. Quantity of Br_2 to react with 2.56 g of Al:

According to the balanced equation 2 mol of Al react with 3 mol of Br_2 .

Calculate the # of moles of Al, then multiply by 3/2 to obtain # mol of Br_2 required.

$$2.56\ g\ Al \cdot \frac{1\ mol\ Al}{26.98\ g\ Al} \cdot \frac{3\ mol\ Br_2}{2\ mol\ Al} \cdot \frac{159.8\ g\ Br_2}{1\ mol\ Br_2} = 22.7\ g\ Br_2$$

Mass of Al_2Br_6 expected:

This could be solved in several ways. The simplest is to recognize that—according to the Law of Conservation of Matter, mass is conserved in a reaction. If 22.7 g of bromine react with exactly 2.56 g of aluminum, the total products would also have a mass of (22.7 g+ 2.56 g) or 25.3 g Al_2Br_6.

16. The reaction of iron with oxygen to given iron(III) oxide:

(a) The balanced equation for the reaction:

$$4\ Fe\ (s) + 3\ O_2\ (g) \rightarrow 2\ Fe_2O_3$$

(b) Mass of Fe_2O_3 produced when 2.68 g Fe react:

$$2.68\ g\ Fe \cdot \frac{1\ mol\ Fe}{55.85\ g\ Fe} \cdot \frac{2\ mol\ Fe_2O_3}{4\ mol\ Fe} \cdot \frac{159.7\ g\ Fe_2O_3}{1\ mol\ Fe_2O_3} = 3.83\ g\ Fe_2O_3$$

(c) Mass of oxygen required:

This could be solved in several ways. The simplest is to recognize that—according to the Law of Conservation of Matter, mass is conserved in a reaction. If 3.83 g Fe_2O_3 are produced when 2.68 g of iron react, the total oxygen required would have a mass of (3.83 g - 2.68 g) or 1.15 g O_2.

18. Removal of SO_2 by $CaCO_3$:

(a) Mass of $CaCO_3$ is required to remove 155 g of SO_2:

$$155 \text{ g } SO_2 \cdot \frac{1 \text{ mol } SO_2}{64.06 \text{ g } SO_2} \cdot \frac{2 \text{ mol } CaCO_3}{2 \text{ mol } SO_2} \cdot \frac{100.1 \text{ g } CaCO_3}{1 \text{ mol } CaCO_3} = 242 \text{ g } CaCO_3$$

(b) Mass of $CaSO_4$ when 155 g of SO_2 is consumed completely:

$$155 \text{ g } SO_2 \cdot \frac{1 \text{ mol } SO_2}{64.06 \text{ g } SO_2} \cdot \frac{2 \text{ mol } CaSO_4}{2 \text{ mol } SO_2} \cdot \frac{136.1 \text{ g } CaSO_4}{1 \text{ mol } CaSO_4} = 329 \text{ g } CaSO_4$$

Limiting Reactants

20. Identify the limiting reactant when 1.6 mol of S_8 and 35 mol of F_2 react:

The balanced equation is: $S_8 + 24 F_2 \rightarrow 8 SF_6$

Determine the required ratio (from the balanced equation): $\dfrac{24 \text{ mol } F_2}{1 \text{ mol } S_8}$

Determine the "available ratio": $\dfrac{35 \text{ mol } F_2}{1.6 \text{ mol } S_8} = \dfrac{21.9 \text{ mol } F_2}{1 \text{ mol } S_8}$

Since the available ratio of fluorine is **less than** the required ratio, **fluorine is the limiting reactant.**

22. For the reaction of methane with water:

$$CH_4 (g) + H_2O(g) \rightarrow CO (g) + 3 H_2(g)$$

(a) Limiting reagent when 995 g of CH_4 react with 2510 g of water:

$$995 \text{ g } CH_4 \cdot \frac{1 \text{ mol } CH_4}{16.04 \text{ g } CH_4} = 62.0 \text{ mol } CH_4$$

$$2510 \text{ g } H_2O \cdot \frac{1 \text{ mol } H_2O}{18.02 \text{ g } H_2O} = 139.3 \text{ mol } H_2O$$

Determine required ratio of methane to water: $\dfrac{1 \text{ mol } H_2O}{1 \text{ mol } CH_4}$

Determine the "available" ratio: $\dfrac{139.3 \text{ mol } H_2O}{62.0 \text{ mol } CH_4} = 2.25$

The "available" ratio indicates that **methane is the limiting reactant.**

(b) Maximum mass of H_2 possible:

$$62.0 \text{ mol } CH_4 \cdot \frac{3 \text{ mol } H_2}{1 \text{ mol } CH_4} \cdot \frac{2.016 \text{ g } H_2}{1 \text{ mol } H_2} = 375 \text{ g } H_2$$

(c) Mass of water remaining:

Since 1 mol of methane reacts with 1 mol of water, we know that 62.0 mol of CH_4 reacts with 62.0 mol of water. The amount of water remaining is: $(139.3 - 62.0) = 77.3$ mol H_2O. The mass of this amount of water is:

$$77.2 \text{ mol } H_2O \cdot \frac{18.02 \text{ g } H_2O}{1 \text{ mol } H_2O} = 1,393 \text{ g } H_2O \text{ or } 1390g \text{ (to 3 sf)}$$

24. The limiting reagent can be determined by calculating the mole-available and mole-needed ratios for the equation:

$$CaO \text{ (s)} + 2 NH_4Cl \text{ (s)} \rightarrow 2 NH_3 \text{ (g)} + H_2O \text{ (g)} + CaCl_2 \text{ (s)}$$

Calculate the moles of CaO and of NH_4Cl :

$$112 \text{ g CaO} \cdot \frac{1 \text{ mol CaO}}{56.08 \text{ g CaO}} = 2.00 \text{ mol CaO}$$

$$224 \text{ g } NH_4Cl \cdot \frac{1 \text{ mol } NH_4Cl}{53.49 \text{ g } NH_4Cl} = 4.19 \text{ mol } NH_4Cl$$

moles-required ratio: $\dfrac{2 \text{ mol } NH_4Cl}{1 \text{ mol CaO}}$

moles-available ratio: $\dfrac{4.19 \text{ mol } NH_4Cl}{2.00 \text{ mol CaO}} = \dfrac{2.10 \text{ mol } NH_4Cl}{1.00 \text{ mol CaO}}$

(a) CaO is the limiting reagent, and will determine the maximum amount of products obtainable:

$$112 \text{ g CaO} \cdot \frac{1 \text{ mol CaO}}{56.08 \text{ g CaO}} \cdot \frac{2 \text{ mol } NH_3}{1 \text{ mol CaO}} \cdot \frac{17.03g \text{ } NH_3}{1 \text{ mol } NH_3} = 68.0 \text{ g } NH_3$$

(b) The balanced equation shows that for each mole of CaO, 2 moles of NH_4Cl are required. So 2.00 mol of CaO would require 4.00 mol of NH_4Cl, leaving $(4.19 - 4.00)$ 0.19 mol of NH_4Cl in excess. This number of moles would have a mass of:

$$0.19 \text{ mol } NH_4Cl \cdot \frac{53.49 \text{ g } NH_4Cl}{1 \text{ mol } NH_4Cl} = 10. \text{ g } NH_4Cl$$

Percent Yield

26. Percent yield of CH_3OH:

$$\frac{\text{actual}}{\text{theoretical}} = \frac{332 \text{ g } CH_3OH}{407 \text{ g } CH_3OH} = 81.6\% \text{ yield}$$

28. In the formation of $Cu(NH_3)_4SO_4$:

    (a) The theoretical yield of $Cu(NH_3)_4SO_4$ from 10.0 g of $CuSO_4$:

$$10.0 \text{ g CuSO}_4 \cdot \frac{1 \text{ mol CuSO}_4}{159.6 \text{ g CuSO}_4} \cdot \frac{1 \text{ mol Cu(NH}_3)_4\text{SO}_4}{1 \text{ mol CuSO}_4} \cdot \frac{227.7 \text{ g Cu(NH}_3)_4\text{SO}_4}{1 \text{ mol Cu(NH}_3)_4\text{SO}_4} = 14.3 \text{ g Cu(NH}_3)_4\text{SO}_4$$

    (b) Percentage yield of the compound:
$$\frac{12.6 \text{ g compound}}{14.3 \text{ g compound}} \cdot 100 = 88.3\%$$

Analysis of Mixtures

30. Mass percent of $CuSO_4 \bullet 5 H_2O$ in the mixture:

    Mass of H_2O = 1.245 g - 0.832 g = 0.413 g H_2O

    Since this water was a part of the hydrated salt, let's calculate the mass of that salt present:
In 1 mol of $CuSO_4 \bullet 5 H_2O$ there are 90.10 g H_2O and 159.61 g $CuSO_4$ or 249.71 g
$CuSO_4 \bullet 5 H_2O$. These masses correspond to the molar masses of anhydrous $CuSO_4$ and
5• mol H_2O.

    So:

$$0.413 \text{ g H}_2\text{O} \cdot \frac{249.71 \text{ g CuSO}_4 \cdot 5\text{H}_2\text{O}}{90.10 \text{ gH}_2\text{O}} = 1.14 \text{ g hydrated salt}$$

$$\% \text{ hydrated salt} = \frac{1.14 \text{ g hydrated salt}}{1.245 \text{ g mixture}} \text{ x } 100 = 91.9\%$$

32. Mass percent of $CaCO_3$ in a limestone sample:

    Note that the ratio of carbon dioxide to calcium carbonate is 1:1 (from balanced equation).
Calculate the amount of carbon dioxide represented by 0.558 g CO_2

$$0.558 \text{ g CO}_2 \cdot \frac{1 \text{ mol CO}_2}{44.01 \text{ g CO}_2} = 0.01268 \text{ mol CO}_2$$

    Since there would have been 0.01268 mol of $CaCO_3$, the mass is :

$$0.01268 \text{ mol CaCO}_3 \cdot \frac{100.1 \text{ g CaCO}_3}{1 \text{ mol CaCO}_3} = 1.269 \text{ g CaCO}_3.$$

    The percent of $CaCO_3$ in the sample is:
$$\frac{1.269 \text{ g CaCO}_3}{1.506 \text{ g sample}} \cdot 100 = 84.3\%$$

34. Mass percent of Tl_2SO_4 in 10.20 g sample:

Begin by balancing the equation:
$$Tl_2SO_4(aq) + 2\ NaI(aq) \rightarrow 2\ TlI(s) + Na_2SO_4(aq)$$
This equation tells us that for each mol of Tl_2SO_4, we expect **two** moles of TlI.

Determine the amount of TlI: $0.1964\ g\ TlI\ \bullet\ \dfrac{1\ mol\ TlI}{331.29\ g\ TlI} = 5.928 \times 10^{-4}\ mol\ TlI$

Use the ratio described by the balanced equation:

$5.928 \times 10^{-4}\ mol\ TlI\ \bullet\ \dfrac{1\ mol\ Tl_2SO_4}{2\ mol\ TlI} = 2.964 \times 10^{-4}\ mol\ Tl_2SO_4$

Using the molar mass of thallium(I) sulfate gives the mass of the sulfate present in the

sample: $2.964 \times 10^{-4}\ mol\ Tl_2SO_4\ \bullet\ \dfrac{504.83\ g\ Tl_2SO_4}{1\ mol\ Tl_2SO_4} = 0.1496\ g\ Tl_2SO_4$

The mass percent of Tl_2SO_4 in the sample is $\dfrac{0.1496\ g\ Tl_2SO_4}{10.20\ g\ sample} \bullet 100 = 1.467\%$

Determination of Empirical and Molecular Formulas

36. The basic equation is:

$$C_xH_y + O_2 \rightarrow x\ CO_2 + \frac{y}{2}\ H_2O$$

Without balancing the equation, one can see that all the C in CO_2 comes from the styrene as does all the H in H_2O. Let's use the percentage of C in CO_2 to determine the mass of C in styrene, and the percentage of H in H_2O to provide the mass of H in styrene.

$1.481\ g\ CO_2\ \bullet\ \dfrac{12.01\ g\ C}{44.01\ g\ CO_2} = 0.404\ g\ C$

Similarly:

$0.303\ g\ H_2O\ \bullet\ \dfrac{2.02\ g\ H}{18.02\ g\ H_2O} = 0.0340\ g\ H$

Alternatively, the mass of H could be determined by subtracting the mass of C from the 0.438 g styrene

Mass H = 0.438 g styrene - 0.404 g C

Establish the ratio of C atoms to H atoms

$0.0340\ g\ H\ \bullet\ \dfrac{1\ mol\ H}{1.008\ g\ H} = 0.0337\ mol\ H$

$0.404\ g\ C\ \bullet\ \dfrac{1\ mol\ C}{12.011\ g\ C} = 0.0336\ mol\ C$

This number of H and C atoms indicates an empirical formula for styrene of 1:1 or CH.

38. Combustion of Cyclopentane:

 The approach is similar to that for study question 36.

 The general equation is $C_xH_y + O_2 \rightarrow x\, CO_2 + \frac{y}{2}\, H_2O$

 All the carbon from cyclopentane eventually resides in the CO_2 produced.

 Calculate the amount of C in cyclopentane:

 $$0.300 \text{ g CO}_2 \cdot \frac{1 \text{ mol CO}_2}{44.01 \text{ g CO}_2} \cdot \frac{1 \text{ mol C}}{1 \text{ mol CO}_2} = 0.00682 \text{ mol C}$$

 Similarly the amount of H in cyclopentane:

 $$0.123 \text{ g H}_2\text{O} \cdot \frac{1 \text{ mol H}_2\text{O}}{18.01 \text{ g H}_2\text{O}} \cdot \frac{2 \text{ mol H}}{1 \text{ mol H}_2\text{O}} = 0.0137 \text{ mol H}$$

 (a) The empirical formula is: $\dfrac{0.0137 \text{ mol H}}{0.00682 \text{ mol C}} = \dfrac{2.00 \text{ mol H}}{1.00 \text{ mol C}}$ or CH_2

 (b) If the molar mass is 70.1 g/mol, the molecular formula is:

 Since the molecular formula represents some **multiple** of the empirical formula,

 calculate the "empirical formula mass".

 For CH_2 that mass is 14.03 g/empirical formula [(1 x 12.0115) + (2 x 1.0079)]

 Calculate the # of empirical formulas in a molecular formula:

 $$\frac{70.1 \text{ g/molecular formula}}{14.03 \text{ g/empirical formula}} = 5 \text{ , giving a molecular formula of } C_5H_{10}.$$

40. Calculate the mass of K and O in the compound, K_xO_y:

 The compound has 0.233 g of K and a total mass of 0.328 g.

 This means that the compound has (0.328 − 0.233) = 0.095 g O.

 Calculate the amount of K and O in the compound:

 $$0.233 \text{ g K} \cdot \frac{1 \text{ mol K}}{39.10 \text{ g K}} = 0.00596 \text{ mol K}$$

 $$0.095 \text{ g O} \cdot \frac{1 \text{ mol O}}{15.9994 \text{ g O}} = 0.00594 \text{ mol O}$$

 The ratio of K:O is 1:1 and the empirical formula is then KO.

42. Formula of the carbonyl compound formed with nickel:

$$Ni_x(CO)_y(s) + O_2 \rightarrow x\, NiO\,(s) + y\, CO_2(g)$$

Given the mass of NiO formed, we can calculate the mass of Ni in NiO (and also in the nickel carbonyl compound)

Quantity of Ni :

$$0.0426\ g\ NiO \bullet \frac{1\ mol\ NiO}{74.6924\ g\ NiO} = 5.70 \times 10^{-4}\ mol\ NiO \quad \text{(and an equal number of moles}$$

of Nickel since the oxide has 1mol Ni:1mol NiO). That number of moles of Ni would have

a mass of: $5.70 \times 10^{-4}\ mol\ Ni \bullet \dfrac{58.693 g\ Ni}{1\ mol\ Ni} = 0.03347\ g\ Ni.$

Quantity of CO contained in the nickel carbonyl compound:

$$0.0973\ g\ compound - 0.03347\ g\ Ni = 0.06383\ g\ CO$$

Amount of CO contained in the compound:

$$0.06383\ g\ CO \bullet \frac{1\ mol\ CO}{28.010\ g\ CO} = 2.279 \times 10^{-3}\ mol\ CO$$

The ratio of Ni: CO is: $\dfrac{2.279 \times 10^{-3}\, molCO}{5.70 \times 10^{-4}\ mol\ Ni} = 4$ and the empirical formula is $Ni(CO)_4$.

General Questions on Stoichiometry

44. Balance:

(a) synthesis of urea:

$$CO_2(g) + 2\, NH_3(g) \rightarrow CO(NH_2)_2(s) + H_2O(\ell)$$

1. Note the need for two NH_3 in each molecule of urea, so multiply NH_3 by 2.
2. $2\, NH_3$ provides the two H atoms for a molecule of H_2O.
3. Each CO_2 provides the O atom for a molecule of H_2O.

(b) synthesis of uranium(VI) fluoride

$$UO_2(s) + 4\, HF(aq) \rightarrow UF_4(s) + H_2O(\ell)$$

$$UF_4(s) + F_2(g) \rightarrow UF_6(s)$$

1. The 4 F atoms in UF_4 requires 4 F atoms from HF.   (equation 1)
2. The H atoms in HF produce 2 molecules of H_2O.   (equation 1)
3. The 1:1 stoichiometry of $UF_6 : UF_4$ provides a simple balance.   (equation 2)

(c) synthesis of titanium metal from TiO_2:

$$TiO_2(s) + 2\, Cl_2(g) + 2\, C(s) \rightarrow TiCl_4(\ell) + 2\, CO(g)$$

$$TiCl_4(\ell) + 2\, Mg(s) \rightarrow Ti(s) + 2\, MgCl_2(s)$$

1. The O balance mandates 2 CO for each TiO_2.   (equation 1)
2. A coefficient of 2 for C provides C balance.   (equation 1)

3. The Ti balance (TiO_2 : $TiCl_4$) requires 4 Cl atoms, hence 2 Cl_2 (equation 1)

4. The Cl balance requires 2 $MgCl_2$, hence 2 Mg.   (equation 2)

46. For the reaction of methane with oxygen:

(a) the products of the reaction:

Combination of CH_4 with O_2 gives the oxide of C and the oxygen of H:

$$CO_2(g) + H_2O(g)$$

(b) the balanced equation for the reaction is:

$$CH_4(g) + 2\,O_2(g) \rightarrow CO_2(g) + 2\,H_2O(g)$$

(c) mass of oxygen , in grams, needed to completely consume the 16.04 g CH_4.

$$16.04 \text{ g } CH_4 \cdot \frac{1 \text{ mol } CH_4}{16.04 \text{ g } CH_4} \cdot \frac{2 \text{ mol } O_2}{1 \text{ mol } CH_4} \cdot \frac{31.999 \text{ g } O_2}{1 \text{ mol } O_2} = 64.00 \text{ g } O_2 \,.$$

(d) total mass of products expected:

This could be solved in several ways. Perhaps the simplest is to recognize that—according to the Law of Conservation of Matter, mass is conserved in a reaction. If 64.00 g of oxygen react with exactly 16.04 g of methane, the total products would also have a mass of (64.00+16.04) g or 80.04 g.

48. Mass of acetone available from decomposition of 125mg of acetoacetic acid(AA)

$$CH_3COCH_2CO_2H \rightarrow CH_3COCH_3 + CO_2$$

molar mass: 102.09 g/mol   58.08 g/mol

$$125 \text{ mg AA} \cdot \frac{1 \text{ mol AA}}{102.09 \text{ g AA}} \cdot \frac{1 \text{ mol acetone}}{1 \text{ mol AA}} \cdot \frac{58.08 \text{ g acetone}}{1 \text{ mol acetone}} = 71.1 \text{ mg acetone}$$

Note that it **is not necessary** to convert milligrams of acetoacetic acid into grams of acetoacetic acid, since the mass of product (acetone) is also expressed in milligrams.

50.   (a) Balanced equation: $2\,Fe(s) + 3\,Cl_2(g) \rightarrow 2\,FeCl_3(s)$

(b) 1. Mass of Cl_2 to react with 10.0 g iron:

$$10.0 \text{ g Fe} \cdot \frac{1 \text{ mol Fe}}{55.85 \text{ g Fe}} \cdot \frac{3 \text{ mol } Cl_2}{2 \text{ mol Fe}} \cdot \frac{70.91 \text{ g } Cl_2}{1 \text{ mol } Cl_2} = 19.0 \text{ g } Cl_2$$

2. Amount of $FeCl_3$ produced :

$$10.0 \text{ g Fe} \cdot \frac{1 \text{ mol Fe}}{55.85 \text{ g Fe}} \cdot \frac{2 \text{ mol } FeCl_3}{2 \text{ mol Fe}} = 0.179 \text{ mol } FeCl_3$$

$$0.179 \text{ mol } FeCl_3 \cdot \frac{162.2 \text{ g } FeCl_3}{1 \text{ mol } FeCl_3} = 29.0 \text{ g } FeCl_3$$

(c) Percent yield of FeCl3:

$$\frac{\text{Actual}}{\text{Theoretical}} \times 100 \quad \text{or} \quad \frac{18.5 \text{ g FeCl}_3}{29.0 \text{ g FeCl}_3} \times 100 = 63.7\%$$

52. The reaction of ammonia with oxygen:

$$4 \text{ NH}_3 \text{ (g)} + 5 \text{ O}_2\text{(g)} \rightarrow 6\text{H}_2\text{O(g)} + 4 \text{ NO(g)}$$

(a) Mass of water produced by reaction of 750. g of NH_3 and 750. g of O_2:

From page 4-12 of the text, we can see that O_2 is the **limiting reactant**.

$$750. \text{ g NH}_3 \cdot \frac{1 \text{ mol NH}_3}{17.02 \text{ g NH}_3} = 44.1 \text{ mol NH}_3$$

$$750. \text{ g O}_2 \cdot \frac{1 \text{ mol O}_2}{32.00 \text{ g O}_2} = 23.4 \text{ mol O}_2$$

Calculating the ratios of amounts required and amounts available:

Amounts required: $\dfrac{5 \text{ mol O}_2}{4 \text{ mol NH}_3} = \dfrac{1.25 \text{ mol O}_2}{1 \text{ mol NH}_3}$

Amounts available: $\dfrac{23.4 \text{ mol O}_2}{44.1 \text{ mol NH}_3} = \dfrac{0.531 \text{ mol O}_2}{1 \text{ mol NH}_3}$

Since the ratio of oxygen available is less than oxygen required, O_2 is the limiting reactant. Since oxygen controls the amount of products possible, we use the ratio between O_2 and H_2O to determine the maximum amount of water possible.

$$23.4 \text{ mol O}_2 \cdot \frac{6 \text{ mol H}_2\text{O}}{5 \text{ mol O}_2} \cdot \frac{18.02 \text{ g H}_2\text{O}}{1 \text{ mol H}_2\text{O}} = 507 \text{ g H}_2\text{O}$$

(b) Quantity of O_2 required to consume 750. g NH_3 :

From above, we know that 750. g of ammonia = 44.1 mol NH3

The stoichiometric amount of O_2 is $44.1 \text{ mol NH}_3 \cdot \dfrac{5 \text{ mol O}_2}{4 \text{ mol NH}_3} = 55.1 \text{ mol O}_2$

The mass associated with that amount of oxygen is:

$$55.1 \text{ mol O}_2 \cdot \frac{32.00 \text{ g O}_2}{1 \text{ mol O}_2} = 1760 \text{ g O}_2 \quad \text{(to 3sf)}$$

54. Empirical formula of B_xH_y

Using the molar mass of B_2O_3, calculate the amount of the compound formed and the amount of B contained in that compound.

$$0.422 \text{ g B}_2\text{O}_3 \cdot \frac{21.62 \text{ g B}}{69.62 \text{ g B}_2\text{O}_3} = 0.131 \text{ g B}$$

The Law of Conservation of matter tells us that since 0.148 g of the compound B_xH_y has 0.131 g B, then it must also contain (0.148 – 0.131) or 0.017g H.

Now calculate the amount of B and H, using the respective atomic masses:

$$0.131 \text{ g B} \cdot \frac{1 \text{ mol B}}{10.811 \text{ g B}} = 1.21 \times 10^{-2} \text{ mol B and}$$

$$0.017 \text{ g H} \cdot \frac{1 \text{ mol H}}{1.0079 \text{ g H}} = 1.7 \times 10^{-2} \text{ mol H (actually } 1.69 \times 10^{-2} \text{ to 3 sf)}$$

The ratios of B:H are then: $\dfrac{1.7 \times 10^{-2} \text{ mol H}}{1.21 \times 10^{-2} \text{ mol B}} = \dfrac{1.39 \text{ mol H}}{1 \text{ mol B}}$

Noting that the numerator is almost 1.4, one might assume that a multiplier of 5 would provide a formula of B_5H_7.

<u>56</u>. The combustion of menthol can be represented: $C_xH_yO_z + O_2 \rightarrow CO_2 + H_2O$

It's important to note that while **all** the C in CO_2 originated in the menthol and **all** the H in H_2O originated in the menthol—**not all** of the oxygen originates in the menthol. So we begin by calculating the masses of C and H , and subtracting those masses from the 95.6 mg of compound to determine the mass of O present in menthol.

$$269 \text{ mg CO}_2 \cdot \frac{12.01 \text{ g C}}{44.02 \text{ g CO}_2} = 73.41 \text{ mg C}$$

$$110 \text{ mg H}_2O \cdot \frac{2.02 \text{ g H}}{18.02 \text{ g H}_2O} = 12.33 \text{ mg H}$$

Mass O = 95.6 mg - (73.41 mg C + 12.33 mg H) = 9.86 mg O

Now we can calculate the # of moles of each of these atoms:

$$73.41 \times 10^{-3} \text{ g C} \cdot \frac{1 \text{ mol C}}{12.01 \text{ g C}} = 6.11 \times 10^{-3} \text{ mol C}$$

$$12.33 \times 10^{-3} \text{ g H} \cdot \frac{1 \text{ mol H}}{1.008 \text{ g H}} = 12.23 \times 10^{-3} \text{ mol H}$$

$$9.86 \times 10^{-3} \text{ g O} \cdot \frac{1 \text{ mol O}}{16.00 \text{ g O}} = 0.616 \times 10^{-3} \text{ mol O}$$

We can express the **ratio** of C,H, and O present in menthol by dividing these three amounts by the smallest (O content).

$$\frac{6.11 \times 10^{-3} \text{ mol C}}{0.616 \times 10^{-3} \text{ mol O}} = 10 \qquad \frac{12.23 \times 10^{-3} \text{ mol H}}{0.616 \times 10^{-3} \text{ mol O}} = 20$$

The empirical formula is then $C_{10}H_{20}O$.

58. 1.056 g MCO3 produced MO + 0.376 g CO2

The MO had a mass of (1.056 - 0.376) 0.680 g

According to the equation given, CO_2 and MO are produced in equimolar amounts.

$$0.376 \text{ g CO}_2 \cdot \frac{1 \text{ mol CO}_2}{44.0 \text{ g CO}_2} = 8.54 \times 10^{-3} \text{ mol CO}_2$$

So the metal oxide (0.680 g) must correspond to 8.54×10^{-3} mol of metal oxide. The

molar mass of metal oxide is then: $\dfrac{0.680 \text{ g}}{8.54 \times 10^{-3} \text{ mol}} = 79.6 \text{ g/ mol}$

If the oxide contains one mol of O atoms per mol of M atoms, we can deduce the molar

mass of M   MO = 79.6 g/mol

79.6 = M g/mol + 16.0 g O/mol

63.6 g/mol = M

The metal with the atomic weight close to 63.6 is (b) Cu.

60. $TiO_2 + H_2 \rightarrow H_2O + Ti_xO_y$

1.598 g   1.438 g

The moles of TiO_2 initially present:

$$1.598 \text{ g TiO}_2 \cdot \frac{1 \text{ mol TiO}_2}{79.879 \text{ g TiO}_2} = 0.02000 \text{ mol TiO}_2 \text{ (and Ti)}$$

The mass of Ti present in the TiO_2 :

$$1.598 \text{ g TiO}_2 \cdot \frac{47.88 \text{ g Ti}}{79.879 \text{ g TiO}_2} = 0.9579 \text{ g Ti}$$

Since all the Ti in the unknown compound, Ti_xO_y, originates in the TiO_2, 0.9579 g of the

1.438 g of the new oxide is Ti, leaving (1.438 - 0.9579) g of O.

The number of moles of O is:

$$0.480 \text{ g O} \cdot \frac{1 \text{ mol O}}{16.00 \text{ g O}} = 0.03000 \text{ mol O}$$

The new oxide is then Ti $_{0.02000}$ O$_{0.03000}$ or Ti_2O_3

62. Thioridazine content per tablet:

$$\frac{0.301 \text{ g BaSO}_4}{12 \text{ tablets}} \cdot \frac{1 \text{ mol BaSO}_4}{233.4 \text{ g BaSO}_4} \cdot \frac{1 \text{ mol S}}{1 \text{ mol BaSO}_4} \cdot \frac{1 \text{ mol thioridazine}}{2 \text{ mol S}} \cdot$$

$$\frac{370.6 \text{ g thioridazine}}{1 \text{ mol thioridazine}} \cdot \frac{1000 \text{ mg thioridazine}}{1 \text{ g thioridazine}} = 19.9 \text{ mg thioridazine/tablet}$$

64. The graph representing the mass of iron consumed with a **fixed** mass of bromine shows that for the mass of bromine used, the amount of iron consumed maximizes at 2 g of Fe.

(a) According to the graph, the mass of Fe consumed maximizes at 2.0 g Fe with a fixed amount of bromine(about 8.6 g). Any excess iron that is added does not result in additional product.

$$2.0 \text{ g Fe} \cdot \frac{1 \text{ mol Fe}}{55.85 \text{ g Fe}} = 0.036 \text{ mol Fe}$$

The amount of bromine would be:

$$8.6 \text{ g Br} \cdot \frac{1 \text{ mol Br}}{79.9 \text{ g Br}} = 0.108 \text{ mol Br}$$

(b) The ratio of $\dfrac{\text{Br}}{\text{Fe}}$ is $\dfrac{0.108}{0.036} = 2.99$ or 3:1 (to 1 sf)

(c) The empirical formula is $FeBr_3$.

(d) The balanced equation: $2 \text{ Fe} + 3 \text{ Br}_2 \rightarrow 2 \text{ FeBr}_3$

(e) The product is $FeBr_3$: *iron(III) bromide*

(f) When 1.00 g of Fe is added to the Br_2, *Fe is the limiting reactant* (the addition of more Fe results in the formation of more product).

Chapter 5
Reactions in Aqueous Solution

Reviewing Important Concepts

1. The difference between a solvent and a solute:

   The solvent is the medium in which the solute is dissolved. As an example, when dissolving table salt in water, the salt is the solute and water is the solvent.

3. What is an electrolyte? What are experimental means for discriminating between weak and strong electrolytes?

   An electrolyte is a substance whose aqueous solution conducts an electric current.

   As to experimental means for discriminating between weak and strong electrolytes, refer to the apparatus in the Chapter 5 Focus. NaCl is a strong electrolyte and would cause the bulb to glow brightly—reflecting a large number of ions in solution while aqueous ammonia or vinegar (an aqueous solution of acetic acid) would cause the bulb to glow only dimly—indicating a smaller number of ions in solution.

5. For the following copper salts:

   Water soluble: $Cu(NO_3)_2$, $CuCl_2$ — nitrates and chlorides are soluble

   Water insoluble: $CuCO_3$, $Cu_3(PO_4)_2$ — carbonates and phosphates are insoluble

7. **Spectator ions** in the following equation and the net ionic equation:

   $$2\,H^+(aq) + 2\,NO_3^-(aq) + Mg(OH)_2(s) \rightarrow 2\,H_2O(\ell) + Mg^{2+}(aq) + 2\,NO_3^-(aq)$$

   The emboldened nitrate ions are the spectator ions. The net ionic equation would be the first equation shown above without the spectator ions:

   $$2\,H^+(aq) + Mg(OH)_2(s) \rightarrow 2\,H_2O(\ell) + Mg^{2+}(aq)$$

9. For the reaction of chlorine with NaBr:

   $$Cl_2(g) + 2\,NaBr(aq) \rightarrow 2\,NaCl(aq) + Br_2(\ell)$$

   Oxidized: **bromine's** oxidation number is changed from -1 to 0

   Reduced: **chlorine's** oxidation number is changed from 0 to –1
   Oxidizing agent: Cl_2 removes the electrons from NaBr

   Reducing agent: **NaBr** provides the electrons to the chlorine.

11. Identify the following as oxidizing or reducing agents:

HNO_3–an oxidizing agent (furnishes oxygen to another substance)

Na—a reducing agent (gives electrons to another substance)

Cl_2–an oxidizing agent (removes electrons from another substance—as it is reduced)

O_2–an oxidizing agent (furnishes oxygen to another substance)

$KMnO_4$–an oxidizing agent (furnishes oxygen to another substance)

13. Greater mass of solute in 1 L of 0.1M NaCl or 1 L of 0.06 M Na_2CO_3

First determine the number of moles of each solute in the amount of solution indicated. Since Molarity is defined as #moles/Liter solution, the product of Molarity • Volume (in L) gives the # of moles. An alternate, but equivalent, question to ask is which has the greater mass: 0.1 mol of NaCl or 0.06 mol of Na_2CO_3 . The molar mass of NaCl is approximately (23 + 35.5) or 58.8 grams, while that for Na_2CO_3 is [2(23) + 12 + 3(16)] or 106 grams. One-tenth mol of NaCl would have a mass of approximately (0.1 • 58.4) or 5.84 g while 0.06 mol of sodium carbonate would have a mass of (0.06 • 106) or 6.36 grams. The **solution of Na_2CO_3** has the greater mass of solute.

15. Reactions described as product or reactant-favored:

(a) The formation of an acid and based from a salt is **reactant-favored.**

(b) The formation of a precipitate during a reaction makes the reaction **product-flavored.**

17. To prepare a 0.500 molar solution of KCl, recall that M = #mol/volume(L). Since the volumetric flask at your disposal is a 250.mL flask, what amount of solute is needed?

$$0.500 \text{ M} = \frac{\text{\# mol KCl}}{0.250 \text{ L}} = 0.125 \text{ mol KCl} \quad \text{(or 9.32 g)}$$

Take 1/4 of the KCl, place it in the volumetric flask, add distilled water until the solute dissolves. Add water until the meniscus of the solution rests at the calibrated mark on the neck of the volumetric flask. Cap the flask and swirl to ensure adequate mixing.

19. The equivalence point in a titration is a mathematical **and** chemical point. It describes the point at which a **stoichiometric** or equivalent amount of one reactant has been added to the second reactant in a chemical process—frequently applied in the case of bases reacting with acids. The function of an indicator is to **change color** when a very slight excess of one reactant is present.

Practicing Skills
Solubility of Compounds

20. Predict water solubility:

 (a) $CuCl_2$ is expected to be soluble, while CuO and $FeCO_3$ are not. Chlorides are generally water soluble, while oxides and carbonates are not.

 (b) $AgNO_3$ is soluble. AgI and Ag_3PO_4 are not soluble. Nitrate salts are soluble. Phosphate salts are generally insoluble. While halides are generally soluble, those of Ag^+ are not.

 (c) K_2CO_3, KI and $KMnO_4$ are soluble. In general, salts of the alkali metals are soluble.

22. Ions produced when the compounds dissolve in water.

Compound	Cation	Anion
(a) KOH	K^+	OH^-
(b) K_2SO_4	$2 K^+$	SO_4^{2-}
(c) $LiNO_3$	Li^+	NO_3^-
(d) $(NH_4)_2SO_4$	$2 NH_4^+$	SO_4^{2-}

24.

Compound	Water Soluble	Cation	Anion
(a) Na_2CO_3	yes	$2 Na^+$	CO_3^{2-}
(b) $CuSO_4$	yes	Cu^{2+}	SO_4^{2-}
(c) NiS	no		
(d) $BaBr_2$	yes	Ba^{2+}	$2 Br^-$

Precipitation Reactions

26. $CdCl_2(aq) + 2 NaOH(aq) \rightarrow Cd(OH)_2(s) + 2 NaCl(aq)$

 Net ionic equation: $Cd^{2+}(aq) + 2 OH^-(aq) \rightarrow Cd(OH)_2(s)$

28. Balanced equations for precipitation reactions:

 (a) $NiCl_2(aq) + (NH_4)_2S(aq) \rightarrow NiS(s) + 2 NH_4Cl(aq)$

 (b) $3 Mn(NO_3)_2(aq) + 2 Na_3PO_4(aq) \rightarrow Mn_3(PO_4)_2(s) + 6 NaNO_3(aq)$

Writing Net Ionic Equations

30.   (a) $(NH_4)_2CO_3(aq) + Cu(NO_3)_2(aq) \rightarrow CuCO_3(s) + 2 NH_4NO_3(aq)$

        (net) $CO_3^{2-}(aq) + Cu^{2+}(aq) \rightarrow CuCO_3(s)$

      (b) $Pb(OH)_2(s) + 2 HCl(aq) \rightarrow PbCl_2(s) + 2 H_2O(\ell)$

        (net) $Pb(OH)_2(s) + 2 H^+(aq) + 2 Cl^-(aq) \rightarrow PbCl_2(s) + 2 H_2O(\ell)$

(c) $BaCO_3(s) + 2 HCl(aq) \rightarrow BaCl_2(aq) + H_2O(\ell) + CO_2(g)$

(net) $BaCO_3(s) + 2 H^+(aq) \rightarrow Ba^{2+}(aq) + H_2O(\ell) + CO_2(g)$

32.   (a) $AgNO_3(aq) + KI(aq) \rightarrow AgI(s) + KNO_3(aq)$

(net) $Ag^+(aq) + I^-(aq) \rightarrow AgI(s)$

(b) $Ba(OH)_2(aq) + 2 HNO_3(aq) \rightarrow 2 H_2O(\ell) + Ba(NO_3)_2(aq)$

(net) $OH^-(aq) + H^+(aq) \rightarrow H_2O(\ell)$

(c) $2 Na_3PO_4(aq) + 3 Ni(NO_3)_2(aq) \rightarrow Ni_3(PO_4)_2(s) + 6 NaNO_3(aq)$

(net) $2 PO_4^{3-}(aq) + 3 Ni^{2+}(aq) \rightarrow Ni_3(PO_4)_2(s)$

Acids and Bases

34. $HNO_3(aq) + H_2O(\ell) \rightarrow H_3O^+(aq) + NO_3^-(aq)$

alternatively $HNO_3(aq) \rightarrow H^+(aq) + NO_3^-(aq)$

36. $H_2C_2O_4(aq) \rightarrow H^+(aq) + HC_2O_4^-(aq)$

$HC_2O_4^-(aq) \rightarrow H^+(aq) + C_2O_4^{2-}(aq)$

38. $MgO(s) + H_2O(\ell) \rightarrow Mg(OH)_2(s)$

Reactions of Acids and Bases

40. Complete and Balance

(a) $2 CH_3CO_2H(aq) + Mg(OH)_2(s) \rightarrow Mg(CH_3CO_2)_2(aq) + 2 H_2O(\ell)$
     acetic   magnesium   magnesium   water
     acid   hydroxide   acetate

(b) $HClO_4(aq) + NH_3(aq) \rightarrow NH_4ClO_4(aq)$
     perchloric   ammonia   ammonium
     acid   perchlorate

42. Write and balance the equation for barium hydroxide reacting with nitric acid:

$Ba(OH)_2(s) + 2 HNO_3(aq) \rightarrow Ba(NO_3)_2(aq) + 2 H_2O(\ell)$
     barium   nitric   barium   water
     hydroxide   acid   nitrate

Gas-Forming Reactions

44. Write and balance the equation for iron(II) carbonate reacting with nitric acid:

$FeCO_3(s) + 2 HNO_3(aq) \rightarrow Fe(NO_3)_2(aq) + H_2O(\ell) + CO_2(g)$
     iron(II)   nitric   iron(II)   water   carbon
     carbonate   acid   nitrate   dioxide

Types of Reactions in Aqueous Solution

46. Acid-Base (AB) , Precipitation (PR), or Gas-Forming (GF)

(a) $Ba(OH)_2(s) + 2\ HCl(aq) \rightarrow BaCl_2(aq) + 2\ H_2O(\ell)$ AB

(b) $2\ HNO_3(aq) + CoCO_3(s) \rightarrow Co(NO_3)_2(aq) + H_2O(\ell) + CO_2(g)$ GF

(c) $2\ Na_3PO_4(aq) + 3\ Cu(NO_3)_2(aq) \rightarrow Cu_3(PO_4)_2(s) + 6\ NaNO_3(aq)$ PR

48. Acid-Base (AB) , Precipitation (PR), or Gas-Forming (GF)

(a) $MnCl_2(aq) + Na_2S(aq)\ \rightarrow\ MnS(s)\ +\ 2\ NaCl(aq)$ PR

(b) $K_2CO_3(aq) + ZnCl_2(aq)\ \rightarrow\ ZnCO_3(s)\ +\ 2\ KCl(aq)$ PR

Net ionic equations:

(a) $Mn^{2+}(aq)\ +\ S^{2-}(aq)\ \rightarrow\ MnS(s)$

(b) $CO_3^{2-}(aq)\ +\ Zn^{2+}(aq)\ \rightarrow\ ZnCO_3(s)$

Product- or Reactant-Favored Reactions

50. Feature causing the reaction to be product-favored:

(a) The formation of the solid CuS "drives" the reaction.

(b) The formation of the liquid H_2O "drives" the reaction

Oxidation Numbers

52. For questions on oxidation number, read the symbol (x) as "the oxidation number of x."

(a) BrO_3^- (Br) + 3(O)   = -1

Since oxygen almost always has an oxidation number of -2, we can substitute this value and solve for the oxidation number of Br.

(Br) + 3(-2)   = -1

(Br)   = +5

(b) $C_2O_4^{2-}$ 2 (C) + 4 (O) = -2

2 (C) + 4 (-2) = -2

2 (C) + -8   = -2

2 (C)   = +6

(C)   = +3

(c) F^- The oxidation number for any monatomic ion is the charge on the ion. So (F) = -1

(d) CaH_2 (Ca) + 2 (H) = 0

(Ca) + 2 (-1) = 0

(Ca)   = +2

(e) H_4SiO_4 $4(H) + (Si) + 4(O) = 0$

                        $4(+1) + (Si) + 4(-2) = 0$

                        $(Si) = +4$

(f) HSO_4^- $(H) + (S) + 4(O) = -1$

                        $(+1) + (S) + 4(-2) = -1$

                        $(S) = +6$

Oxidation-Reduction Reactions

54.   (a) Oxidation-Reduction: $Zn(s)$ has an oxidation number of 0, while $Zn^{2+}(aq)$ has an
oxidation number of +2—hence Zn is being oxidized.
N in NO_3^- has an oxidation number of +5, while N in NO_2
has an oxidation number of +4—hence N is being reduced.

      (b) Acid-Base reaction: There is no change in oxidation number for any of the
elements in this reaction—hence it is NOT an oxidation-
reduction reaction.
H_2SO_4 is an acid, and $Zn(OH)_2$ acts as a base.

      (c) Oxidation-Reduction: $Ca(s)$ has an oxidation number of 0, while $Ca^{2+}(aq)$ has an
oxidation number of +2—hence Ca is being oxidized.
H in H_2O has an oxidation number of +1, while H in H_2
has an oxidation number of 0—hence H is being reduced.

56. Determine which reactant is oxidized and which is reduced:

     (a) $C_2H_4(g) + 3 O_2(g) \rightarrow 2 CO_2(g) + 2 H_2O(g)$

	ox. number			
specie	before	after	has experienced	functions as the
C	-2	+4	oxidation	(C_2H_4) reducing agent
H	+1	+1	no change	
O	0	-2	reduction	(O_2) oxidizing agent

     (b) $Si(s) + 2 Cl_2(g) \rightarrow SiCl_4(\ell)$

	ox. number			
specie	before	after	has experienced	functions as the
Si	0	+4	oxidation	(Si) reducing agent
Cl	0	-1	reduction	(Cl_2) oxidizing agent

Solution Concentration

58. Molarity of Na_2CO_3 solution:

$$6.73 \text{ g } Na_2CO_3 \cdot \frac{1 \text{ mol } Na_2CO_3}{106.0 \text{ g } Na_2CO_3} = 0.0635 \text{ mol } Na_2CO_3$$

$$\text{Molarity} \equiv \frac{\# \text{ mol}}{L} = \frac{0.0635 \text{ mol } Na_2CO_3}{0.250 \text{ L}} = 0.254 \text{ M } Na_2CO_3$$

Concentration of Na^+ and CO_3^{2-} ions:

$$\frac{0.254 \text{ mol } Na_2CO_3}{L} \cdot \frac{2 \text{ mol } Na^+}{1 \text{ mol } Na_2CO_3} = 0.508 \text{ M } Na^+$$

$$\frac{0.254 \text{ mol } Na_2CO_3}{L} \cdot \frac{1 \text{ mol } CO_3^{2-}}{1 \text{ mol } Na_2CO_3} = 0.254 \text{ M } CO_3^{2-}$$

60. Mass of $KMnO_4$:

$$\frac{0.0125 \text{ mol } KMnO_4}{L} \cdot \frac{0.250 \text{ L}}{1} \cdot \frac{158.0 \text{ g } KMnO_4}{1 \text{ mol } KMnO_4} = 0.494 \text{ g } KMnO_4$$

62. Volume of 0.123 M NaOH to contain 25.0 g NaOH:

Calculate moles of NaOH in 25.0 g:

$$\frac{25.0 \text{ g NaOH}}{1} \cdot \frac{1 \text{ mol NaOH}}{40.00 \text{ g NaOH}} = 0.625 \text{ mol NaOH}$$

The volume of 0.123 M NaOH that contains 0.625 mol NaOH:

$$0.625 \text{ mol NaOH} \cdot \frac{1 \text{ L}}{0.123 \text{ mol NaOH}} \cdot \frac{1 \times 10^3 \text{ mL}}{1L} = 5.08 \times 10^3 \text{ mL}$$

64. Identity and concentration of ions in each of the following solutions:
    (a) 0.25 M $(NH_4)_2SO_4$ gives rise to (2 x 0.25) 0.50 M NH_4^+ ions & 0.25 M SO_4^{2-} ions
    (b) 0.123 M Na_2CO_3 gives rise to (2 x 0.123) 0.246 M Na^+ ions & 0.123 M CO_3^{2-}.
    (c) 0.056 M HNO_3 gives rise to 0.056 M H^+ ions & 0.056 M NO_3^- ions.

Preparing Solutions

66. Prepare 500. mL of 0.0200 M solution of Na_2CO_3:

Decide what amount (#moles) of sodium carbonate are needed:

$$\frac{0.0200 \text{ M } Na_2CO_3}{1 \text{ L}} \cdot \frac{0.500 \text{ L}}{1} = 0.0100 \text{ mol } Na_2CO_3$$

What mass does 0.0100 mol sodium carbonate have?

MW = 106.0 g Na_2CO_3/ mol Na_2CO_3 and using 1/100 mol would require:

0.0100 mol Na_2CO_3 • 106.06 g/mol Na_2CO_3 or *1.06 g Na_2CO_3.*

To prepare the solution, take 1.06 g of Na_2CO_3, and transfer it to the volumetric flaks. Add a bit of distilled water and stir carefully until all the solid Na_2CO_3 has dissolved. Once the solid has dissolved, and distilled water to the calibrated mark on the neck of the volumetric flask. Stopper, stir to assure complete mixing!

68. Molarity of HCl in the diluted solution:

We can calculate the molarity if we know the number of moles of HCl in the 25.0 mL solution

1. Moles of HCl in 25.0 mL of 1.50 M HCl :

$$M \times V \ = \ \frac{1.50 \text{ mol HCl}}{L} • \frac{25.00 \times 10^{-3} L}{1} = 0.0375 \text{ mol HCl}$$

2. When that number of moles is distributed in 500. mL:

$$\frac{0.0375 \text{ mol HCl}}{0.500 \text{ L}} = 0.0750 \text{ M HCl}$$

Perhaps a shorter way to solve this problem is to note the number of moles (found by multiplying the original molarity times the volume) is distributed in a given volume, resulting in the diluted molarity. Mathematically: $M_1 \times V_1 \ = \ M_2 \times V_2$

70. Using the definition of molarity we can calculate the amount of H_2SO_4 in the following:

$$\frac{0.125 \text{mol} H_2SO_4}{L} • 1.00 \text{ L} \ = \ 0.125 \text{ moles of } H_2SO_4$$

If we dilute 20.8 mL of 6.00 M H_2SO_4 to one liter we obtain:

$$\frac{6.00 \text{ M } H_2S O_4}{1 \text{ L}} • 0.0208 L = 0.1248 \text{ mol } H_2SO_4 \text{ or a } 0.125 \text{ M solution of } H_2SO_4$$

Method (a) is correct!

pH

72. The hydrogen ion concentration of a wine whose pH = 3.40:

Since pH is defined as $-\log[H^+]$; $[H^+] = 10^{-3.40} \ = \ 4.0 \times 10^{-4}$ M H^+

The solution has a pH<7.00, so it is **acidic**.

74. The $[H^+]$ and pH of a solution of 0.0013 M HNO_3:

Since nitric acid is a strong acid (and strong electrolyte), we can state that:

$[H^+]$= 0.0013 M; the pH = - log (0.0013) or 2.89

76. Make the interconversions and decide if the solution is acidic or basic:

	pH	$[H^+]$	Acidic/Basic
(a)	**1.00**	0.10	Acidic
(b)	**10.50**	3.2×10^{-11} M	Basic
(c)	4.89	**1.3×10^{-5} M**	Acidic
(d)	7.64	**2.3×10^{-8} M**	Basic

pH values less than 7 indicate an acidic solution while those greater than indicate a basic solution. Similarly solutions for which the $[H^+]$ is greater than 1.0×10^{-7} are acidic while those with hydrogen ion concentrations LESS THAN 1.0×10^{-7} are basic.

Stoichiometry of Reactions in Solution

78. Volume of 0.109 M HNO_3 to react with 2.50 g of $Ba(OH)_2$:

Need several steps:

1. calculate mol of barium hydroxide in 2.50 g

$$2.50 \text{ g } Ba(OH)_2 \cdot \frac{1 \text{ mol } Ba(OH)_2}{171.3 \text{ g } Ba(OH)_2}$$

2. calculate mole of HNO_3 needed to react with that # of mol of barium hydroxide

$$2.50 \text{ g } Ba(OH)_2 \cdot \frac{1 \text{ mol } Ba(OH)_2}{171.3 \text{ g } Ba(OH)_2} \cdot \frac{2 \text{ mol } HNO_3}{1 \text{ mol } Ba(OH)_2}$$

3. calculate volume of 0.109 M HNO_3 that contains that # of mol of nitric acid.

$$2.50 \text{ g } Ba(OH)_2 \cdot \frac{1 \text{ mol } Ba(OH)_2}{171.3 \text{ g } Ba(OH)_2} \cdot \frac{2 \text{ mol } HNO_3}{1 \text{ mol } Ba(OH)_2} \cdot \frac{1 \text{ L}}{0.109 \text{ M } HNO_3} = 0.268 \text{ L}$$

or 268 mL.

80. Mass of NaOH formed from 15.0 L of 0.35 M NaCl:

$$\frac{0.35 \text{ mol NaCl}}{1 \text{L}} \cdot \frac{15.0 \text{ L}}{1} \cdot \frac{2 \text{ mol NaOH}}{2 \text{ mol NaCl}} \cdot \frac{40.0 \text{ g NaOH}}{1 \text{ mol NaOH}} = 210 \text{ g NaOH}$$

Mass of Cl_2 obtainable:

$$\frac{0.35 \text{ mol NaCl}}{1 \text{ L}} \cdot \frac{15.0 \text{ L}}{1} \cdot \frac{1 \text{ mol } Cl_2}{2 \text{ mol NaCl}} \cdot \frac{70.9 \text{ g } Cl_2}{1 \text{ mol } Cl_2} = 186 \text{ g } Cl_2 \text{ or } 190 \text{ mL (2 sf)}$$

82. Volume of 0.0138M $Na_2S_3O_3$ to dissolve 0.225 g of AgBr:

Calculate: 1. mol of AgBr

2. mol of $Na_2S_2O_3$ needed to react (balanced equation)

3. volume of 0.0138 M $Na_2S_2O_3$ containing that number of moles.

$$0.225 \text{ g AgBr} \cdot \frac{1 \text{ mol AgBr}}{187.8 \text{ g AgBr}} \cdot \frac{2 \text{ mol } Na_2S_2O_3}{1 \text{ mol AgBr}} \cdot \frac{1 \text{ L}}{0.0138 \text{ mol } Na_2S_2O_3} \cdot \frac{1000 \text{ mL}}{1 \text{ L}} = 174 \text{ mL } Na_2S_2O_3$$

84. The balanced equation:

$$Pb(NO_3)_2(aq) + 2\,NaCl(aq) \rightarrow PbCl_2(s) + 2\,NaNO_3(aq)$$

Volume of 0.750 M $Pb(NO_3)_2$ needed:

$$\frac{2.25\text{ mol NaCl}}{1\text{ L}} \cdot \frac{1.00\text{ L}}{1} \cdot \frac{1\text{ mol Pb(NO}_3)_2}{2\text{ mol NaCl}} \cdot \frac{1\text{ L}}{0.750\text{ mol Pb(NO}_3)_2} \cdot \frac{1000\text{ mL}}{1\text{ L}}$$

$$= 1500\text{ mL or } 1.50 \times 10^3\text{ mL}$$

Titrations

86. To calculate the volume of 0.812 M HCl needed, we calculate the moles of NaOH in 1.45 g, then use the stoichiometry of the balanced equation:

$$HCl(aq) + NaOH(aq) \rightarrow NaCl(aq) + H_2O(\ell)$$

$$1.45\text{ g NaOH} \cdot \frac{1\text{ mol NaOH}}{40.00\text{ g NaOH}} \cdot \frac{1\text{ mol HCl}}{1\text{ mol NaOH}} \cdot \frac{1\text{ L}}{0.812\text{ mol HCl}} \cdot \frac{1000\text{ mL}}{1\text{ L}}$$

$$= 44.6\text{ mL HCl}$$

88. Calculate:

1. moles of Na_2CO_3 corresponding to 2.150 g Na_2CO_3

2. moles of HCl that react with that number of moles (using the balanced equation)

3. the molarity of HCl containing that number of moles of HCl in 38.55 mL.

The balanced equation is:

$$Na_2CO_3(aq) + 2\,HCl(aq) \rightarrow 2\,NaCl(aq) + H_2O(\ell) + CO_2(g)$$

$$2.150\text{ g Na}_2\text{CO}_3 \cdot \frac{1\text{ mol Na}_2\text{CO}_3}{106.0\text{ g Na}_2\text{CO}_3} \cdot \frac{2\text{ mol HCl}}{1\text{ mol Na}_2\text{CO}_3} \cdot \frac{1}{0.03855\text{ L sol}}$$

$$= 1.052\text{ M HCl}$$

90. Molar Mass of an acid if 36.04 mL of 0.509 M NaOH will titrate 0.954 g of an acid H_2A:

Note that the acid is a diprotic acid, and will require 2 mol of NaOH for each mol of the acid.

Calculate the # moles of NaOH: $\dfrac{0.509\text{ mol NaOH}}{1\text{ L}} \cdot 0.03604\text{ L} = 0.0183\text{ mol NaOH}$

The number of moles of acid will be **half** of the number of moles of NaOH:

$$0.0183\text{ mol NaOH} \cdot \frac{1\text{ mol H}_2\text{A}}{2\text{ mol NaOH}} = 0.00917\text{ mol H}_2\text{A}$$

Since we know the mass corresponding to this number of moles of acid, we can calculate the molar mass (# g/mol): $\dfrac{0.954\text{ g H}_2\text{A}}{0.00917\text{ mol H}_2\text{A}} = 104\text{ g/mol}$

92. Mass percent of iron in a 0.598 gram sample that requires 22.24 mL of 0.0123 M $KMnO_4$:

Note from the balanced, net ionic equation that 1 mol MnO_4^- requires 5 mol of Fe^{2+}.

The # mol of MnO_4^- is: $\dfrac{0.0123 \text{ M } MnO_4^-}{1 \text{ L}} \cdot 0.02224 \text{ L} = 2.74 \times 10^{-4} \text{ mol } MnO_4^-$

Using the ratio of permanganate ion to iron(II) ion we obtain:

$2.74 \times 10^{-4} \text{ mol } MnO_4^- \cdot \dfrac{5 \text{ mol } Fe^{2+}}{1 \text{ mol } MnO_4^-} = 1.37 \times 10^{-3} \text{ mol } Fe^{2+}$

Using the mass of the iron(II) ion, we can calculate the mass of iron present:

$1.37 \times 10^{-3} \text{ mol } Fe^{2+} \cdot \dfrac{55.847 \text{ g } Fe^{2+}}{1 \text{ mol } Fe^{2+}} = 0.0764 \text{ g } Fe^{2+}$

The mass percent of iron is then: $\dfrac{0.0764 \text{ g } Fe^{2+}}{0.598 \text{ g sample}} \cdot 100 = 12.8 \% \text{ } Fe^{2+}$

General Questions

94. Volume of 0.054 M H_2SO_4 required to react completely with 1.56 g KOH:

The balanced equation indicates that 2 moles of KOH react with 1 mol of H_2SO_4.

$$2 \text{ KOH} + H_2SO_4(aq) \rightarrow K_2SO_4(aq) + H_2O(\ell)$$

The amount of KOH present in 1.56 g KOH:

$1.56 \text{ g KOH} \cdot \dfrac{1 \text{ mol KOH}}{56.11 \text{ g KOH}} = 2.78 \times 10^{-2} \text{ mol KOH}$

The amount of H_2SO_4 required will be 1/2 that amount:

$2.78 \times 10^{-2} \text{ mol KOH} \cdot \dfrac{1 \text{ mol } H_2SO_4}{2 \text{ mol KOH}} = 1.39 \times 10^{-2} \text{ mol } H_2SO_4$

The volume of 0.054 M H_2SO_4 that contains that number of moles is:

$1.39 \times 10^{-2} \text{ mol } H_2SO_4 \cdot \dfrac{1 \text{ L}}{0.054 \text{ mol } H_2SO_4} = 0.26 \text{ L or 260 mL}$

96.   (a) A balanced equation:

$$Mg(s) + 4 \text{ HNO3}(aq) \rightarrow Mg(NO_3)_2(aq) + 2 \text{ NO}_2(g) + 2 H_2O(\ell)$$

Magnesium changes oxidation state from 0 to +2, reflecting a 2 electron change per magnesium atom. N in nitric acid changes oxidation state from +5 (in nitric acid) to +4 in nitrogen dioxide, reflecting a 1 electron change. [Note that **not all** the nitric acid participates in the redox reaction.] The coefficient of HNO3 needs to be 2 times the coefficient of Mg (to equalize electron gain with loss. Since magnesium nitrate **also** contains 2 nitrate ions per formula unit, 2 additional HNO3 molecules are needed as reactants.

(b) compounds named:

$$Mg(s) + 4\ HNO_3(aq) \rightarrow Mg(NO_3)_2(aq) + 2\ NO_2(g) + 2\ H_2O(\ell)$$

Magnesium nitric
acid

magnesium
nitrate

nitrogen   water
dioxide

(c) net ionic equation:

$$Mg\ + 4\ H^+ + 2\ NO_3^- \rightarrow Mg^{2+} + 2\ NO_2 + 2\ H_2O$$

(d) oxidizing and reducing agents:

Mg loses electrons to nitric acid, serving as the **reducing agent.**

Nitric acid removes electrons from Mg, serving as the **oxidizing agent.**

98.   Species present in aqueous solutions of:

	compound	types of species	species present
(a)	NH_3	molecules (weak base)	NH_3, NH_4^+, OH^-
(b)	CH_3CO_2H	molecules (weak acid)	$CH_3CO_2H, CH_3CO_2^-, H^+$
(c)	NaOH	ions (strong base)	Na^+ and OH^-
(d)	HBr	ions (strong acid)	H^+ and Br^-

In every case, H_2O will be present (but omitted in this list)

100. Limiting reagent between 125 mL of 0.15 M CH_3CO_2H and 15.0 g $NaHCO_3$:

moles CH_3CO_2H: $\dfrac{0.15\ mol\ CH_3CO_2H}{1L} \cdot 0.125\ L = 0.01875\ mol\ CH_3CO_2H$

moles $NaHCO_3$: 15.0 g $NaHCO_3 \cdot \dfrac{1\ mol\ NaHCO_3}{84.01\ g\ NaHCO_3} = 0.179\ mol\ NaHCO_3$

Since the reaction proceeds with a 1:1 stoichiometry, *CH_3CO_2H is the limiting reagent.*

102. Weight percent of oxalic acid in a 4.554 g sample:

1. Determine the moles of NaOH in the solution (V x M)

2. Use the balanced equation to determine the # of moles of oxalic acid present
   The equation shows 1 mol $H_2C_2O_4$ reacts with 2 mol NaOH

3. Determine the mass of oxalic acid present, and from the total mass of sample—the
   percent of oxalic acid in the sample.

$$\dfrac{0.550\ mol\ NaOH}{1\ L} \cdot 0.02958\ L \cdot \dfrac{1\ mol\ H_2C_2O_4}{2\ mol\ NaOH} \cdot \dfrac{90.04\ g\ H_2C_2O_4}{1\ mol\ H_2C_2O_4} = 0.7324\ g\ H_2C_2O_4$$

% oxalic acid $= \dfrac{0.7234\ g H_2C_2O_4}{4.554\ g\ sample} \cdot 100 = 16.1\%\ H_2C_2O_4$ (to 3 sf)

104. Solubility:

   (a) Two soluble copper(II) salts:

   copper(II) nitrate: $Cu(NO_3)_2$ and copper(II) halides: CuF_2, $CuCl_2$, $CuBr_2$, CuI_2

   Two insoluble copper(II) salts:

   copper(II) sulfide, CuS and copper(II) carbonate: $CuCO_3$

   (b) Two soluble barium salts:

   barium nitrate: $Ba(NO_3)_2$ and $BaCl_2$

   Two insoluble barium salts: $BaSO_4$ and $Ba_3(PO_4)_2$

106. Balance and classify the following equations:

   (a) $K_2CO_3(aq) + 2\ HClO_4(aq) \rightarrow 2\ KClO_4(aq) + CO_2(g) + H_2O(\ell)$

   TYPE: **gas-forming**
   NET IONIC: $CO_3^{2-}(aq) + 2\ H^+(aq) \rightarrow CO_2(g) + H_2O(\ell)$

   (b) $FeCl_2(aq) + (NH_4)_2S(aq) \rightarrow FeS(s) + 2\ NH_4Cl(aq)$

   TYPE: **precipitation**
   NET IONIC: $Fe^{2+}(aq) + S^{2-}(aq) \rightarrow FeS(s)$

   (c) $Fe(NO_3)_2(aq) + Na_2CO_3(aq) \rightarrow FeCO_3(s) + 2\ NaNO_3(aq)$

   TYPE: **precipitation**
   NET IONIC: $Fe^{2+}(aq) + CO_3^{2-}(aq) \rightarrow FeCO_3(s)$

108. pH measurements:

   (a) pH of 0.105 M HCl solution: Since HCl is a strong acid (and electrolyte), the
   $[H^+]$=0.105 M. pH = $-\log[0.105]$ = 0.979

   (b) $[H^+]$ in a solution whose pH = 2.56: $[H^+]$ = $10^{-2.56}$ = 2.8 x 10^{-3}

   Solutions with a pH < 7 are considered **acidic**.

   (c) pH = 9.67; $[H^+]$ = $10^{-9.67}$ = 2.1 x 10^{-10}

   Solutions with a pH > 7 are considered **basic**.

110. Concentration of hydrochloric acid:

   Calculate the # of moles of HCl in both of the two solutions:

   $\dfrac{2.50\ M\ HCl}{1\ L}$ • 0.500 L = 1.25 mol HCl

   $\dfrac{3.75\ M\ HCl}{1\ L}$ • 0.250 L = 0.938 mol HCl

   This number of moles is now contained in 0.750 L, so the concentration is:

   $\dfrac{2.1875\ mol\ HCl}{0.750\ L}$ = 2.92 M HCl. Since HCl is a strong acid (and electrolyte), the

   concentration of H^+ = 2.92 M and pH = $-\log[2.92]$ = - 0.465

112. pH of solution resulting when 250. mL of 0.0105 M NaOH is added to 250. mL of HCl solution with pH = 1.92:

HCl will react with NaOH on a 1 mol:1 mol basis. First determine the # mol of HCl in the original solution.

pH = 1.92 so $[H^+]$ = $10^{-1.92}$ = 1.20 x 10^{-2} M HCl and since the volume is 250. mL:

The product of M • V = 3.01 x 10^{-3} mol HCl.

The number of moles of HCl that will react is equal to the # mol NaOH:

$$\frac{0.0105 \text{ mol NaOH}}{1 \text{ L}} \bullet 0.250 \text{ L} = 0.00263 \text{ mol NaOH} = 0.00263 \text{ mol HCl react}$$

The amount of HCl remaining is then:

3.01 x 10^{-3} mol HCl - 2.63 x 10^{-3} mol HCl = 3.8 x 10^{-4} mol HCl

and the concentration is: $\dfrac{3.8 \times 10^{-4}\text{mol}}{0.500 \text{ L}}$ = 7.6 x 10^{-4}M

and the pH = -log[7.6 x 10^{-4}] = 3.13

114. Preparation of iodine:

(a) Oxidation numbers:

$2 \text{ NaI} + 2 \text{ H}_2\text{SO}_4 + \text{MnO}_2 \rightarrow \text{Na}_2\text{SO}_4 + \text{MnSO}_4 + \text{I}_2 + \text{H}_2\text{O (l)}$

Na = +1; H = +1   Mn = +4   Na = +1   Mn = +2   I = 0   H = +1

I = -1   S = +6   O = -2   S = +6   S = +6   O = -2

            O = -2   O = -2   O = -2

(b) Oxidizing agent = MnO_2; Oxidized = NaI

Reducing agent = I^-; Reduced = MnO_2

(c) Mass of I_2 if 20.0 g of NaI mixed with 10.0 g MnO_2

Moles of NaI: 20.0 g NaI • $\dfrac{1 \text{ mol NaI}}{149.9 \text{g NaI}}$ = 0.133 mol

Moles of MnO_2: 20.0 g MnO_2 • $\dfrac{1 \text{ mol MnO}_2}{86.94 \text{g MnO}_2}$ = 0.230 mol

Determine the limiting reactant: Ratio of moles needed: $\dfrac{2 \text{ mol NaI}}{1 \text{ mol MnO}_2}$

The ratio of moles available is: $\dfrac{0.133 \text{ mol NaI}}{0.230 \text{ mol MnO}_2}$ = 0.5

Since the moles available is **less than** the moles needed, NaI will determine the maximum amount of iodine produced.

From the balanced equation, we see that 1mol of I_2 results from 2 mol of NaI, so we get:

$$0.133 \text{ mol NaI} \cdot \frac{1 \text{ mol } I_2}{2 \text{ mol NaI}} \cdot \frac{253.809 \text{ g } I_2}{1 \text{ mol } I_2} = 16.9 \text{ g } I_2$$

116. Preparation of $BaSO_4$ by:

    a precipitation reaction:

$$BaCl_2(aq) + Na_2SO_4(aq) \rightarrow BaSO_4(s) + 2 \text{ NaCl (aq)}$$

    a gas-forming reaction:

$$BaCO_3(s) + H_2SO_4(aq) \rightarrow BaSO_4(s) + H_2O(\ell) + CO_2(g)$$

118.   Prepare zinc chloride by:

    (a) an acid-base reaction:

$$Zn(OH)_2(s) + 2 \text{ HCl}(aq) \rightarrow ZnCl_2(aq) + 2 H_2O(\ell)$$

    (b) a gas-forming reaction

$$ZnCO_3(s) + 2 \text{ HCl}(aq) \rightarrow ZnCl_2(aq) + H_2O(\ell) + CO_2(g)$$

    (c) an oxidation-reduction reaction

$$Zn(s) + 2 \text{ HCl}(aq) \rightarrow ZnCl_2(aq) + H_2(g)$$

To obtain the salt, you would need to heat the solution to drive off the water, leaving solid zinc chloride.

120. To decide the relative concentrations, calculate the dilutions. Let's assume that the HCl has a concentration of 0.100 M. Then calculate the diluted concentrations in each case.

Student 1 :

20.0 mL of 0.100 M HCl is diluted to 40.0 mL total

The diluted molarity is then : $\frac{0.100 \text{ mol HCl}}{1 \text{ L}} \cdot 0.0200 \text{ L} = 0.400 \text{ L} \cdot M$

$$0.050 \frac{\text{mol HCl}}{\text{L}} = M$$

Student 2 :

20.0 mL of 0.100 M HCl is diluted to 80.0 mL total

The diluted molarity is : $\frac{0.100 \text{ mol HCl}}{1\text{L}} \cdot 0.0200 \text{ L} = 0.0800 \text{ L} \cdot M$

$$0.025 \frac{\text{mol HCl}}{\text{L}} = M$$

So the second student's prepared HCl solution is (c) **half the concentration of the first student's. However,** since the total number of moles of HCl in both solutions *is identical,* they will calculate **the same concentration for the original HCl solution.**

122. For the reaction in which Au is dissolved by treatment with sodium cyanide:
      (a) Oxidizing agent: elemental O_2 Reducing agent: elemental Au

      Substance oxidized: elemental Au (oxidation state changes from 0 to +1)
      Substance reduced: elemental O_2 (oxidation state changes from 0 to -2)

      (b) Volume of 0.075 M NaCN to extract gold from 1000 kg of rock:

      Mass of gold in the rock: $\dfrac{0.019 \text{ g gold}}{100 \text{ g rock}} \cdot \dfrac{10^3 \text{ g rock}}{1 \text{ kg rock}} \cdot 10^3 \text{ kg rock} = 190 \text{ g gold}$

      Amount of NaCN needed (from the balanced equation):

      $190 \text{ g Au} \cdot \dfrac{1 \text{ mol Au}}{196.9 \text{ g Au}} \cdot \dfrac{8 \text{ mol NaCN}}{4 \text{ mol Au}} = 1.9 \text{ mol NaCN}$

      Volume of 0.075 M NaCN that contains that amount of NaCN

      $1.9 \text{ mol NaCN} \cdot \dfrac{1 \text{ L}}{0.075 \text{ mol NaCN}} = 26 \text{ L}$

124. Weight percent of Cu in 0.251 g of a copper containing alloy:
      Several steps in this problem:
      (1) Note the relationship between Cu (Cu^{2+}) and I_3^- formed (2 mol Cu^{2+} : 1 mol I_3^-)
      (2) Excess I_3^- reacts with $S_2O_3^{2-}$ in a 1 mol I_3^- to 2 mol $S_2O_3^{2-}$ ratio)
      (3) Calculate the amount of thiosulfate in 26.32 mL of 0.101 M solution:

      $\dfrac{0.101 \text{ mol } S_2O_3^{2-}}{1 \text{ L}} \cdot 0.02632 \text{ L} = 0.00266 \text{ mol } S_2O_3^{2-}$ (to 3 sf)

      (4) This amount can now be used to calculate I_3^- produced (step 2), and the amount of I_3^-
      related to the amount of Cu present (step 1):

      $0.00266 \text{ mol } S_2O_3^{2-} \cdot \dfrac{1 \text{ mol } I_3^-}{2 \text{ mol } S_2O_3^{2-}} \cdot \dfrac{2 \text{ mol } Cu^{2+}}{1 \text{ mol } I_3^-} \cdot \dfrac{1 \text{ mol Cu}}{1 \text{ mol } Cu^{2+}} \cdot \dfrac{63.546 \text{ g Cu}}{1 \text{ mol Cu}} = 0.169 \text{ g Cu}$

      The weight percent is then: $\dfrac{0.169 \text{ g Cu}}{0.251 \text{ g alloy}} \cdot 100 = 67.3 \text{ \% Cu}$

126. (a) The balanced equation for the formation of cisplatin:

      $(NH_4)_2PtCl_4(aq) + 2 NH_3(aq) \rightarrow Pt(NH_3)_2Cl_2(aq) + 2 NH_4Cl(aq)$

      (b) Mass of $(NH_4)_2PtCl_4$ to make 12.50 g cis-platin:

      Noting that the ratio between the two compounds of interest is a 1:1 mol ratio, we
      can calculate the mass of $(NH_4)_2PtCl_4$ needed using the ratio of the masses of the two
      compounds:

      $12.50 \text{ g Pt(NH}_3)_2\text{Cl}_2 \cdot \dfrac{372.97 \text{ g Pt(NH}_4)_2\text{Cl}_4}{300.05 \text{ g Pt(NH}_3)_2\text{Cl}_2} = 15.54 \text{ g } (NH_4)_2PtCl_4$

Volume of 0.125 M NH_3 required to make 12.50 g of cisplatin:

$$12.50 \text{ g Pt(NH}_3)_2\text{Cl}_2 \cdot \frac{1 \text{mol Pt(NH}_3)_2\text{Cl}_2}{300.05 \text{ g Pt(NH}_3)_2\text{Cl}_2} \cdot \frac{2 \text{ mol NH}_3}{1 \text{ mol Pt(NH}_3)_2\text{Cl}_2}$$

or 0.08332 mol NH_3

Calculating the volume that contains that number of moles of NH_3

$$0.08332 \text{ mol NH}_3 \cdot \frac{1 \text{ L}}{0.125 \text{ mol NH}_3} = 0.667 \text{ L or } 667 \text{ mL (3 sf)}$$

(c) Determine the value of x in the compound $Pt(NH_3)_2Cl_2(C_5H_5N)_x$

Determine the # of moles of cisplatin in 0.150 g of the compound:

$$0.150 \text{ g cisplatin} \cdot \frac{1 \text{mol Pt(NH}_3)_2\text{Cl}_2}{300.05 \text{ g Pt(NH}_3)_2\text{Cl}_2} = 0.000500 \text{ mol cisplatin}$$

Determine the **total** number of moles of pyridine:

$$1.50 \text{ mL pyridine} \cdot \frac{0.979 \text{ g pyridine}}{1 \text{ mL pyridine}} \cdot \frac{1 \text{ mol pyridine}}{79.10 \text{ g pyridine}} = 0.0186 \text{ mol}$$

pyridine

Once some of the pyridine has reacted, the amount of pyridine **remaining** can be calculated by titrating with HCl. The balanced equation shows that 1 mol of HCl reacts with 1 mol pyridine. The # moles HCl (and therefore the # moles pyridine) is:

$$\frac{0.475 \text{ mol HCl}}{1 \text{ L}} \cdot 0.0370 \text{ L} = 0.0176 \text{ mol HCl} = 0.0176 \text{ mol pyridine remaining}$$

The pyridine that reacted is then (0.0186 mol - 0.0176 mol) = 0.0010 mol pyridine.

The # of moles of pyridine PER mol of cisplatin is then : $\dfrac{0.0010 \text{ mol pyridine}}{0.000500 \text{ mol cisplatin}} = 2$

The formula is then: $Pt(NH_3)_2Cl_2(C_5H_5N)_2$.

128.   The amount of HCl contained in all 3 flasks is identical. The amount of Zn in flask 2 is a stroichiometric amount (1 mol Zn: 2 mol HCl). Flask 1 has excess Zn—but HCl limits the amount of hydrogen gas. Flask 3 contains an insufficient amount of Zn to react with all the HCl present—and a smaller volume of hydrogen gas results.

Chapter 6
Principles of Reactivity:
Energy and Chemical Reactions

Reviewing Important Concepts

2. State whether the process is exothermic or endothermic:
   (a) The conversion of liquid water to solid water (we normally call it "ice") requires that heat be removed from the liquid (an exothermic process).
   (b) When elemental oxygen and hydrogen are combined to form water, heat is released to the surroundings. The process is exothermic.
   (c) To cool water from 25 °C to 15 °C, heat must be removed from the water. This process is exothermic.
   (d) To convert liquid water to gaseous water, we must add heat to reach the "higher energy" state. This process is endothermic.

4. Which of the following are state functions?
   (a) The volume of a balloon **is a state function**. The balloon can be filled to a specific volume in many ways, achieving some finite volume which is larger than the initial volume—before inflation.
   (b) The time to drive from your home to your college or university will **not be a state function**. The time will definitely change as a function of the route you choose. If you're caught in a traffic jam on a city street, it might take longer than the time needed if you chose an "expressway". So the time is NOT a state function.
   (c) The temperature of water in a coffee cup is **a state function**. The water could be at 30°C as a result of adding hot water to the cup, then pouring in some cold water. Alternatively, you could reach a temperature of 30°C by starting with water at 25°C and pouring in very hot water.
   (d) The potential energy of a ball in your hand **is a state function**, depending on how high the ball is held above some "ground" or zero energy level. You might hold the ball seven feet off the ground by lifting it above your head. Alternatively you might wish to stand on a short ladder, and hold the ball at waist level. As long as the ball is seven feet of the ground (the zero energy level), the potential energy of the ball is the same—without regard to how the ball reached that height.

10. How do product- and reactant-favored reactions differ? What is the relationship between these terms and the enthalpy change for a reaction?

Product-favored reactions are those in which some substance precipitates, an acid-base reaction occurs, or some gas is formed. These "driving forces" result in the transformation of reactants to products. Reactant-favored reactions are those in which reactants remain largely unchanged, with only some products being formed. **In general**, reactions which are **product-favored** are **exothermic,** while reactions which are **reactant-favored** are typically **endothermic.**

Practicing Skills

Energy Units

12. Express 1200 Calories/day in Joules:

$$\frac{1200\ Cal}{day} \bullet \frac{1000\ calorie}{1\ Cal} \bullet \frac{4.184\ J}{1\ cal} = 5.0 \times 10^6\ Joules/day$$

Specific Heat Capacity

14. What is the specific heat capacity of mercury, if the molar heat capacity is 28.1 J/mol • K?

Note that the difference in **units** of these two quantities is in the amount of substance. In one case, moles, while in the other grams.

$$28.1 \frac{J}{mol \bullet K} \bullet \frac{1\ mol}{200.59\ g} = 0.140 \frac{J}{g \bullet K}$$

16. Heat energy to warm 168g copper from -12.2 °C to 25.6 °C:

Heat = mass x heat capacity x ΔT

For copper = $(168\ g)(\frac{0.385\ J}{g \bullet K})[25.6°C - (-12.2)\ °C] \bullet \frac{1\ K}{1\ °C}$

$= 2.44 \times 10^3\ J$ or 2.44 kJ

18. The final temperature of a 344 g sample of iron when 2.25 kJ of heat are added to a sample originally at 18.2 °C: The energy added is:

q_{Fe} = (mass)(heat capacity)(ΔT)

$2.25 \times 10^3\ J = (344\ g)(0.449 \frac{J}{g \bullet K})(x)$ and solving for x we get:

14.57 K = x. and since 1K = 1°C, ΔT = 14.57 °C.

The final temperature is (14.57 + 18.2)°C or 32.8°C.

20. Final T of copper-water mixture:

We must **assume** that **no energy** will be transferred to or from the beaker containing the water. Then the **magnitude** of energy lost by the hot copper and the energy gained by the cold water will be equal (but opposite in sign).

$$q_{copper} = -q_{water}$$

Using the heat capacities of H_2O and copper, and expressing the temperatures in Kelvin $(K = °C + 273.15)$ we can write:

$$(45.5 \text{ g})(0.385 \frac{J}{g \cdot K})(T_{final} - 372.95 \text{ K}) = -(152. \text{ g})(4.184 \frac{J}{g \cdot K})(T_{final} - 291.65 K)$$

Simplifying each side gives:

$$17.52 \frac{J}{K} \cdot T_{final} - 6533 \text{ J} = -636.0 \frac{J}{K} \cdot T_{final} + 185,480 \text{ J}$$

$$653.52 \frac{J}{K} \cdot T_{final} = 192013 \text{ J}$$

$$T_{final} = 293.81 \text{ K or } (293.81 - 273.15) \text{ or } 20.7 °C$$

Don't forget: **Round numbers only at the end.**

22. Final temperature of water mixture:

This problem is solved almost exactly like question 20. The difference is that both samples are samples of water. From a mechanical standpoint, the heat capacity of both samples will be identical—and can be omitted from both sides of the equation:

$$q_{water} \text{ (at 95 °C)} = -q_{water} \text{ (at 22 °C)}$$

$$(85.2 \text{ g})(4.184 \frac{J}{g \cdot K})(T_{final} - 368.15 \text{ K}) = -(156 \text{ g})(4.184 \frac{J}{g \cdot K})(T_{final} - 295.15 \text{ K}) \text{ or}$$

$$(85.2 \text{ g})(T_{final} - 368.15 \text{ K}) = -(156 \text{ g})(T_{final} - 295.15 \text{ K}) \text{ or}$$

$$85.2 \text{ J/K } T_{final} - 31366.38 \text{ J} = -156 \text{ J/K } T_{final} + 46043.4 \text{ J}$$

rearranging: $241.2 \text{ J/K} \cdot T_{final} = 77409.78 \text{ J}$

$$321.0 \text{ K} = T_{final}$$

$$\text{or } 47.8 °C$$

24. Exercise 6.4 in your text is another good example for this problem.

$$q_{metal} = -q_{water}$$

Remembering that $\Delta T = T_{final} - T_{initial}$, we can calculate the change in temperature for the water and the metal. Further, since we know the final and initial for both the metal and the water, we can calculate the temperature difference in units of Celsius degrees, since the **change** in temperature on the **Kelvin** scale would be numerically identical.

For the metal : $\Delta T = T_{final} - T_{initial} = (27.1 - 98.8)$ or $-71.7°C$ or -71.7 K.

For the water: $\Delta T = T_{final} - T_{initial} = (27.1 - 25.0)$ or $2.1°C$ or 2.1 K (recalling that a Celsius degrees and a Kelvin are the same "size".

$$(13.8 \text{ g})(C_{metal})(-71.7 \text{ K}) = -(45.0 \text{ g})(4.184 \frac{J}{g \cdot K})(2.1 \text{ K})$$

$$-989.46 \text{ g} \cdot K(C_{metal}) \quad = -395 \text{ J}$$

$$C_{metal} \quad = 0.40 \frac{J}{g \cdot K} \quad \text{(to 2 significant figures)}$$

Changes of State

26. Quantity of energy evolved when 1.0 L of water at 0 °C solidifies to ice:.

The mass of water involved: If we assume a density of liquid water of 1.000 g/mL, 1.0 L of water (1000 mL) would have a mass of 1000 g.

To freeze 1000 g water: $1000 \text{ g ice} \cdot \dfrac{333 \text{ J}}{1.000 \text{ g ice}} = 333. \times 10^3$ J or 330 kJ (to 2sf)

28. Heat required to vaporize (convert liquid to solid) 125 g C_6H_6:

The heat of vaporization of benzene is 30.8 kJ/mol.

Convert mass of benzene to moles of benzene: $125 \text{ g} \cdot 78.11 \text{ g/mol} = 1.60$ mol
Heat required: $1.60 \text{ mol } C_6H_6 \cdot 30.8 \text{ kJ/mol} = 49.3$ kJ

NOTE: As in question 26, no sign has been attached to the amount of heat, since we wanted to know the **amount**. If we want to assign a **direction** of heat flow in this question, then we would add a (+) to 49.3 kJ to indicate that heat is being **added** to the liquid benzene.

30. To calculate the quantity of heat for the process described, think of the problem in two steps:

       1. cool liquid from 23.0 °C to liquid at – 38.8 °C

       2. freeze the liquid at it's freezing point (– 38.8 °C)

Note that the specific heat capacity is expressed in units of mass, so convert the volume of liquid mercury to **mass**. $1.00 \text{ mL} \cdot 13.6 \text{ g/mL} = 13.6$ g Hg (Recall:1 cm^3 = 1 mL)

1. The energy to cool 13.6 g of Hg from 23.0 °C to liquid at – 38.8 °C is:

$$\Delta T = (234.35 \text{ K} - 296.15 \text{ K}) \text{ or } - 61.8 \text{ K}$$

$$13.6 \text{ g Hg} \cdot 0.140 \frac{J}{g \cdot K} \cdot -61.8 \text{ K} = -118 \text{ J}$$

2. To convert liquid mercury to solid Hg at this temperature:

$$-11.4 \text{ J/g} \cdot 13.6 \text{ g} = -155 \text{ J} \text{ (The (-) sign indicates that heat is being removed from the Hg.}$$

The total energy released by the Hg is: $[-118 \text{ J} + -155 \text{ J}] = -273$ J and since $q_{mercury} = -q_{surroundings}$ the amount released to the surroundings is 273 J.

32. To accomplish the process, one must:

1. heat the ethanol from 20.0 °C to 78.29 °C (ΔT = 58.29 K)

2. boil the ethanol (convert from liquid to gas) at 78.29 °C

Using the specific heat for ethanol, the energy for the first step is:

$$(2.44 \frac{J}{g \bullet K})(1000 \text{ g})(58.29 \text{ K}) = 142,227.6 \text{ J} (142,000 \text{ J to 3 sf})$$

To boil the ethanol at 78.29 °C, we need:

$$855 \frac{J}{g} \bullet 1000 \text{ g} = 855,000 \text{ J}$$

The total heat energy needed (in J) is (142,000 + 855,000) = 997,000 or 9.97×10^5 J

Enthalpy

Note that in this chapter, I have left negative signs with the value for heat released

(heat released = - ; heat absorbed = +)

34. For a process in which the $\Delta H°$ is negative, that process is **exothermic**.

To calculate heat released when 1.25 g NO react, note that the energy shown (-114.1 kJ) is released when **2** moles of NO react, so we'll need to account for that:

$$1.25 \text{ g NO} \bullet \frac{1 \text{ mol NO}}{30.01 \text{ g NO}} \bullet \frac{-114.1 \text{ kJ}}{2 \text{ mol NO}} = -2.38 \text{ kJ}$$

36. The combustion of isooctane (IO) is **exothermic**. The molar mass of IO is: 114.2 g/mol.

The heat evolved is:

$$1.00 \text{ L of IO} \bullet \frac{0.69 \text{ g IO}}{1 \text{mL}} \bullet \frac{1 \times 10^3 \text{ mL}}{1 \text{ L}} \bullet \frac{1 \text{ mol IO}}{114.2 \text{ g IO}} \bullet \frac{-10922 \text{ kJ}}{2 \text{ mol IO}} = -3.3 \times 10^4 \text{ kJ}$$

Calorimetry

38. 100.0 mL of 0.200 M CsOH and 50.0 mL of 0.400 M HCl each supply 0.0200 moles of base and acid respectively. If we assume the specific heat capacities of the solutions are 4.2 J/g • K, the **heat evolved** for 0.200 moles of CsOH is:

$$q = (4.2 \text{ J/g} \bullet \text{K})(150. \text{ g})(24.28 \text{ °C} - 22.50\text{°C}) \text{ [and since } 1.78°\text{C} = 1.78 \text{ K]}$$
$$q = (4.2 \text{ J/g} \bullet \text{K})(150. \text{ g})(1.78 \text{ K})$$
$$q = 1120 \text{ J}$$

The molar enthalpy of neutralization is: $\dfrac{-1120 \text{ J}}{0.0200 \text{ mol CsOH}} = -56000$ J/mol (to 2 sf)

or -56 kJ/mol

70

40. For the problem, we'll assume that the coffee-cup calorimeter absorbs **no** heat.

$$\text{Since } q_{metal} = -q_{water}$$

Remembering that $\Delta T = T_{final} - T_{initial}$, we can calculate the change in temperature for the water and the metal. Further, since we know the final and initial for both the metal and the water, we can calculate the temperature difference in units of Celsius degrees, since the **change** in temperature on the **Kelvin** scale would be numerically identical.

For the metal : $\Delta T = T_{final} - T_{initial} = (24.3 - 99.5)$ or $-75.2°C$ or $-75.2 K$.

For the water: $\Delta T = T_{final} - T_{initial} = (24.3 - 21.7)$ or $2.6°C$ or $2.6 K$ (recalling that a Celsius degrees and a Kelvin are the same "size".

$$(20.8 \text{ g})(C_{metal})(-75.2 \text{ K}) = -(75.0 \text{ g})(4.184 \frac{J}{g \cdot K})(2.6 \text{ K})$$

$$-1564.16 \text{ g} \cdot K(C_{metal}) = -816 \text{ J}$$

$$C_{metal} = 0.52 \frac{J}{g \cdot K} \text{ (to 2 significant figures)}$$

42. Enthalpy change when 5.44 g of NH_4NO_3 is dissolved in 150.0 g water at 18.6 °C..

Calculate the heat released by the solution: $\Delta T = (16.2 - 18.6)$ or $-2.4 °C$ or $-2.4 K$

$$(155.4 \text{ g})(4.2 \frac{J}{g \cdot K})(-2.4 \text{ K}) = -1566 \text{ J or } -1600 \text{ J(to 2 sf)}$$

Calculate the amount of NH_4NO_3: $5.44 \text{ g NH}_4NO_3 \cdot \frac{1 \text{ mol NH}_4NO_3}{80.04 \text{ g NH}_4NO_3} = 0.0680 \text{ mol}$

Recall that the energy that was released by the solution is **absorbed** by the ammonium nitrate, so we change the sign from (-) to (+). The enthalpy change has been requested in units of kJ, so divide the energy (in J) by 1000:

Enthalpy of dissolving $= \dfrac{1.566 \text{ kJ}}{0.0680 \text{ mol}} = 23.0 \text{ kJmol or 23 kJ/mol (to 2 sf)}$

44. Calculate the heat evolved (per mol SO_2) for the reaction of sulfur with oxygen to form SO_2
There are several steps:

1) Calculate the heat transferred to the water:

$$815 \text{ g} \cdot 4.184 \frac{J}{g \cdot K} \cdot (26.72 - 21.25)°C \cdot 1K/1°C = 18{,}700 \text{ J}$$

2) Calculate the heat transferred to the bomb calorimeter

$$923 \text{ J/K} \cdot (26.72 - 21.25)°C \cdot 1K/1°C = 5{,}050. \text{J}$$

3) Amount of sulfur present: $2.56 \text{ g} \cdot \dfrac{1 \text{ mol S}_8}{256.536 \text{ g S}_8} = 0.010 \text{ mol S}_8$

Note from the equation that 8 mol of SO_2 form from each mole of S_8

4) Calculating the quantity of heat related per mol of SO_2 yields:

$$\frac{(18{,}700 \text{ J} + 5{,}050 \text{ J})}{0.08 \text{ mol SO}_2} = 297{,}000 \text{ J/mol SO}_2 \text{ or 297 kJ/mol SO}_2$$

46. Quantity of heat evolved in the combustion of benzoic acid:

    As in study question 44, we can approach this in several steps:

    1) Calculate the heat transferred to the water:

    $$775 \text{ g} \bullet 4.184 \frac{J}{g \bullet K} \bullet (31.69 - 22.50)°C \bullet 1K/1°C = 29,800 \text{ J}$$

    2) Calculate the heat transferred to the bomb calorimeter

    $$893 \text{ J/K} \bullet (31.69 - 22.50)°C \bullet 1K/1°C = 8,210 \text{ J}$$

    3) Amount of benzoic acid:

    $$1.500 \text{ g benzoic acid} \bullet \frac{1 \text{ mol benzoic acid}}{122.1 \text{ g benzoic acid}} = 1.229 \times 10^{-2} \text{ mol benzoic acid}$$

    4) Heat evolved per mol of benzoic acid is:

    $$\frac{(29,800 \text{ J} + 8,210 \text{ J})}{1.229 \times 10^{-2} \text{ mol}} = 3,090 \text{ J/mol}$$

48. Heat absorbed by the ice : $\frac{333 \text{ J}}{1.00 \text{ g ice}} \bullet 3.54 \text{ g ice} = 1,180 \text{ J}$ (to 3 sf)

    Since this energy (1180 J) is released by the metal, we can calculate the heat capacity of the

    metal: heat = heat capacity x mass x ΔT

    -1180 J = C x 50.0 g x (273.2 K - 373 K) [Note that ΔT is negative!]

    $0.236 \frac{J}{g \bullet K}$ = C

    Note that the heat released (left side of equation) has a negative sign to indicate the

    directional flow of the energy.

Hess's Law

50.   (a) Hess' Law allows us to calculate the overall enthalpy change by the appropriate

        combination of several equations. In this case we add the two equations, reversing the

        second one (with the concommitant reversal of sign)

$CH_4 (g) + 2 O_2 (g) \rightarrow CO_2 (g) + 2 H_2O (g)$	$\Delta H° = - 802.4 \text{ kJ}$
$CO_2 (g) + 2 H_2O (g) \rightarrow CH_3OH(g) + 3/2 O_2 (g)$	$\Delta H° = + 676 \text{ kJ}$
$CH_4 (g) + 1/2 O_2 (g) \rightarrow CH_3OH(g)$	$\Delta H° = - 126 \text{ kJ}$

(b) A graphic description of the energy change:

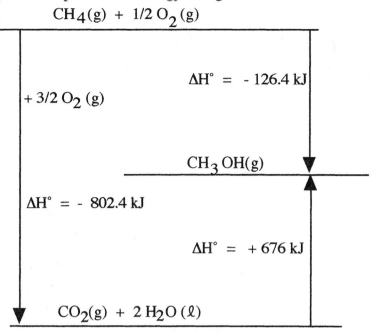

$$CH_4(g) + 1/2\,O_2(g)$$

$\Delta H° = -126.4$ kJ

$+ 3/2\,O_2(g)$

$CH_3OH(g)$

$\Delta H° = -802.4$ kJ

$\Delta H° = +676$ kJ

$CO_2(g) + 2\,H_2O(\ell)$

52. The overall enthalpy change for $1/2\,N_2(g) + 1/2\,O_2(g) \rightarrow NO(g)$

For the overall equation, note that elemental nitrogen and oxygen are on the "left" side of the equation, and NO on the "right" side of the equation. Noting that equation 2 has 4 ammonia molecules consumed, let's multiply equation 1 by 2:

$2\,N_2(g) + 6\,H_2(g) \rightarrow 4\,NH_3(g)$ $\qquad\qquad \Delta H = (2)(-91.8\text{ kJ})$

The second equation has NO on the right side :

$4\,NH_3(g) + 5\,O_2(g) \rightarrow 4\,NO(g) + 6\,H_2O(g)$ $\qquad \Delta H = -906.2\text{kJ}$

The third equation has water as a product, and we need to "consume" the water formed in equation two, so let's reverse equation 3—changing the sign--AND multiply it by 6

$6\,H_2O(g) \rightarrow 6\,H_2(g) + 3\,O_2(g)$ $\qquad\qquad \Delta H = (+241.8)(6)\text{ kJ}$

Adding these 3 equations gives

$2\,N_2(g) + 2\,O_2(g) \rightarrow 4\,NO(g)$ $\qquad\qquad \Delta H = 361\text{ kJ}$

So, dividing all the coefficients by 4 provides the desired equation with a $\Delta H = +361 \cdot 0.25$ kJ or 90.3 kJ

Standard Enthalpies of Formation
54. The equation requested requires that we form **one** mol of product liquid CH_3OH from its elements—each in their standard state.

Begin by writing a balanced equation:

$2\,C\,(s,graphite) + O_2(g) + 2\,H_2(g) \rightarrow \qquad 2\,CH_3OH(\ell)$

Now express the reaction so that you form one mole of CH3OH--divide coefficients by 2

$$C \text{ (s) } + 1/2 \text{ } O_2\text{(g) } + H_2\text{(g) } \rightarrow \quad CH_3OH(\ell)$$

And from Appendix L, the $\Delta H_f°$ is reported as -238.4 kJ/mol

56.   (a) The equation of the formation of Cr_2O_3 (s) from the elements:

$$2 \text{ Cr (s) } + 3/2 \text{ } O_2 \text{ (g) } \rightarrow Cr_2O_3 \text{ (s)}$$

from Appendix L $\Delta H_f°$ is reported as: -1134.7 kJ/mol for the oxide.

(b) The enthalpy change if 2.4 g of Cr is oxidized to Cr_2O_3 (g) is:

$$2.4 \text{ g Cr} \bullet \frac{1 \text{ mol Cr}}{52.0 \text{ g Cr}} \bullet \frac{-1134.7 \text{ kJ}}{1 \text{ mol Cr}} = -52 \text{ kJ} \quad \text{(to 2 sf)}$$

58.   Calculate $\Delta H_{rx}°$ for the following processes:

(a) 1.0 g of white phosphorus burns:

$$P_4 \text{ (s) } + 5 \text{ } O_2 \text{ (g) } \rightarrow P_4O_{10} \text{ (s)} \quad \text{from Appendix L: } \Delta H_f° \text{ -2984.0 kJ/mol}$$

$$1.0 \text{ g } P_4 \bullet \frac{1.0 \text{ mol } P_4}{123.89 \text{ g } P_4} \bullet \frac{-2984.0 \text{ kJ}}{1 \text{ mol } P_4} = -24 \text{ kJ}$$

(b) 0.20 mol NO (g) decomposes to N_2 (g) and O_2 (g):

From Appendix L: $\Delta H_f°$ for NO = 90.29 kJ/mol

Since the reaction requested is the **reverse** of $\Delta H_f°$, we change the sign to

-90.29 kJ/mol. The enthalpy change is then $\dfrac{-90.29 \text{ kJ}}{1 \text{ mol}} \bullet 0.20 \text{ mol} = -18 \text{ kJ}$

(c) 2.40 g NaCl is formed from elemental Na and elemental Cl_2:

From Appendix L: $\Delta H_f°$ for NaCl (s) = -411.12 kJ/mol

The amount of NaCl is: $2.40 \text{ g NaCl} \bullet \dfrac{1 \text{ mol NaCl}}{58.44 \text{ g NaCl}} = 0.0411 \text{ mol}$

The overall energy change is: -411.12 kJ/mol $\bullet$ 0.0411 mol = -16.9 kJ

(d) 250 g of Fe oxidized to Fe_2O_3 (s):

From Appendix L: $\Delta H_f°$ for Fe_2O_3 (s) = -824.2 kJ/mol

The overall energy change is:

$$250 \text{ g Fe} \bullet \frac{1 \text{ mol Fe}}{55.8847 \text{ g Fe}} \bullet \frac{1 \text{ mol Fe}_2O_3}{2 \text{ mol Fe}} \frac{-824.2 \text{ kJ}}{1 \text{ mol Fe}_2O_3} = -1.8 \times 10^3 \text{ kJ}$$

60. (a) The enthalpy change for the reaction:

$$4 \text{ NH}_3\text{(g) } + 5 \text{ } O_2\text{(g) } \quad \rightarrow \quad 4 \text{ NO(g) } + 6 \text{ } H_2O\text{(g)}$$

$\Delta H°_f\text{(kJ/mol)}$ -45.90   0   +90.29   -241.83

$$\Delta H°_{rxn} = [(4 \text{ mol})(+90.29 \tfrac{\text{kJ}}{\text{mol}}) + (6 \text{ mol})(-241.83 \tfrac{\text{kJ}}{\text{mol}})] -$$

$$[(4 \text{ mol})(-45.90 \tfrac{\text{kJ}}{\text{mol}}) + (5 \text{ mol})(0)]$$

$$= (-1089.82 \text{ kJ}) - (-183.6 \text{ kJ})$$

$$= -906.2 \text{ kJ} \quad \textbf{The reaction is exothermic.}$$

(b) Heat evolved when 10.0 g NH3 react:

The balanced equation shows that 4 mol NH3 result in the release of 906.2 kJ.

$$10.0 \text{ g NH3} \cdot \frac{1 \text{ mol NH3}}{17.03 \text{ g NH3}} \cdot \frac{-906.2 \text{ kJ}}{4 \text{ mol NH}_3} = -133 \text{ kJ}$$

62. (a) The enthalpy change for the reaction:

$$\text{BaO}_2 \text{ (s)} \quad \rightarrow \quad \text{BaO (s)} + 1/2 \text{ O}_2 \text{ (g)}$$

Given $\Delta H°_f$ for BaO is: - 553.5 kJ/mol and $\Delta H°_f$ for BaO2 is: -634.3 kJ/mol

This equation can be seen as the summation of the two equations:

(1)   Ba (s) + 1/2 O2 (g) →   BaO (s)

(2) $\underline{\text{BaO}_2 \text{ (s)} \quad \rightarrow \quad \text{Ba (s)} + \text{O}_2 \text{ (g)}}$

Equation (1) corresponds to the formation of BaO while equation(2) corresponds to the **reverse** of the formation of BaO2

$$\Delta H°_{rxn} = (\Delta H°_f \text{ for BaO}) + -1(\Delta H°_f \text{ for BaO2}) =$$

$$\Delta H°_{rxn} = 80.8 \text{ kJ} \quad \text{and the reaction is } \textbf{endothermic.}$$

(b) Energy level diagram for the equations in question:

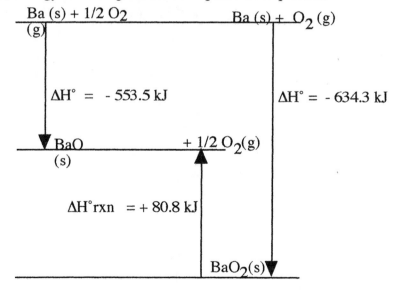

64. The molar enthalpy of formation of naphthalene can be calculated since we're given the enthalpic change for the reaction:

$$\text{C}_{10}\text{H}_8 \text{ (s)} + 12 \text{ O}_2\text{(g)} \quad \rightarrow \quad 10 \text{ CO}_2\text{(g)} + 4 \text{ H}_2\text{O(}\ell\text{)}$$

| $\Delta H°_f$(kJ/mol) | ? | 0 | -393.509 | -285.83 |

$$\Delta H°_{rxn} \quad = \sum \Delta H°_f \text{ products } - \sum \Delta H°_f \text{ reactants}$$

$$-5156.1 \text{ kJ} \quad = [(10 \text{ mol})(-393.509 \frac{kJ}{mol}) + (4 \text{ mol})(-285.83 \frac{kJ}{mol})] - [\Delta H°_f C_{10}H_8]$$

$$-5156.1 \text{ kJ} \quad = (-5078.41 \text{ kJ}) - \Delta H°_f C_{10}H_8$$

$$-77.7 \text{ kJ} \quad = -\Delta H°_f C_{10}H_8$$

$$77.7 \text{ kJ} \quad = \Delta H°_f C_{10}H_8$$

Product- and Reactant-Favored Reactions

66. Calculate $\Delta H°_{rxn}$ for the equations and draw an energy level diagram for:

(a) Aluminum and chlorine to form $AlCl_3$ (s): Al (s) $+ 3/2 \, Cl_2$ (g) $\rightarrow AlCl_3$ (s)

This equation represents the formation of $AlCl_3$ from it's elements, so we can

calculate the $\Delta H°_{rxn}$ by looking up the $\Delta H°_f$ for $AlCl_3$ in Appendix L which is:

–705.63 kJ/mol   (Reaction is product-favored.)

(b) Decomposition of HgO to form elemental liquid mercury and oxygen gas:
HgO (s) $\rightarrow Hg$ (ℓ) $+ 1/2 \, O_2$ (g)

This equation represents the decomposition of HgO into its component elements,
so we can calculate the $\Delta H°_{rxn}$ by looking up the $\Delta H°_f$ for HgO in Appendix L

(which is: –90.83 kJ/mol) and **changing the sign to +90.83 kJ/mol.**

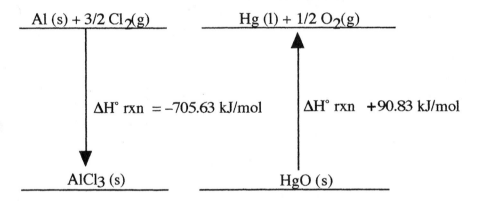

General Questions on Thermochemistry

68. $q_{metal} = $ heat capacity x mass x ΔT

$q_{metal} = C_{metal} \cdot 27.3 \text{ g} \cdot (299.47 \text{ K} - 372.05 \text{ K})$

Note that ΔT is negative, since T_{final} of the metal is LESS THAN $T_{initial}$]

and $q_{water} = 15.0 \text{ g} \cdot \dfrac{4.184 \text{ J}}{g \cdot K} \cdot (299.47 \text{ K} - 295.65 \text{K}) = 239.7 \text{ J}$

Setting $q_{metal} = - q_{water}$

$C_{metal} \cdot 27.3 \text{ g} \cdot (-72.58 \text{ K}) = - 239.7 \text{ J}$ and solving for C gives:

$C_{metal} = 0.121 \dfrac{J}{g \cdot K}$

70. Final T of copper-water mixture:

We must **assume** that **no energy** will be transferred to or from the beaker containing the water. Then the **magnitude** of energy lost by the hot copper and the energy gained by the cold water will be equal (but opposite in sign).

$$q_{copper} = -q_{water}$$

Using the heat capacities of H_2O and copper, and expressing the temperatures in Kelvin we can write:

mass of water

$$(192 \text{ g})(0.385 \frac{J}{g \cdot K})(T_{final} - 373.2 \text{ K}) = -(750. \text{ mL})(1.00 \frac{g}{mL})(4.184 \frac{J}{g \cdot K})(T_{final} - 277.2 \text{ K})$$

Simplifying each side gives:

$$73.92 \frac{J}{K} \cdot T_{final} - 27{,}600 \text{ J} = -3138 \frac{J}{K} \cdot T_{final} + 870{,}000 \text{ J}$$

$$3212 \frac{J}{K} \cdot T_{final} = 897600 \text{ J}$$

$$T_{final} = 279.4 \text{ K} \text{ or } 6.2 \text{ °C}$$

Don't forget: **Round numbers only at the end.** To show the steps, I have rounded some of these intermediate numbers.

72. To calculate the quantity of heat for the process described, think of the problem in three steps:

1. melt ice at 0°C to liquid water at 0°C
2. warm liquid water from 0°C to 100°C
3. convert liquid water at 100 °C to gaseous water at 100 °C

1. The energy to melt 60.1 g of ice at 0 °C is:

$$60.1 \text{ g ice} \cdot \frac{333 \text{ J}}{\text{g ice}} = 2.00 \times 10^4 \text{ J}$$

2. The energy required to warm the liquid water from 0°C to 100 °C ($\Delta T = 100$ K) is:

$$60.1 \text{ g} \cdot 4.18 \frac{J}{g \cdot K} \cdot 100 \text{ K} = 2.51 \times 10^4 \text{ J}$$

3. To convert liquid water at 100 °C to gaseous water at 100 °C:

$$2260 \text{ J/g} \cdot 60.1 \text{ g} = 13.6 \times 10^4 \text{ J}$$

The total energy required is: $[2.00 \times 10^4 + 2.51 \times 10^4 \text{ J} + 13.6 \times 10^4 \text{J}] = 180$ kJ

<u>74</u>. To calculate the quantity of heat for the process described, think of the problem in two steps:

    1. calculate energy required to cool the tea to 0 °C from 20.0 °C.

    2. calculate amount of ice that absorbs that amount of energy while it melts.

Assume that the density of tea at this temperature is 1.00 g/ml, so 5.00×10^2 mL will have a mass of 5.00×10^2 g.

1. The energy to cool the tea to 0 °C is:

$$5.00 \times 10^2 \, g \bullet 4.18\frac{J}{g \bullet K} \bullet 20 \, K = 4.18 \times 10^4 \, J$$

2. The amount of ice that absorbs that amount of energy:

$$x \, g \, ice \bullet \frac{333 \, J}{1 \, g \, ice} = 4.18 \times 10^4 \, J \text{ so we solve for x.}$$

$$\frac{4.18 \times 10^4 \, J}{1} \bullet \frac{1 \, g \, ice}{333 \, J} = 125.5 \, g \, ice \; (\text{ or } 126 \text{ to 3 sf})$$

The total amount of ice present (initially) is 3 cubes • 45g/cube = 135 g ice

The amount left unmelted (and floating) would be: (135 g ice - 125.5 g ice) or 9 g (to 1 sf)

76. Let's assume that the cola is chilled from 10.5 °C to 0 °C. Further we'll assume that the density of the cola is 1.0g/mL, so 240 mL of cola will have a mass of 240 g.

The energy required to do that would be:

$$240 \, g \bullet 4.18\frac{J}{g \bullet K} \bullet 10.5 \, K = 1.05 \times 10^4 \, J$$

Calculate the feasibility of this happening! Do we have enough ice?

$$\frac{1.05 \times 10^4 \, J}{1} \bullet \frac{1 \, g \, ice}{333 \, J} = 31.6 \, g \, ice \text{ required to provide that energy.}$$

So **we have enough ice**! We should have the cola chilled to 0°C and (45-31.6)g or 13 g of ice should remain (to 2 sf).

78. Calculate the enthalpy change for the precipitation of AgCl (in kJ/mol):

1) How much AgCl is being formed?

250. mL of 0.16M $AgNO_3$ will contain (0.250L • 0.16 mol/L) 0.040 mol of $AgNO_3$

125 mL of 0.36M NaCl will contain (0.125L • 0.36 mol/L) 0.045 mol of NaCl.

Given the stoichiometry of the process, we anticipate the formation of 0.040 mol of AgCl.

2) How much energy is evolved?

$$375 \, g \bullet 4.2\frac{J}{g \bullet K} \bullet (296.05 \, K - 294.30) \, K = 2{,}800 \, J \text{ (to 2 sf)}$$

The enthalpy change is then - 2800 J (since the reaction **releases** heat).

The change in kJ/mol is $\dfrac{2800 \, J}{0.040 \, mol} \bullet \dfrac{1 \, kJ}{1000 \, J} = -69 \, kJ/mol$

80. Heat evolved when ammonium nitrate is decomposed:
    $\Delta T = (20.72-18.90) = 1.82\ °C$ (or 1.82 K).
    Heat absorbed by the calorimeter: 155 J/K • 1.82K = 282 J

    Heat absorbed by the water : 415 g • $4.18\dfrac{J}{g\cdot K}$ • 1.82 K = 3160 J

    Total heat **released** by the decomposition: 3160 J + 282 J = 3,440 J (to 3 sf)

    With 7.647 g NH4NO3 = 0.09554 mol, heat released $= \dfrac{3440\ J}{0.09554\ mol} = 36.0$ kJ/mol

82. One can arrive at the desired answer if you recall the **definition** of $\Delta H°_f$. The definition is the
    enthalpy change associated with the formation of **one mole** of the substance (in this case
    B_2H_6) from its elements—each in their standard state (s for boron and g for hydrogen).

    (a) Note that the 1st equation given uses **four** moles of B as a reactant — and we'll need
    only 2, so divide the first equation by 2 to give:
    $$2\ B\ (s) + 3/2\ O_2\ (g)\quad\rightarrow\quad B_2O_3\ (s)\qquad \Delta H° = 1/2(-2543.8\ kJ) = -1271.9\ kJ$$
    The formation of 1 mole of B_2H_6 will require the use of 6 moles of H (or 3 moles of H_2),
    so multiply the second equation by 3 to give:
    $$3\ H_2\ (g) + 3/2\ O_2\ (g)\quad\rightarrow\quad 3\ H_2O\ (g)\qquad \Delta H° = 3(-241.8\ kJ) = -725.4\ kJ$$
    Finally the third equation given has B_2H_6 as a **reactant and not a product**. So reverse
    the third equation to give:
    $$B_2O_3\ (s) + 3\ H_2O\ (g)\ \rightarrow B_2H_6\ (g) + 3\ O_2\ (g)\ \Delta H° = -(-2032.9\ kJ) = +2032.9kJ$$
    (b) Adding the three equations gives the equation:
    $$2\ B\ (s) + 3\ H_2\ (g) \rightarrow\ B_2H_6\ (g)\ \text{with a }\Delta H° = (-1271.9 + -725.4 + 2032.9)kJ$$
    $$\text{or a }\Delta H°\text{ for }B_2H_6\ (g)\text{ of } + 35.6\ kJ$$

    (c) Energy level diagram for the reactions:

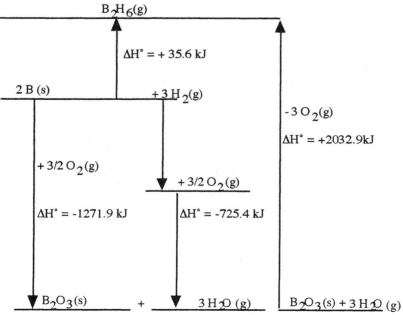

(d) Formation of B_2H_6 (g) is reactant-favored.

84. The enthalpy change for the reaction:

$$Mg(s) + 2 H_2O(\ell) \rightarrow Mg(OH)_2 (s) + H_2(g)$$

$\Delta H°_f$(kJ/mol)   0   -285.83   -924.54   0

$$\Delta H°_{rxn} = (1\ mol)(-924.54\ \frac{kJ}{mol}) - (2\ mol)(-285.83\ \frac{kJ}{mol})$$

$$= -352.88\ kJ\ or\ -3.5288 \times 10^5\ J$$

Each mole of magnesium releases 352.88 kJ of heat energy.

Calculate the heat required to warm 250 mL of water from 25 to 85 °C.

$$heat = heat\ capacity\ x\ mass\ x\ \Delta T$$

$$= (4.184\ \frac{kJ}{mol})(250\ mL)(\frac{1.00\ g}{1\ mL})(60\ K)$$

$$= 62760\ or\ 63000\ J\ or\ 63\ kJ\ (to\ 2\ sf)$$

Magnesium required:

$$63\ kJ \cdot \frac{1\ mol\ Mg}{352.88\ kJ} \cdot \frac{24.3\ g\ Mg}{1\ mol\ Mg} = 4.3\ g\ Mg$$

86. Enthalpy change for:

$$C(s) + H_2O(g) \rightarrow CO(g) + H_2(g)$$

$\Delta H°_f$(kJ/mol)   0   -241.83   -110.525   0

$$\Delta H°_{rxn} = [(1\ mol)(-110.525\ \frac{kJ}{mol}) + 0] - [0 + (1\ mol)(-241.83\ \frac{kJ}{mol})]$$

$$= +131.31\ kJ$$

(b) The process is **endothermic,** so the reaction is **reactant-favored**.

(c) Heat involved when 1.0 metric ton (1000.0 kg) of C is converted to coal gas:

$$1000.0\ kg\ C \cdot \frac{1000\ g\ C}{1\ kg\ C} \cdot \frac{1\ mol\ C}{12.011\ g\ C} \cdot \frac{+131.31\ kJ}{1\ mol\ C} = 1.0932 \times 10^7\ kJ$$

88.   For the combustion of C_8H_{18}:

$$C_8H_{18}(\ell) + 25/2\ O_2(g) \rightarrow 8\ CO_2(g) + 9\ H_2O(\ell)$$

$$\Delta H°_{rxn} = [(8\ mol)(-393.509\ \frac{kJ}{mol}) + (9\ mol)(-285.83\ \frac{kJ}{mol})] - [(1\ mol)(-259.2\ \frac{kJ}{mol}) + 0]$$

$$\Delta H°_{rxn} = -5461.3\ kJ$$

Expressed on a gram basis: $-5461.3\ \frac{kJ}{mol} \cdot \frac{1\ mol\ C_8H_{18}}{114.2\ g\ C_8H_{18}} = -47.81\ kJ/g$

For the combustion of CH_3OH:

$$2\ CH_3OH(\ell)\ +\ 3\ O_2\ (g)\ \rightarrow\ 2\ CO_2\ (g)\ +\ 4\ H_2O(\ell)$$

$$\Delta H^\circ_{rxn}\ =\ [(2 mol)(-393.509\ kJ/mol)\ +\ (4\ mol)(-285.83\ kJ/mol)]$$
$$-\ [(2mol)(-238.4\ kJ/mol)\ +\ 0]$$

$$=\ [(-787.0)\ +\ (-967.2)]\ +\ 477.4\ kJ$$

$$=\ -1453.5\ kJ\ or\ -726.77\ kJ/mol$$

Express this on a per mol and per gram basis:

$$\frac{-1453.5\ kJ}{2\ mol\ CH_3OH}\ \bullet\ \frac{1\ mol\ CH_3OH}{32.04\ g\ CH_3OH}\ =\ -22.682\ kJ/g$$

On a per gram basis, **octane liberates the greater amount** of heat energy.

90.   The molar heat capacities for Al, Fe, Cu, and Au are:

$$0.897\ \frac{J}{g\bullet K}\ \bullet\ \frac{26.98\ g\ Al}{1\ mol\ Al}\ =\ 24.2\ \frac{J}{mol\bullet K}$$

$$0.449\ \frac{J}{g\bullet K}\ \bullet\ \frac{55.85\ gFe}{1\ mol\ Fe}\ =\ 25.1\ \frac{J}{mol\bullet K}$$

$$0.385\ \frac{J}{g\bullet K}\ \bullet\ \frac{63.55\ g\ Cu}{1\ mol\ Cu}\ =\ 24.5\ \frac{J}{mol\bullet K}$$

$$0.129\ \frac{J}{g\bullet K}\ \bullet\ \frac{197.0\ g\ Au}{1\ mol\ Au}\ =\ 25.4\ \frac{J}{mol\bullet K}$$

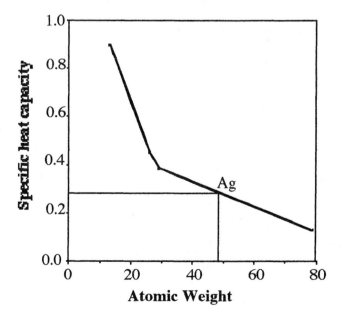

The graph shown is a plot of specific heat capacity versus atomic weight. As you can see, no simple linear relationship exists for these metals. The plot of the specific heat of Cu (atomic weight 63.55) and Au (atomic weight 197) does show a **decreasing** value of specific heat capacity as the atomic weight of the element increases. If you estimate the atomic weight to be about 100 (exact value is about 108), one could **estimate** a value of approximately 0.28 as the specific heat (compared to the experimental value of 0.236.

Alternatively, a quick examination of the values for the four metals above indicates that they are **quite similar**, with an average of 24.8 J/mol • K. This translates into:

$$24.8 \frac{J}{mol \bullet K} \bullet \frac{1 \ mol \ Au}{107.9 \ g \ Au} = 0.230 \ J/g \bullet K$$

<u>92</u>. This is a losing battle. To extract heat from the inside of the refrigerator, work has to be done. That work (by the condenser and motor) releases heat to the environment (your room). So while the temporary relief of cool air from the inside of the refrigerator is pleasant, the motor has to do work—and heats your room.

94.   (a) Energy used to raise temperature of 350mL of soda from 5°C to 37 °C:

Assume the density of the soda is 1.0 g/mL and a specific heat capacity of 4.18 J/g•K.

Heat absorbed (**gained ; q=+**) by the soda (and therefore released by the body) is:

$$350 \ g \ soda \bullet 4.18 \frac{J}{g \bullet K} \bullet 32 \ K = 46816 \ J \ or \ 47 \ kJ \ (\ to \ 2 \ sf)$$

Converting this energy to kcal (the food Calorie):

$$47 \ kJ \bullet \frac{1 \ kcal}{4.184 \ kJ} \bullet \frac{1 \ Calorie}{1 \ kcal} = 11 \ Calorie$$

(b) The soda has 1 Calorie. The body **consumed** 11 Calories—a net **loss** of 10 Calories.

96. Mass of methane needed to heat the air from 15.0 to 22.0 °C:

Calculate the volume of air, then with the density and average molar mass, the moles of air present:

$$275 \ m^2 \bullet 2.50 \ m \bullet \frac{1000 \ L}{1 \ m^3} \bullet \frac{1.22 \ g \ air}{1 \ L \ air} \bullet \frac{1 \ mol \ air}{28.9 \ g \ air} = 2.90. \times 10^4 \ mol \ air$$

The energy needed to change the temperature of that amount of air by (22.0 − 15.0)°C:

$$2.90. \times 10^4 \ mol \ air \bullet 29.1 \frac{J}{mol \bullet K} \bullet 7.0 \ K = 5.9 \times 10^6 \ J$$

What quantity of energy does the combustion of methane provide?

The reaction may be written: $CH_4 (g) + 2\ O_2 (g) \rightarrow 2\ H_2O (g) + CO_2 (g)$

Using data from Appendix L:

$$\Delta H_{rxn} = [(2\ mol)(-241.83\ kJ/mol) + (1mol)(-393.509\ kJ/mol)]$$

$$- [(1\ mol)(-74.87\ kJ/mol) + (2\ mol)(0)] = -802.3\ kJ$$

The amount of methane necessary is:

$$5.9 \times 10^6\ J \cdot \frac{1\ kJ}{1000\ J} \cdot \frac{1\ mol\ CH_4}{802.3\ kJ} \cdot \frac{16.0\ g\ CH_4}{1\ mol\ CH_4} = 120\ g\ CH_4\ \text{(to 2 sf)}$$

98. The Energy level diagram for the isomers of butene:

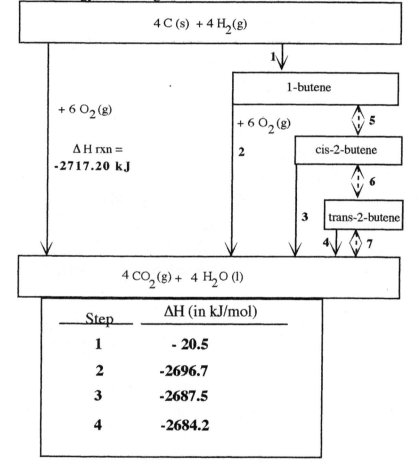

Step	ΔH (in kJ/mol)
1	- 20.5
2	-2696.7
3	-2687.5
4	-2684.2

The enthalpy change from *cis*-2-butene to *trans*-2-butene:

This energy difference corresponds to step 6 in the diagram. Note that this difference corresponds to the difference in the Enthalpies of Combustion of steps 3-4 or: $\Delta H = (-2684.2) - (-2687.5) = -3.3\ kJ$

The enthalpies of formation for both *cis*-2-butene and *trans*-2-butene:

> Step 1 corresponds to the ΔH_f for 1-butene. Step 1 + step 5 corresponds to the ΔH_f for *cis*-2-butene. Step 5 corresponds to the difference in the Enthalpies of Combustion of steps 2-3 or:
>
> $\Delta H = (-2687.5) - (-2696.7) = -9.2$ kJ
>
> Step 1 + Step 5 = $(-20.5) + (-9.2) = -29.7$ kJ (ΔH_f for *cis*-2-butene)

> Using the same logic as above, step 1 + step 5 + step 6 corresponds to the ΔH_f for *trans*-2-butene. The magnitude for step 6 corresponds to the difference in the Enthalpies of Combustion of steps 3-4, as calculated in part b above, or:
>
> $\Delta H = (-2684.2) - (-2687.5) = -3.3$ kJ
>
> Step 1 + Step 5 + Step 6 = $(-20.5) + (-9.2) + (-3.3) = -33.0$ kJ
>
> (ΔH_f for *trans*2-butene)

100.   The desired equation is: CH_4 (g) + 3 Cl_2 (g) $\rightarrow$ 3 HCl (g) + $CHCl_3$ (g)

> Begin with equation 1 (the combustion of methane)
>
> CH_4 (g) + 2 O_2 (g) $\rightarrow$ 2 H_2O (l) + CO_2 (g) $\Delta H = -890.3$ kJ = -890.3 kJ
>
> Noting that we form HCl as one of the products, using the second equation, we need to **reverse** it and (to adjust the coefficient of HCl to 3), multiply by 3/2 to give:
>
> 3/2 H_2 (g) + 3/2 Cl_2 (g) $\rightarrow$ 3 HCl $\Delta H = -3/2(+184.6)$ kJ = -276.9 kJ
>
> Note that CO_2 formed in equation 1 doesn't appear in the overall equation so let's use equation 3 (reversed) to "consume" the CO_2:
>
> CO_2 (g) $\rightarrow$ C (graphite) + O_2 (g) $\Delta H = -1(-393.5)$ kJ = + 393.5 kJ
>
> Noting also that equation 1 produces 2 water molecules, let's "consume" them by using equation 4 (reversed) multiplied by 2:
>
> 2 H_2O (l) $\rightarrow$ 2 H_2 (g) + O_2 (g) $\Delta H = -2(-285.8)$ kJ = + 571.6 kJ
>
> and finally we need to produce $CHCl_3$ which we can do with the equation that represents the ΔH_f for $CHCl_3$:
>
> C(graphite) + 1/2 H_2 (g) + 3/2 Cl_2 (g) $\rightarrow$ $CHCl_3$ (g) $\Delta H = -103.1$ kJ
>
> The overall enthalpy change would then be:
>
> $\Delta H = -890.3$ kJ - 276.9 kJ + 393.5 kJ + 571.6 kJ -103.1 kJ = -305.3 kJ

102. The parking lot is 325m long and 50.0 m wide, or a surface area of 16250 m^2

> If the solar radiation is 2.6 x 10^7 J/m^2 (per day) , the parking lot would receive:
>
> is 2.6 x 10^7 J/m^2 • 16250 m^2 = 4.2 x 10^{11} J (per day).

Chapter 7
Atomic Structure

Reviewing Important Concepts

15. Number of nodal surfaces for the following orbital types:

orbital type	nodal surfaces
s	0 (because $\ell = 0$)
p	1 (because $\ell = 1$)
d	2 (because $\ell = 2$)
f	3 (because $\ell = 3$)

17. Orbital types associated with values of "ℓ"

orbital type	values of "ℓ"
f	$\ell = 3$
s	$\ell = 0$
p	$\ell = 1$
d	$\ell = 2$

19.

Orbital Type	Number of orbitals in a Given Subshell	Number of Nodal Surfaces
s	1	0
p	3	1
d	5	2
f	7	3

Practicing Skills
Electromagnetic Radiation

20. Using Figure 7.3:

(a) Microwave radiation is less energetic than X-ray radiation.

(b) Red light uses higher frequency light than Radar.

(c) Infrared radiation is of longer wavelength than ultraviolet.

22. (a) The higher frequency light is the green 500 nm light.

Recall that frequency and wavelength are inversely related.

(b) The frequency of amber light (595 nm) is:

$$\text{frequency} = \frac{\text{speed of light}}{\text{wavelength}} = \frac{2.9979 \times 10^8 \text{m/s}}{595 \text{ nm}} \cdot \frac{1.00 \times 10^9 \text{nm}}{1.00 \text{ m}}$$

$$= 5.04 \times 10^{14} \text{ s}^{-1}$$

24. To calculate the energy of one photon of light with 500 nm wavelength, we need to first calculate the frequency of the radiation:

$$\text{frequency} = \frac{\text{speed of light}}{\text{wavelength}} = \frac{2.9979 \times 10^8 \text{ m/s}}{5.0 \times 10^2 \text{ nm}} \cdot \frac{1.00 \times 10^9 \text{ nm}}{1.00 \text{ m}}$$

$$= 6.0 \times 10^{14} \text{ s}^{-1}$$

And the energy is then E = hυ or $(6.626 \times 10^{-34} \text{ J} \cdot \text{s} \cdot \text{photons}^{-1})(6.0 \times 10^{14} \text{ s}^{-1})$

$$= 4.0 \times 10^{-19} \text{ J} \cdot \text{photons}^{-1}$$

Energy of 1.00 mol of photons $= 4.0 \times 10^{-19} \text{ J} \cdot \text{photon}^{-1} \cdot \dfrac{6.0221 \times 10^{23} \text{ photons}}{1.00 \text{ mol photons}}$

$$= 2.4 \times 10^5 \text{ J/mol photon}$$

26.   The frequency of the line at 396.15 nm:

$$\text{frequency} = \frac{\text{speed of light}}{\text{wavelength}} = \frac{2.9979 \times 10^8 \text{ m/s}}{3.9615 \times 10^2 \text{ nm}} \cdot \frac{1.00 \times 10^9 \text{ nm}}{1.00 \text{ m}}$$

$$= 7.5676 \times 10^{14} \text{ s}^{-1}$$

The energy of a photon of this light may be determined :   E = hυ

Planck's constant, h, has a value of $6.626 \times 10^{-34} \text{ J} \cdot \text{s} \cdot \text{photons}^{-1}$

$$E = (6.626 \times 10^{-34} \text{ J} \cdot \text{s} \cdot \text{photons}^{-1})(7.5676 \times 10^{14} \text{ s}^{-1})$$

$$= 5.0143 \times 10^{-19} \text{ J} \cdot \text{photon}^{-1}$$

Energy of 1.00 mol of photons $= 5.0143 \times 10^{-19} \text{ J} \cdot \text{photon}^{-1} \cdot \dfrac{6.0221 \times 10^{23} \text{ photons}}{1.00 \text{ mol photons}}$

$$= 3.02 \times 10^5 \text{ J/mol photon} \text{ or } 302 \text{ kJ/mol photon.}$$

28.   Since energy is proportional to frequency (E = hυ), we can arrange the radiation in order of increasing energy per photon by listing the types of radiation in increasing frequency (or decreasing wavelength).

→	Energy increasing	→	
FM music	microwave	yellow light	x-rays

$$\rightarrow \text{ Frequency } (\upsilon) \text{ increasing } \rightarrow$$
$$\leftarrow \text{ Wavelength } (\lambda) \text{ increasing } \leftarrow$$

Photoelectric Effect

30.   Energy $= 2.0 \times 10^2$ kJ/mol $\bullet \dfrac{1\ mol}{6.0221 \times 10^{23}\ photons} \bullet \dfrac{1.00 \times 10^3\ J}{1.00\ kJ}$

$= 3.3 \times 10^{-19}$ J $\bullet$ photons^{-1}

What wavelength of light would provide this energy ?

$E = h\upsilon = \dfrac{hc}{\lambda}$ or $\lambda = \dfrac{hc}{E} = \dfrac{(6.626 \times 10^{-34}\ J \bullet s \bullet photons^{-1})(2.9979 \times 10^8\ m \bullet s^{-1})}{3.3 \times 10^{-19}\ J \bullet photons^{-1}}$

$= 6.0 \times 10^{-7}$ m or 6.0×10^2 nm

Radiation of this wavelength--in the **visible** region of the electromagnetic spectrum-- would appear **orange**.

Atomic Spectra

32.   (a) The **most energetic light** would be represented by the light of **shortest wavelength** (253.652 nm).

(b) The frequency of this light is :

$\dfrac{2.9979 \times 10^8\ m/s}{253.652\ nm} \bullet \dfrac{1.00 \times 10^9\ nm}{1.00\ m} = 1.18190 \times 10^{15}\ s^{-1}$

The energy of 1 photon with this wavelength is:

$E = h\upsilon = (6.62608 \times 10^{-34} \dfrac{J \bullet s}{photon})(1.18190 \times 10^{15}\ s^{-1})$

$= 7.83139 \times 10^{-19} \dfrac{J}{photon}$

(c) The line emission spectrum of mercury shows the visible region between $\approx$ 400 and 750 nm. The lines at 404 and 436 nm are present while the lines at 253 nm, 365 nm and 1013 nm lie outside the visible region. The 404 nm line is violet, while the 436 nm line is blue.

34. The Balmer series of lines terminates with $n_f = 2$. According to Figure 7.12, the transition originates at $n_i = 6$. Light of wavelength 410.2 nm would be violet.

36.   (a)   <u>Transitions from</u>   <u>to</u>

n = 4   n = 3, 2, or 1

n = 3   n = 2 or 1

n = 2   n = 1

Six transitions are possible from these four quantum levels, providing 6 emission lines.

(b) Photons of the lowest energy will be emitted in a transition from the level with **n = 4** to the level **n = 3**. This is easily seen with the aid of the equation

$$\Delta E = Rhc(\frac{1}{n^2_f} - \frac{1}{n^2_i}).$$

Since R, h, and c are constant for any transition, inspection shows that the smallest change in energy results if $n_f = 3$ and $n_i = 4$.

(c) The emission line having the **shortest wavelength** also has the **highest frequency**. A transition from $n_i = 4$ to $n_f = 1$ would provide the shortest wavelength line.

38. Regarding energy emitted when an electron changes energy levels:

(a) The energy levels 2 and 3 are much closer to one another than 4 and 2, so less energy is involved in moving an electron from n = 3 to n = 2.

(b) Since the energy levels are more closely "spaced" the farther one goes from the nucleus, the levels n = 5 and n = 2 are closer in energy than the levels n = 4 and n = 1, so the n = 4 to n = 1 transition would be the one of greater energy.

40. The wavelength of emitted light for the transition n = 3 to n = 1.

$$\Delta E = -Rhc(\frac{1}{1^2} - \frac{1}{3^2})$$ and the value of Rhc = 1312 kJ/mol, so

$$\Delta E = -1312 \text{ kJ/mol} (\frac{1}{1^2} - \frac{1}{3^2}) \text{ or } = -1312 \text{ kJ/mol} (8/9) = -1166 \text{ kJ/mol}$$

To calculate the frequency and wavelength, we use E = $h\upsilon$. Recall that we must first express the energy **per photon** (as opposed to a mole of photons).

$$\Delta E = \frac{-1166 \text{ kJ/mol photons}}{6.022 x 10^{23} \text{ photons}/1 \text{ mol photons}} \cdot \frac{10^3 J}{1 \text{ kJ}} = 1.936 \times 10^{-18} \text{ J/photon}$$

and solving for frequency, $\upsilon = \dfrac{1.936 \times 10^{-18} \text{ J/photon}}{6.626 \times 10^{-34} \text{ J} \cdot \text{s}/\text{photon}} = 2.923 \times 10^{15} \text{ s}^{-1}$

Substituting into the relationship $\lambda\upsilon = c$ we get $\dfrac{2.998 \times 10^8 \text{ m}}{2.923 \times 10^{15} \text{ s}^{-1}} = 1.0257 \times 10^{-7} \text{m}$

$\lambda = 1.0257 \times 10^{-7}$ m or 102.6 nm (far ultraviolet)

DeBroglie and Matter Waves

42.   Mass of an electron: 9.11×10^{-31} kg

Planck's constant: 6.626×10^{-34} J $\bullet$ s $\bullet$ photon^{-1}

Velocity of the electron: 2.5×10^8 cm $\bullet$ s^{-1} or 2.5×10^6 m $\bullet$ s^{-1}

$$\lambda = \frac{h}{m \bullet v} = \frac{6.626 \times 10^{-34} \text{ J} \bullet \text{s}}{(9.11 \times 10^{-31} \text{ kg} \bullet 2.5 \times 10^6 \text{ m} \bullet \text{s}^{-1})}$$

$$= 2.9 \times 10^{-10} \text{ m} = 2.9 \text{ Angstroms} = 0.29 \text{ nm}$$

44. The wavelength can be determined exactly as in Question 42:

$$\lambda = \frac{h}{m \bullet v} = \frac{6.626 \times 10^{-34} \text{ J} \bullet \text{s}}{(1.0 \times 10^{-1} \text{ kg} \bullet 30. \text{ m} \bullet \text{s}^{-1})}$$

$$= 2.2 \times 10^{-34} \text{ m or } 2.2 \times 10^{-25} \text{ nm}$$

Velocity to have a wavelength of 5.6×10^{-3} nm

(First convert the wavelength to units of meters)

$$5.6 \times 10^{-3} \text{ nm} \bullet \frac{1 \text{ m}}{1 \times 10^{9} \text{ nm}} = 5.6 \times 10^{-12} \text{ m}$$

Then rewriting the above equation:

$$v = \frac{h}{m \bullet \lambda} = \frac{6.626 \times 10^{-34} \text{ J} \bullet \text{s}}{(1.0 \times 10^{-1} \text{ kg} \bullet 5.6 \times 10^{-12} \text{ m})} = 1.2 \times 10^{-21} \frac{m}{s}$$

Quantum Mechanics

46.   (a) n = 4   possible ℓ values = 0,1,2,3 $(\ell = 0,1,... (n-1))$

      (b) $\ell = 2$ possible m_ℓ values = -2,-1,0,+1,+2 $(-\ell ..., 0,....+ \ell)$

      (c) orbital = 4s   n = 4; ℓ = 0; m_ℓ = 0

      (d) orbital = 4f   n = 4; ℓ = 3; m_ℓ = -3,-2,-1,0,+1,+2,+3

48. An electron in a 4p orbital must have n = 4 and ℓ = 1. The possible m_l values give rise to the following sets of n, ℓ, and m_ℓ

n	ℓ	m_ℓ	
4	1	-1	Note that the **three values** of m describe
4	1	0	**three orbital orientations**.
4	1	+1	

50. Subshells in the electron shell with n = 4 :

There are 4: s, p, d, and f sublevels corresponding to ℓ = 0, 1, 2, and 3 respectively.

Recall that values of ℓ from 0 to a maximum of (n-1) are possible.

52. Explain why each of the following is not a possible set of quantum numbers for an electron in an atom.

(a) n = 2, ℓ = 2, m_ℓ = 0   For n = 2, maximum value of ℓ is one (1).

(b) n = 3, ℓ = 0, m_ℓ = -2   For ℓ = 0, possible value of m_l is 0.

(c) n = 6, ℓ = 0, m_ℓ = 1   For ℓ = 0, possible value of m_l is 0.

54.   quantum number designation   maximum number of orbitals
        (a) $n = 3$; $\ell = 0$; $m_\ell = +1$ none; for $\ell = 0$, the only possible value of $m_\ell = 0$
        (b) $n = 5$; $\ell = 1$ 3 ("**p**" orbitals)
        (c) $n = 7$; $\ell = 5$ eleven; the # of orbitals is "$2\ell + 1$"
        (d) $n = 4$; $\ell = 2$; $m_\ell = -2$ 1 (one of the three 4 "**p**" orbitals)

56.   Which of the following orbitals cannot exist and why:

        2s exists $n = 2$ permits ℓ values as large as 1 ($\ell = 0$ is an s sublevel)

        2d cannot exist $\ell = 2$ is not permitted for $n < 3$ ($\ell = 2$ is a d sublevel)

        3p exists $n = 3$ permits ℓ values as large as 2 ($\ell = 1$ is a p sublevel)

        3f cannot exist $\ell = 3$ is not permitted for $n < 4$ ($\ell = 3$ is an f sublevel)

        4f exists $\ell = 4$ permits ℓ values as large as 3 ($\ell = 3$ is an f sublevel)

        5s exists $n = 5$ permits ℓ values as large as 4 ($\ell = 0$ is an s sublevel)

58.   The complete set of quantum numbers for :

		$\underline{n}$	$\underline{\ell}$	$\underline{m_\ell}$	
(a)	2p	2	1	-1, 0, +1	(3 orbitals)
(b)	3d	3	2	-2, -1, 0, +1, +2	(5 orbitals)
(c)	4f	4	3	-3, -2, -1, 0, +1, +2, +3	(7 orbitals)

60. With an $n = 4$, and $\ell = 2$, this orbital belongs in the 4th level ($n = 4$) and with an $\ell = 2$, this
    must be a "d" type orbital.

62. The number of nodal surfaces possessed by an orbital is equal to the value of the "ℓ"
    quantum number, so for each of the following:

	orbital		number of planar nodes
(a)	2s	($\ell = 0$)	0
(b)	5d	($\ell = 2$)	2
(c)	5f	($\ell = 3$)	3

General Questions on Atomic Structure

64. The energy needed to move an electron from $n = 1$ to $n = 5$ will be the same as the amount
    emitted as the electron relaxed from $n = 5$ to $n = 1$, i.e. 2.093×10^{-18} J.

66. Regarding red and green light emitted by a sign:
    (a) Green light has the shorter wavelength and therefore higher energy photons.

(b) Green light has higher energy photons than red light, the shorter wavelength (500 nm) is green

(c) Since frequency and wavelength are inversely related, the shorter wavelength (green) must have the higher frequency.

68. For radiation of 850 MHz:

(a) The wavelength: $\dfrac{3.00 \times 10^8 \text{ m/s}}{850 \times 10^6 \text{ Hz}} \cdot \dfrac{1 \text{ Hz}}{1 \text{ s}^{-1}} = 0.35 \text{ m}$

(b) The energy of 1.0 mol of photons with $\upsilon = 850$ MHz:

$$E = \frac{hc}{\lambda} = \frac{(6.626 \times 10^{-34} \text{ J} \bullet \text{ s} \bullet \text{ photons}^{-1})(2.9979 \times 10^8 \text{ m} \bullet \text{ s}^{-1})}{0.35 \text{ m}} \bullet \frac{6.0221 \times 10^{23} \text{ photons}}{1 \text{ mol photons}}$$

$$= 0.34 \text{ J/mol}$$

(c) The energy of a mole of photons of 420 nm light:

$$E = \frac{hc}{\lambda} = \frac{(6.626 \times 10^{-34} \text{ J} \bullet \text{ s} \bullet \text{ photons}^{-1})(2.9979 \times 10^8 \text{ m} \bullet \text{ s}^{-1})}{420 \times 10^{-9} \text{ m}} \bullet \frac{6.0221 \times 10^{23} \text{ photons}}{1 \text{ mol photons}}$$

$$= 2.9 \times 10^5 \text{ J/mol}$$

(d) The energy of a mole of photons of blue light is much greater than that of the corresponding photons from a cell phone.

70. "A Closer Look" following Example 7.3 in your text illustrates the calculation of the ionization energy for H's electron

$$E = \frac{-Z^2 Rhc}{n^2} = -2.179 \times 10^{-18} \text{ J/atom} \implies -1312 \text{ kJ/mol}$$

For He$^+$ the calculation yields

$$E = \frac{-(2)^2(1.097 \times 10^7 \text{ m}^{-1})(6.626 \times 10^{-34} \text{ J} \bullet \text{s})(2.998 \times 10^8 \text{ m} \bullet \text{s}^{-1})}{(1)^2}$$

$= -8.717 \times 10^{-18}$ J/ion and expressing this energy for a mol of ions

$$= \frac{-8.717 \times 10^{-18} \text{ J}}{\text{ion}} \bullet \frac{6.0221 \times 10^{23} \text{ atoms}}{\text{mol}} \bullet \frac{1 \text{ kJ}}{1000 \text{ J}} = -5248 \frac{\text{kJ}}{\text{mol}}$$

The energy to remove the electron is then 5248 kJ/mol of ions. Note that this energy is four times that for H.

72.   (i) The photon with the smallest energy would be produced by a transition between the
           closest levels, or (b) n = 7 to n=6.
       (ii) The photon with the highest frequency (and with the highest energy) would be
           produced by a transition between the **two most distant levels**, (a) n = 7 to n= 1.
       (iii) The photon with the shortest wavelength (and therefore with the highest energy)
           would be produced by a transition between the **two most distant levels**,
           (a) n = 7 to n= 1.

74. The number of orbitals that correspond to:
       (a) 3p   3 orbitals (in a p sublevel)
       (b) 4p   3 orbitals
       (c) 4 p_x 1 (of the 3 p orbitals)
       (d) 6d   5 orbitals (in a d sublevel)
       (e) 5d   5 orbitals
       (f) 5f   7 orbitals (in an f sublevel)
       (g) n = 5   25 orbitals in the level n = 5 (1-s, 3-p, 5-d, 7-f, 9-g for a total of 25)
       (h) 7s   1 orbital (in an s sublevel)

76. Bohr's model violates the uncertainty principle by theorizing that we can know an exact
       distance of an electron from the nucleus—and hence it's energy. The uncertainty principle
       restricts the "certainty" with which we can *simultaneously* know the position and the
       energy of the electron.

78.   Observables:
       (b),(g)—energies of light absorbed or emitted can be measured.
       (e),(f)-- diffraction patterns produced by electrons and light
       (h),(i)-- atoms and molecules can now be observed with scanning tunneling microscopes.
               (though certainly not with the unaided eye)
       (j) --water waves are visible

80.   For the first three electron shells:
               N = 1   N = 2   N = 3
               L = 1   L = 2   L = 3
               M = ±1,0   M = ±1,0   M = ±1,0
       There would be 3 orbitals in each of the 3 shells for a **total of 9 orbitals**.

82. Wavelength and frequency for a photon with E = 1.173MeV:

Using the energy relationship: $E = h\upsilon$, we solve for frequency:

$$\upsilon = \frac{1.172 \times 10^6 \text{ ev}}{6.626 \times 10^{-34} \text{J} \bullet \text{s}} \bullet \frac{9.6485 \times 10^4 \text{ J/mol}}{1 \text{ ev}} \bullet \frac{1 \text{ mol}}{6.022 \times 10^{23} \text{ photons}}$$

$$= 2.836 \times 10^{20} \text{ s}^{-1}$$

Substituting into the wavelength-frequency relationship, we have:

$$\lambda = c/\upsilon = \frac{2.998 \times 10^8 \text{ m/s}}{2.836 \times 10^{20} \text{ s}^{-1}} = 1.057 \times 10^{-12} \text{m}$$

84. Time for Sojourner's signal to travel 7.8×10^7 km:

If light travels at 2.9979×10^8 m $\bullet$ s^{-1}, we can calculate the time:

$$\frac{7.8 \times 10^7 \text{ km}}{1} \bullet \frac{1 \times 10^3 \text{ m}}{1 \text{ km}} \bullet \frac{1 \text{ s}}{2.9979 \times 10^8 \text{ m}} = 260 \text{ s}$$

Given that there are 60 s in 1 minute: $260 \text{ s} \bullet \dfrac{1 \text{min}}{60 \text{s}} = 4.3$ minutes

86. (a) The quantum number n describes the **size (and energy)** of an atomic orbital .

(b) The shape of an atomic orbitals is given by the quantum number ℓ.

(c) A photon of green light has **more energy** than a photon of orange light.

(d) The maximum number of orbitals that may be associated with the quantum numbers n= **4**, ℓ = **3** is **seven.** (corresponding to m_l values of $\pm 3, \pm 2, \pm 1$, and 0)

(e) The maximum number of orbitals that may be associated with the quantum numbers n= **3**, ℓ = **2**, and m_l = -2 is **one.**

(f) The orbital on the left is a **d orbital** and the one on the right is a **p orbital**, while the orbital in the middle is an **s orbital**.

(g) When n = 5, the possible values of ℓ **are 0,1,2,3, and 4**. (Range is 0 ..... (n-1))

(h) The maximum number of orbitals that can be assigned to the n = 4 shell is **16**.

n = 4	l = 0	m_ℓ = 0	1 orbital
n = 4	ℓ = 1	m_ℓ = -1,0,+1	3 orbitals
n = 4	l = 2	m_ℓ = -2,-1,0,+1,+2	5 orbitals
n = 4	ℓ = 3	m_ℓ = -3,-2,-1,0,+1,+2,+3	<u>7 orbitals</u>
			16 orbitals

88. Regarding Technetium:

(a) Technetium is in group 7B of the fifth period.

(b) Quantum numbers for an electron in the 5s subshell:
$$n = 5, \; \ell = 0, \; m_l = 0$$

(c) Wavelength and frequency of a photon with energy of 0.141 MeV:

Using $E = h\upsilon$, we solve for frequency:

$$\upsilon = \frac{0.141 \times 10^6 \, ev}{6.626 \times 10^{-34} J \bullet s} \bullet \frac{9.6485 \times 10^4 \, J/mol}{1 \, ev} \bullet \frac{1 \, mol}{6.022 \times 10^{23} \, photons}$$
$$= 3.41 \times 10^{19} \, s^{-1}$$

Substituting into the wavelength-frequency relationship, we have:

$$\lambda = c/\upsilon = \frac{2.998 \times 10^8 \, m/s}{3.41 \times 10^{19} \, s^{-1}} = 8.79 \times 10^{-12} m$$

(d) In the preparation of $NaTcO_4$:

(i) The balanced equation: $HTcO_4(aq) + NaOH(aq) \rightarrow NaTcO_4(aq) + H_2O(\ell)$

(ii) Mass of $NaTcO_4$ from 4.5 mg of Tc:

$$4.5 \times 10^{-3} \, g \; Tc \bullet \frac{1 \, mol \; Tc}{98 \, g \; Tc} \bullet \frac{1 \, mol \; NaTcO_4}{1 \, mol \; Tc} \bullet \frac{185 \, g \; NaTcO_4}{1 \, mol \; NaTcO_4} = 8.5 \times 10^{-3} \, g \; NaTcO_4$$

Mass of NaOH required:

$$4.5 \times 10^{-3} \, g \; Tc \bullet \frac{1 \, mol \; Tc}{98 \, g \; Tc} \bullet \frac{1 \, mol \; HTcO_4}{1 \, mol \; Tc} \bullet \frac{1 \, mol \; NaOH}{1 \, mol \; HTcO_4} \bullet \frac{40.0 \, g \; NaOH}{1 \, mol \; NaOH} = 1.8 \times 10^{-3} \, g \; NaOH$$

Using Electronic Resources

90. The pickle glows since the materials in the pickle are being "excited" by the addition of the energy (electric current). Since the pickle has been soaked in brine (NaCl), the electrons in the sodium atom are excited and release energy as they "return" to lower energy states, providing "yellow" light. The same kind of light is visible in many street lamps.

Chapter 8
Atomic Electron Configurations and Chemical Periodicity

Reviewing Important Concepts

3. Electron configuration for Li in both orbital box diagram and spdf notation:

    (See page 298 of your text)

	Orbital box notation	Spectroscopic notation

    Li   [↑↓] [↑] $1s^2 2s^1$

8. The element in the fourth period in Group 4A is: Germanium (Ge). The location tells us that (a) the outermost electrons are in the "4th" shell, with a total of 4 electrons in the outer shell—2 in the 4s sublevel and 2 in the 4p sublevel.

    Total configuration: $1s^2 2s^2 2p^6 3s^2 3p^6 3d^{10} 4s^2 4p^2$

Practicing Skills
Writing Electron Configurations of Atoms

10. The orbital box and spdf notation for P and Cl

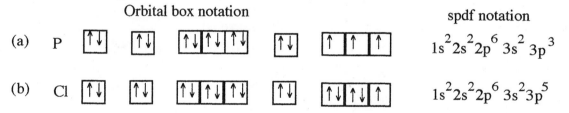

Orbital box notation   spdf notation

(a) P $1s^2 2s^2 2p^6 3s^2 3p^3$

(b) Cl $1s^2 2s^2 2p^6 3s^2 3p^5$

Note that Cl is in group 7A (17) indicating that there are SEVEN electrons in the outer shell, while P is in group 5A (15) indicating that there are FIVE electrons in the outer shell. Both Cl and P are on the "right side" of the periodic table—where elements have their "outermost" electrons in p subshells.

12. Electron configuration of chromium and iron:

    (a) Cr: $1s^2 2s^2 2p^6 3s^2 3p^6 3d^5 4s^1$

    (b) Fe: $1s^2 2s^2 2p^6 3s^2 3p^6 3d^6 4s^2$

    The "surprising" configuration of elemental chromium—compared to the electron configuration of the preceding element, vanadium, arises from the stability of the "half-filled 3d sublevel" (which can be visualized as having a 4s electron occupy a 3d orbital—with the resultant $4s^1$ configuration.)

14.   (a) Arsenic's electron configuration: (33 electrons)

spdf notation: $1s^2 2s^2 2p^6 3s^2 3p^6 3d^{10} 4s^2 4p^3$

noble gas notation: $[Ar] 3d^{10} 4s^2 4p^3$

(b) Krypton's electron configuration: (36 electrons)

spdf notation: $1s^2 2s^2 2p^6 3s^2 3p^6 3d^{10} 4s^2 4p^6$

noble gas notation: $[Kr]$

16.   (a) Tantalum's noble gas and spdf notation (73 electrons)

spdf notation: $1s^2 2s^2 2p^6 3s^2 3p^6 3d^{10} 4s^2 4p^6 4d^{10} 4f^{14} 5s^2 5p^6 5d^3 6s^2$

noble gas notation: $[Xe] 4f^{14} 5d^3 6s^2$

(b) Platinum's noble gas and spdf notation (78 electrons)

spdf notation: $1s^2 2s^2 2p^6 3s^2 3p^6 3d^{10} 4s^2 4p^6 4d^{10} 4f^{14} 5s^2 5p^6 5d^9 6s^1$

noble gas notation: $[Xe] 4f^{14} 5d^9 6s^1$

Tantalum's configuration is expected, with Ta in period 5. and in group 5B. Platinum's configuration is a bit unexpected, but like other transition metals, it attempts to fill that d sublevel, resulting in a $5d^9 6s^1$ configuration rather than the expected $5d^8 6s^2$. Transition elements in the levels past period 4, do have several exceptions to the Aufbau principle.

18. Americium's noble gas and spdf notation (95 electrons)

spdf notation: $1s^2 2s^2 2p^6 3s^2 3p^6 3d^{10} 4s^2 4p^6 4d^{10} 4f^{14} 5s^2 5p^6$
$5d^{10} 5f^7 6s^2 6p^6 7s^2$

noble gas notation: $[Rn] 5f^7 7s^2$

Electron Configurations of Atoms and Ions and Magnetic Behavior

20. The orbital box representations for the following ions:

Orbital box notation

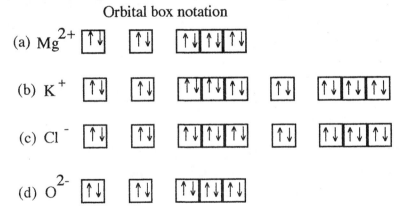

22. Electron configurations of:

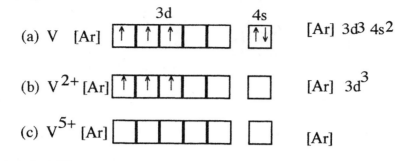

(a) V [Ar]   [Ar] $3d^3 4s^2$

(b) V^{2+} [Ar]   [Ar] $3d^3$

(c) V^{5+} [Ar]   [Ar]

Note that the V^{2+} ion contains unpaired electrons, and is therefore paramagnetic.

24. Manganese's orbital box and noble gas diagrams:

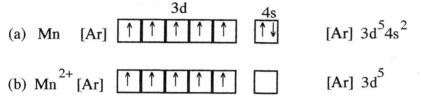

(a) Mn   [Ar]   [Ar] $3d^5 4s^2$

(b) Mn^{2+} [Ar]   [Ar] $3d^5$

(c) Having unpaired electrons, Mn $^{2+}$ is **paramagnetic.**

(d) Mn $^{2+}$ has five (5) unpaired electrons.

Quantum Numbers and Electron Configurations

26. Explain why the following sets of quantum numbers are not valid:

(a) $n = 4$, $\ell = 2$, $m_\ell = 0$, $m_s = 0$:

The possible values of m_s can only be +1/2 or -1/2

(b) $n = 3$, $\ell = 1$, $m_\ell = -3$, $m_s = -1/2$:

The possible values for m_ℓ is - ℓ......0.....+ℓ. Changing m_ℓ to -1,0,+1 would give a valid set of quantum numbers.

(c) $n = 3$, $\ell = 3$, $m_\ell = -1$, $m_s = +1/2$

The maximum value of ℓ is (n-1). Changing ℓ to 2 would provide a valid set of quantum numbers.

28. Maximum number of electrons associated with the following sets of quantum numbers:

	Characterized as	Maximum number of electrons
(a) $n = 4$ and $\ell = 3$	4f electrons	14 (an "f" sublevel)
(b) $n = 6$, $\ell = 1$, $m_\ell = -1$	6p electrons	2 (a "p" orbital)
(c) $n = 3$, $\ell = 1$, $m_\ell = 2$, and $m_s = +1/2$	NONE	With $\ell = 1$, the maximum value of m_ℓ can be +1.

30. The electron configuration for Mg using the orbital box method:

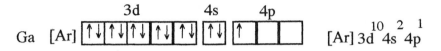

               1s   2s   2p   3s

Mg:

Electron number   11 12

The noble gas notation: $[Ne]3s^2$

Electron

number:	n	ℓ	m_ℓ	m_s
11	3	0	0	+ 1/2
12	3	0	0	− 1/2

32. The electron configuration for Gallium using the orbital box method:

                                         Noble gas configuration

               3d   4s   4p

Ga [Ar]   [Ar] $3d^{10} 4s^2 4p^1$

A possible set of quantum numbers for the highest energy electron:

n	ℓ	m_ℓ	m_s
4	1	−1	+ 1/2

Periodic Properties

34. Elements arranged in order of increasing size: C < B < Al < Na < K

    Radii from Figure 8.9 (in pm)   77 < 83 < 143 < 186 < 227

       Since K is in period 4, we anticipate it being larger than Na, its analog in 1A (period 3).

       Al is to the right of Na, so we expect it to be smaller than Na. B and C are in period 2,

       with B to the right of C, and therefore larger than C.

36. The specie in each pair with the larger radius:

       (a) Cl⁻ is larger than Cl -- The ion has more electrons/proton than the atom.

       (b) Al is larger than O -- Al is in period 3, while O is in period 2.

       (c) In is larger than I -- Atomic radii decrease, in general, across a period.

38. The group of elements with correctly ordered increasing ionization energy (IE):

                     (c) Li < Si < C < Ne.

       Neon would have the greatest IE. Silicon, being slightly larger in atomic radius than carbon,

       has a lesser IE. Lithium, the largest atom of this group, would have the smallest IE.

40. For the elements Na, Mg, O, and P:

(a) The largest atomic radius: Na

The greater the period number, the larger the atom. Radius also decreases to the right in a given period.

(b) The largest (most negative) electron affinity: O

Nonmetals have a more negative EA than metals. Down a group, the EA becomes more positive (as the "metallic" character increases).

(c) Increasing ionization energy: Na < Mg < P < O

The ionization energy varies inversely with the atomic radius. The smaller the atom, the greater the IE.

42.   (a) Increasing ionization energy: S < O < F

Ionization energy is inversely proportional to atomic size.

(b) Largest ionization energy of O, S, or Se: O

Oxygen is the smallest of these Group 6A elements, and hence has the largest IE.

(c) Most negative electron affinity of Se, Cl, or Br: Cl

Chlorine is the smallest of these three elements. EA tends to increase on a diagonal from the lower left of the periodic table to the upper right.(See Figure 8.12).

(d) Largest radius of O^{2-}, F^-, F: O^{2-}

The oxide ion has the largest electron : proton ratio. If one considers the attraction of the nuclear species (protons) for the extranuclear species (electrons), the greater the number of protons/electron the smaller the specie—owing to an increased electron-proton attraction. So the oxide ion (with 10 electrons and 8 protons) has the fewest electrons (of these three species) per proton.

General Questions on Electron Configurations and Periodic Trends

44. Electron configuration of Rutherfordium(104 electrons) with spdf and noble gas notations:

spdf notation: $1s^2\ 2s^2\ 2p^6\ 3s^2\ 3p^6\ 3d^{10}\ 4s^2\ 4p^6 4d^{10}\ 4f^{14}\ 5s^2\ 5p^6$
$5d^{10}5f^{14}6s^26p^66d^27s^2$

noble gas notation: [Rn] $5f^{14}6d^27s^2$

Note that Rutherfordium is in group 4B, so we anticipate 4 electrons ($6d^27s^2$). It immediately follows Lawrencium (which would have the filled 5f sublevel).

46. The orbital box diagram and noble gas configurations for:

(a) Ce and Ce $^{3+}$

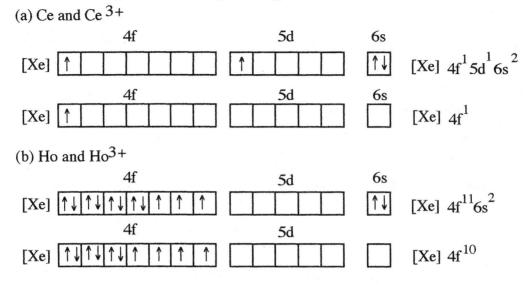

Note that both ions are formed by the loss of 3 electrons from the highest energy orbitals. (6s and 5d for Cerium, and 6s and 4f for Holmium.)

48. For Meitnerium (109 electrons), its spdf and noble gas configuration is:

spdf notation: $1s^2\ 2s^2\ 2p^6\ 3s^2\ 3p^6\ 3d^{10}\ 4s^2\ 4p^6 4d^{10}\ 4f^{14}\ 5s^2\ 5p^6$
$5d^{10}5f^{14}6s^26p^66d^77s^2$

noble gas notation: $[Rn]\ 5f^{14}6d^77s^2$

It resides in the group with Cobalt, Rhodium, and Iridium.

50. An electron in a 4p orbital must have $n = 4$ and $\ell = 1$. The possible m_ℓ and m_s values give

rise to the following **six** sets of quantum numbers. (For brevity, I have included the +/- values of m_s on one line for each of the 3 possible values of m_ℓ.

n	l	m_ℓ	m_s
4	1	-1	+1/2, and −1/2
4	1	0	+1/2, and −1/2
4	1	+1	+1/2, and −1/2

52. The element with the characteristics:

(a) The element with electron configuration $1s^2\ 2s^2\ 2p^6\ 3s^2\ 3p^3$ is: P (atomic # = 15)

(b) The alkaline earth element with smallest atomic radius is Be; atomic radius increases as one goes down the group.

(c) The element with the largest IE in Group 5A would also be the element with the smallest atomic radius—N

(d) The element whose 2+ ion would have the configuration: [Kr] $4d^5$: Tc

(e) The element with the most negative EA in Group 7A : Cl

(f) The element with electron configuration: [Ar] $3d^{10} 4s^2$ is: Zn (atomic # = 30)

54. Relative ionization energies for Cl, Ca^{2+} and Cl^-: Examining the number of protons per electron, we see that both Ca^{2+} and Cl^- are isoelectronic, with Ca^{2+} having 20 protons and 18 electrons and Cl^- having 17 protons and 18 electrons. Since the calcium ion has a 2+ charge, removing another electron from that ion would take "more work"—a larger IE. Removing an electron from the chloride ion would require less energy than the removal of an electron from the Cl atom, so the order is proposed to be: $Cl^- < Cl < Ca^{2+}$

56. For the elements A and B with electron configurations as shown:

    A: [Ar]$4s^2$ and B: [Ar] $3d^{10}4s^24p^5$

   (a) Element A is a metal (with only 2 electrons in the valence shell).

   (b) Element B is a nonmetal (with 7 electrons in the valence shell).

   (c) B would have a larger IE, since B will have an atomic radius smaller than A and be in the same period as A.

   (d) B has the smaller radius, since it would lie to the right of A in the period.

58. Ions in order of decreasing size:

	K^+	Cl^-	S^{2-}	Ca^{2+}
# protons	19	17	16	20
# electrons	18	18	18	18

   Note that the four ions are isoelectronic (same # of electrons). So we anticipate that the greater number of protons/electron would result in a smaller ionic radius. Therefore the sizes range: $S^{2-} > Cl^- > K^+ > Ca^{2+}$

60. For the ions: Cl^-, K^+, Ca^{2+}

   (a) In order of increasing size: $Ca^{2+} < K^+ < Cl^-$. See question 58 for the explanation.

   (b) Increasing IE: Given the increasing + charge, we anticipate
       IE increasing: $Cl^- < K^+ < Ca^{2+}$

   (c) Increasing electron affinity: $Cl^- < K^+ < Ca^{2+}$ We anticipate that the calcium ion will happily gain an electron (reducing its' overall charge), and that the potassium ion will follow the same trend. The chloride ion is not likely to gain another electron (increasing it's overall negative charge—hence a smaller (less negative) EA.

62. Examine the electron configurations for the elements in the 2nd transition series, Y thru Cd. Note that Tc has the configuration, [Kr]4d^5 5s^2 and Rh the configuration, [Kr]4d^{8}5s^1. The loss of three electrons would give, for Tc^{3+} [Kr]4d^4 and for Rh^{3+} the configuration, [Kr]4d^6. Each of these two species would have 4 unpaired electrons.

64. (a) The element (containing 27 electrons) is Cobalt

     (b) The sample contains unpaired electrons—so it is paramagnetic

     (c) The 3+ ion of cobalt would be formed by the loss of the two 4s electrons and 1 of the d electrons— leaving four unpaired electrons.

66. Explain the trends in size:

     (a) The decrease in atomic size across a period is attributable to the increasing nuclear charge with an increasing number of protons. Given that the electrons are in the same outer energy level, the nuclear attraction for the electrons results in a diminishing atomic radius.

     (b) The slight decrease in atomic radius of the transition metals is a result of increased repulsions of (n-1)d electrons for (n)s electrons. Thus repulsion reduces the effects of the increasing nuclear charge across a period.

68. The element with the greatest difference between the first and second IE's is lithium. Of the four elements shown, Li is in the first group. The loss of the first electron results in an ion with a filled outer shell—a very stable configuration. Removal of a second electron would require a much larger amount of energy.

70. Ions not likely to be formed from the list: K^{2+}, Cs$^+$, Al^{4+}, F^{2-}, Se^{2-}

     Cs$^+$ and Se^{2-}are likely to be formed, since these ions have the noble gas configuration. The other species are **not likely to be formed.**

72. Since <u>Ca</u> is smaller than <u>K</u>, we would expect the first IE of <u>Ca</u> to be greater than that of <u>K</u>. Once <u>K</u> has lost its "first" electron, it possess an [Ar] core. Removal of an additional electron (the second) requires much energy. <u>Ca</u>, on the other hand, can lose a "second" electron to obtain the stable noble gas configuration with a much smaller amount of energy (smaller IE).

74. For the reaction to be: 2 Ca (s) + 3 F$_2$ (g) → 2 CaF$_3$ (s), calcium would have to form a 3+ cation. Since Calcium (in Group 2A) forms 2+ cations (and produces an ion that has the noble gas configuration), the formation of the 3+ cation is highly unlikely.

76. A plot of the atomic radii of the elements K—V shows a decrease. With the mass of these elements increasing from K through V (as more protons, neutrons, and electrons are added), the density is expected to increase.

<u>78.</u> Since effective nuclear charge increases across a period, we anticipate the ionization energies of the elements to be in the order: Li<Be<B<C. Experiment indicates that the order is: Li<Be>B<C. This is understood if you consider that removing an electron from Be would be destroying the filled 2s sublevel. The atomic radius of Boron (83 pm) is **much** smaller than that of Be (113 pm) , so we anticipate that it would be **much** more difficult (i.e. take more energy) to remove an electron. The difference in radius between B and C is not that great (C radius is 77 pm), so while the IE would be higher for C, we don't anticipate it being that much higher.

80. Following the general trend that atomic radii increase **down** a group, and decrease **to the right** in a period, we anticipate the order to be:

Element	Radius (pm)
On	90
M	120
E	140
Ch	180

Using Electronic Resources

82. For the reaction between elemental sodium and chlorine gas:
   (a) Na is the reducing agent (as are most metals—with their tendencies to lose electrons in the process of attaining a noble gas configuration.
   (b) Chlorine is the oxidizing agent, and does so in the process of gaining electrons—to attain a noble gas configuration.
   (c) The reaction produces NaCl since as each Na atom loses 1e$^-$ (to attain the noble gas configuration of Ne) and each Cl atom gains 1e$^-$ (to attain the noble gas configuration of Ar), monopositive (Na$^+$) and mononegative ions (Cl$^-$) are formed.

84.   (a) Is Effective nuclear charge ever the same as the nuclear charge? Since the nuclear charge and effective nuclear charge differ owing to the presence of electrons that shield other electrons from the charge of the nucleus, H (a 1 electron specie) would have no

other electrons shielding the charge, and hence would have an effective nuclear charge that is identical to the nuclear charge.

(b) Trend in effective nuclear charge proceeding across the periodic table. As a general trend, the effective nuclear charge increases across the periodic table.(See Table 8.2 as an example.) One result of this increasing nuclear charge is the trend of atoms to have smaller radii as one crosses a given period in the table.

(c) Change in effective nuclear charge going from Ne to Na. As electrons occupy the 3s sublevel (for Na), the 10 electrons in the inner shell shield the "11th" electron on Na. The result is that the effective nuclear charge is **less**($Z^* = 5.85$ for Ne, and 2.20 for Na), and the atomic radius of Na is **larger**. This will affect the chemical and physical properties of Ne and Na. The increased atomic radius of Na will make it easier to remove an electron (making Na a great reducing agent) while Ne (with its filled shell) does not add or remove an electron easily. So the chemical property of reactivity (towards loss or gain of electrons) is affected by this difference in effective nuclear charge.

86. (a) Al has 2 electrons in the 2s sublevel and 1 electron in the 2p sublevel. Mg has only 2 electrons in the 2s sublevel (a filled s sublevel). Removal of the "2p" electron from Al (IE approximately 578 kJ/mol) would result in an ion (effectively a Mg atom in terms of electrons). Removal of a "2s" electron from Mg would partially empty the "stable" filled 2s sublevel—and require a greater energy(IE for Mg approximately 738 kJ/mol). The highest energy occupied orbital in Mg is lower in energy than the comparable orbital in Al, and therefore more difficult to remove the electron.

(b) As one proceeds from Al to Ar, the orbital energies decrease. One result of this change in orbital energies is that the ionization energies increase from Al to Ar.

Chapter 9:
Bonding and Molecular Structure: Fundamental Concepts

Reviewing Important Concepts

4. Prediction on the bonding in:

KI	ionic	bonding between metal and nonmetal
MgS	ionic	bonding between metal and nonmetal
CS_2	covalent	bonding between two nonmetals
P_4O_{10}	covalent	bonding between two nonmetals

6. $CaCl_4$ is not likely to exist. Calcium (Group 2A) forms a 2+ cation and chlorine (Group 7A) forms a 1- anion, making $CaCl_2$ the likely formula of the compound between these two elements.

8. The dot structure for BCl_3 is shown on the left. The coordinate covalent bond between B and N is shown as a pair of electrons in the dot structure on the right.

The bond between B-N and the three B-Cl bonds provide an octet configuration for Boron.

10. Which of the following are odd-electron species:

Specie	electrons	total	electron number
NO_2	$5 + 2(6)$	17	**odd**
SF_4	$6 + 4(7)$	34	even
SO_3	$6 + 3(6)$	24	even
O_2^-	$2(6) + 1$	13	**odd**

12. Bond order in acetylene and phosgene:

$$H - C \equiv C - H$$

The C-H bonds are of bond order =1 (single bond); the C-C bond is of bond order 3. The Cl-C bonds (in phosgene) are bond order = 1; the C-O bond is of bond order 2.

14. The N-O bond order in the nitrate ion:

$$\left[\ :\overset{..}{\underset{..}{O}}-\overset{\overset{\overset{..}{O}\;}{\underset{1}{\|}}}{\underset{1}{N}}\overset{2}{=}\overset{..}{\underset{..}{O}}\ \right]^{-}$$

The N-O bonds labeled (1) are of bond order 1, the N-O bond labeled (2) is of bonder order 2. This is only one structure that we could draw for the nitrate ion. The others would show a similar electron distribution. We have **4 pairs** of electrons making 3 bonds so the "average" bond order is 4/3 of 1 1/3.

16. Bond dissociation energy—The energy (enthalpy) required to break a bond in a molecule with the reactants and products (in the gas phase) under standard conditions. Bond-breaking reactions **always** require the input of energy, and so always have a + sign.

18. To estimate the enthalpy change for the formation of water (g) from hydrogen and oxygen one needs: O=O (double bond) energy   H-H bond energy

    H-O bond energy

    ΔHreaction $= \Sigma$ E(bonds broken) - Σ E (bonds made)

    $= $ [E (O=O) bond + 2 E (H-H)bond] - 4[E (O-H) bond]

    $=$[498 kJ + 2•436 kJ] – [4 • 463 kJ] = - 482 kJ

20. Difference between electronegativities and electron affinities:

    Electronegativity is the ability of an atom in a molecule to attract electrons to itself.

    Electron affinity is the energy change when an atom of an element (in the gas phase) gains an electron.

22. The principle of **electroneutrality** state that atoms in molecules or ions should have **formal charges as small as possible**.

    The resonance structure for carbon dioxide shown below can be eliminated since—using this structure the formal charges on the atoms are:

    (1+   0   1-).

    $:O\equiv C-\overset{..}{\underset{..}{O}}:$

    The resonance structure showing 2 C=O double bonds has a formal charge on **all 3 atoms**— a much better "picture".

24. The **electron pair geometry** shows the *relative positions of all the valence electrons* around a central atom in a molecule or ion. The **molecular geometry** is concerned with the *relative*

*positions of the **atoms*** in a molecule or ion. For water, the electron pair geometry is tetrahedral, while the molecular geometry is bent.

26. Four electron pairs form a **pyramidal molecule if** one of the electron pairs is a non–bonding electron pair. A **bent molecule** is obtained if **two** electron pairs are non-bonding pairs. In either case, the angle between two electron pairs and the central atom is approximately 109°.

Practicing Skills
Valence Electrons and the Octet Rule

28.

Element		Group Number	Number of Valence Electrons
(a)	O	6A	6
(b)	B	3A	3
(c)	Na	1A	1
(d)	Mg	2A	2
(e)	F	7A	7
(f)	S	6A	6

30.

Group Number	Number of Bonds
3A	3
4A	4
5A	3 (or 4 in species such as NH_4^+)
6A	2 (or 3 as in H_3O^+)
7A	1

Ionic Compounds

32. Compound with the most negative energy of ion pair formation? least negative?

Coulomb's Law tells us that the most negative IP energy will result from (a) increased charges on the ions and (b) decreased distance between the ions. Compiling those data for the ions involved we get:

ion	charge	ionic radius (pm)
Na^+	1	116
Mg^{2+}	2	86
Cl^-	1	167
F^-	1	119
S^{2-}	2	170

Ignoring for the moment the charges on the electrons (since they are the same on all electrons, we can calculate terms for the numerators (# of + and - charges) and denominators (sum of ionic radii) for the three compounds:

$$NaCl \quad \frac{(1 \cdot 1)}{(116+167)} = \frac{1}{283} = \frac{1}{283}$$

$$MgF_2 \quad \frac{(2 \cdot 1)}{(86 + 119)} = \frac{2}{205} = \frac{1}{103}$$

$$MgS \frac{(2 \bullet 2)}{(86 + 170)} = \frac{4}{256} = \frac{1}{64}$$

The third column represents the reduction of all numerators to unity. The result is that **MgS would have the largest negative energy of ion pair formation, and NaCl would have the least negative value.**

34. Arrange lattice energies from least negative to most negative:

Since CaO involves 2+ and 2- ions, the lattice energy for this compound is greater than the other compounds. Given that lattice energy is **inversely** related to the distance between the ions, the lattice energy for LiI is greater(more negative) than that for RbI ($Rb^+ > Li^+$). The small diameter of the fluoride ion (compared to iodide) indicates that the lattice energy for LiF would be more negative than for either LiI or RbI. The lattice energies are then:

least ----- RbI ----- LiI ----- LiF ----- CaO ----- **most**
negative   **negative**

36. Since melting a solid involves disassembling the crystal lattice of cations and anions, the lesser the distance between cations and anions—the greater the attraction between the cation and anion, and the harder it becomes to disassemble the lattice, and hence the **higher the melting point.**

Lewis Electron Dot Structures

38. (a) NF_3 : $[1(5) + 3(7)] = 26$ valence electrons

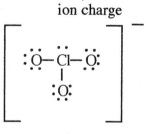

(b) ClO_3^- : $[1(7) + 3(6) + 1] = 26$ valence electrons

↑
ion charge

(c) HOBr: $[1(1) + 1(6) + 1(7)] = 14$ valence electrons

(d) SO_3^{2-} : $[1(6) + 3(6) + 2] = 26$ valence electrons

↑
ion charge

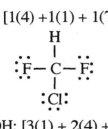

40.   (a) $CHClF_2$: $[1(4) + 1(1) + 1(7) + 2(7)] = 26$ valence electrons

$$
\begin{array}{c}
\text{H} \\
| \\
:\ddot{F}-\underset{|}{\overset{|}{C}}-\ddot{F}: \\
:\ddot{Cl}:
\end{array}
$$

(b) CH_3COOH: $[3(1) + 2(4) + 2(6) + 1(1)] = 24$ valence electrons

$$
\begin{array}{c}
\text{H}\quad :\ddot{O}: \\
|\qquad || \\
\text{H}-\underset{|}{\overset{|}{C}}-\underset{}{C}-\ddot{O}-\text{H} \\
\text{H}
\end{array}
$$

(c) H_3CCN: $[3(1) + 2(4) + 1(5)] = 16$ valence electrons

$$
\begin{array}{c}
\text{H} \\
| \\
\text{H}-\underset{|}{\overset{|}{C}}-C\equiv N: \\
\text{H}
\end{array}
$$

(d) F_2CCF_2: $[4(7) + 2(4)] = 36$ valence electrons

$$
\begin{array}{c}
:\ddot{F}:\ :\ddot{F}: \\
|\quad\ | \\
:\ddot{F}-C=C-\ddot{F}:
\end{array}
$$

42.   Resonance structures for:
       (a) SO_2:

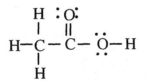

       (b) NO_2^-:

$$\left[\ddot{O}=\ddot{N}-\ddot{O}:\right]^{-} \longleftrightarrow \left[:\ddot{O}-\ddot{N}=\ddot{O}\right]^{-}$$

       (c) SCN^- :

$$\left[:\ddot{N}-C\equiv S:\right]^{-} \longleftrightarrow \left[\ddot{N}=C=\ddot{S}\right]^{-} \longleftrightarrow \left[:N\equiv C-\ddot{S}:\right]^{-}$$

44.   (a) BrF_3 : $[1(7) + 3(7)] = 28$ valence electrons

$$
\begin{array}{c}
\ddot{\mathrm{F}}\!-\!\mathrm{Br}\!-\!\ddot{\mathrm{F}} \\
| \\
\ddot{\mathrm{F}}
\end{array}
$$

(b) I_3^- : $[3(7) + 1] = 22$ valence electrons

$$
\ddot{\mathrm{I}}\!-\!\ddot{\mathrm{I}}\!-\!\ddot{\mathrm{I}}
$$

(c) XeO_2F_2 : $[1(8) + 2(6) + 2(7)] = 34$ valence electrons

$$
\begin{array}{c}
\ddot{\mathrm{O}} \qquad \ddot{\mathrm{F}} \\
\diagdown \qquad \diagup \\
\mathrm{Xe} \\
\diagup \qquad \diagdown \\
\ddot{\mathrm{F}} \qquad \ddot{\mathrm{O}}
\end{array}
$$

(d) XeF_3^+ : $[1(8) + 3(7) - 1] = 28$ valence electrons

$$
\left[\begin{array}{c}
\ddot{\mathrm{F}}\!-\!\mathrm{Xe}\!-\!\ddot{\mathrm{F}} \\
| \\
\ddot{\mathrm{F}}
\end{array} \right]^{+}
$$

Formal Charges

46. Formal charge on each atom in the following:

(a) N_2H_4

Atom	Formal Charge
H	$1 - 1/2(2) = 0$
N	$5 - 2 - 1/2(6) = 0$

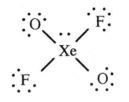

(b) PO_4^{3-}

Atom	Formal Charge
P	$5 - 1/2(8) = +1$
O	$6 - 6 - 1/2(2) = -1$
	Sum = -3 (charge on ion)

(c) BH_4^-

Atom	Formal Charge
B	$3 - 1/2(8) = -1$
H	$1 - 1/2(2) = 0$
	Sum = -1 (charge on ion)

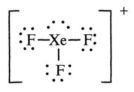

(d) NH_2OH

Atom	Formal Charge
N	5 - 2 - 1/2(6) = 0
H	1 - 1/2(2) = 0
O	6 - 4 - 1/2(4) = 0

$$H - \overset{\displaystyle H}{\underset{\displaystyle |}{N}} - \ddot{\underset{..}{O}} - H$$

48.   Formal charge on each atom in the following:

(a) NO_2^+

Atom	Formal Charge
O	6 - 4 - 1/2(4) = 0
N	5 - 0 - 1/2(8) = +1

$$\left[\ddot{\underset{..}{O}} = N = \ddot{\underset{..}{O}}\right]^+$$

(b) NO_2^-

Atom	Formal Charge
O1	6 - 4 - 1/2(4) = 0
O2	6 - 6 - 1/2(2) = -1
N	5 - 2 - 1/2(6) = 0

$$\left[\ddot{\underset{..}{O}} = N - \ddot{\underset{..}{\ddot{O}}}\right]^-$$
$$\quad 1 \qquad\qquad 2$$

(c) NF_3

Atom	Formal Charge
F	7 - 6 - 1/2(2) = 0
N	5 - 2 - 1/2(6) = 0

$$\ddot{\underset{..}{:}F} - \overset{..}{N} - F\ddot{\underset{..}{:}}$$
$$\quad\;\; | $$
$$\quad\; :F:$$

(d) HNO_3

Atom	Formal Charge
O1	6 - 4 - 1/2(4) = 0
O2	6 - 6 - 1/2(2) = -1
O3	6 - 4 - 1/2(4) = 0
N	5 - 0 - 1/2(8) = +1
H	1 - 1/2(2) = 0

$$H : \ddot{\underset{..}{O}} \underset{1}{-} N \underset{3}{=} \ddot{O}$$

[Note: This is only 1 possible structure]

Bond Polarity and Electronegativity

50.   Indicate the more polar bond (Arrow points toward the more negative atom in the dipole).

(a) C-O > C-N
$\;\;\; \rightarrow \qquad\quad \rightarrow$

(b) P-Cl > P-Br
$\;\;\;\; \rightarrow \qquad\quad \rightarrow$

(c) B - O > B - S
$\;\;\;\; \rightarrow \qquad\qquad \rightarrow$

(d) B-F > B-I
$\;\;\; \rightarrow \qquad\quad \rightarrow$

52.   For the bonds in acrolein the polarities are as follows:

	H-C	C-C	C=O
$\dfrac{\Delta\chi}{\Sigma\chi}$	0.09	0	0.16

(Note that χ represents electronegativity)

(a) The C-C bonds are nonpolar, the C-H bonds are slightly polar, and the C=O bond is polar.

(b) The most polar bond in the molecule is the C=O bond, with the oxygen atom being the negative end of the dipole.

Bond Polarity and Formal Charge

54.   Atom(s) on which the negative charge resides in:

(a) BF_4^- Formal charges : B = -1; F = 0   but Fluorine is MUCH more electronegative than B, so the negative charge (predicted to reside on B) resides on the fluorines--and is indeed distributed over the molecule

(b) BH_4^- Formal charges: B = -1: H = 0   Hydrogen is only slightly more electronegative than B (2.1 compared to 2.0), so the negative charge would reside on the H atoms (although the B-H bonds are **not very polar**).

(c) OH^- Formal charges: O = -1 ; H = 0   Oxygen is much more electronegative than H so the negative charge resides on the Oxygen atom.

(d) $CH_3CO_2^-$

Formal charges : C = 0; O1 = 0, O2 = -1   Oxygen is more electronegative than C, so the charge would reside on the oxygens as opposed to the C. The picture shown here is a bit misleading, since **either oxygen** could have the double bond to the C, so **two resonance structures are available** with the negative charge distributed (delocalized) over both C-O bonds.

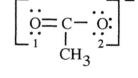

56.   (a) Resonance structures of N_2O :

$:N{\equiv}N-\ddot{O}:$ ⟷ $\ddot{N}{=}N{=}\ddot{O}$ ⟷ $:\ddot{N}-N{\equiv}O:$
   1   2   1   2   1   2

(b) Formal Charges:
      N_1 5 - 2 - 1/2(6) = 0   5 - 4 - 1/2(4) = -1   5 - 6 - 1/2(2) = -2
      N_2 5 - 0 - 1/2(8) = +1   5 - 0 - 1/2(8) = +1   5 - 0 - 1/2(8) = +1
      O   6 - 6 - 1/2(2) = -1   6 - 4 - 1/2(4) = 0   6 - 2 - 1/2(6) = +1

(c) Of these three structures, the first is the most reasonable in that the most electronegative atom, O, bears a formal charge of -1.

58. Resonance structures for NO_2^- :

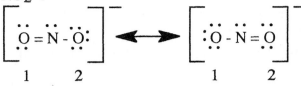

1   2   1   2

Formal charges:

O_1 6 - 4 - 1/2(4) = 0	O_1 6 - 6 - 1/2(2) = -1
N 5 - 2 - 1/2(6) = 0	N 5 - 2 - 1/2(6) = 0
O_2 6 - 6 - 1/2(2) = -1	O_2 6 - 4 - 1/2(4) = 0

Given that the formal charge on the oxygen atoms is –1, an H^+ ion will attach to the more negative oxygen atom.

Bond Order and Bond Length

60.

	Specie	Number of bonds	Bond Order : Bonded Atoms
(a)	H_2CO	3	1 : CH 2: C = O
(b)	SO_3^{2-}	3	1 : SO
(c)	NO_2^+	2	2 : NO
(d)	NOCl	2	1: N-Cl 2: N=O

Bond order is calculated: (number of shared electron pairs between atoms/number of links)

62. In each case the shorter bond length should be between the atoms with smaller radii--if we assume that the bond orders are equal.

(a) B-Cl	B is smaller than Ga		(b) C-O	C is smaller than Sn
(c) P-O	O is smaller than S		(d) C=O	O is smaller than N

64. The bond order for NO_2^+ is 2, for NO_2^- is 3/2 while the bond order for NO_3^- is 4/3. The Lewis dot structure for the NO_2^+ ion indicates that both NO bonds are double, while in the nitrate ion, any resonance structure (there are three) shows one double bond and two single bonds. The nitrite ion has—in either resonance structure (there are two)—one double and one single bond. Hence the **NO bonds in the nitrate ion will be longest** while those in the **NO_2^+ ion will be shortest**.

Bond Energy

66. The CO bond in carbon monoxide is shorter. The CO bond in carbon monoxide is a **triple bond**, thus it requires more energy to break than the CO double bond in H_2CO.

68.

$$H_3C\text{-}CH_2\text{-}\overset{\overset{H}{|}}{C}=\overset{\overset{H}{|}}{C}\text{-}H \quad + \quad H_2 \quad \longrightarrow \quad H_3C\text{-}CH_2\text{-}\overset{\overset{H}{|}}{\underset{\underset{H}{|}}{C}}\text{-}\overset{\overset{H}{|}}{\underset{\underset{H}{|}}{C}}\text{-}H$$

Energy input: 1 mol C=C = 1 mol • 610 kJ/mol =   610 kJ

                     1 mol H-H = 1 mol • 436 kJ/mol =   436 kJ

                                                Total input   =   1046 kJ

Energy release: 1 mol C-C = 1 mol • 346 kJ/mol =   346 kJ

                     2 mol C-H = 2 mol • 413 kJ/mol =   826 kJ

                                                Total released = 1172 kJ

Energy change: 1046 kJ - 1172 kJ = - 126 kJ

70. OF_2 (g) + H_2O (g) → O_2 (g) + 2 HF (g)   ΔH = - 318 kJ

Energy input :   2 mol O-F = 2 x (where x = O-F bond energy)

                     2 mol O-H = 2 mol • 463 kJ/mol = 926 kJ

                                                Total input   = (926 + 2x) kJ

Energy release: 1 mol O=O = 1 mol • 498 kJ/mol =   498 kJ

                     2 mol H-F = 2 mol • 565 kJ/mol = 1130 kJ

                                                Total release = 1628 kJ

          - 318 kJ   = 926 kJ + 2x - 1628 kJ

          384 kJ   = 2x

          192 kJ/mol = O-F bond energy

Molecular Geometry

72. Using the Lewis structure describe the Electron-pair and molecular geometry:

   (a).

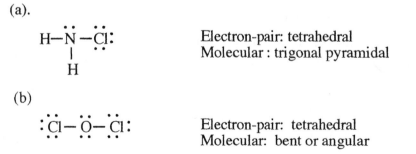

   H—N̈—C̈l: Electron-pair: tetrahedral
      | Molecular : trigonal pyramidal
      H

   (b)

   :C̈l— Ö— C̈l: Electron-pair: tetrahedral
                       Molecular: bent or angular

114

(c)

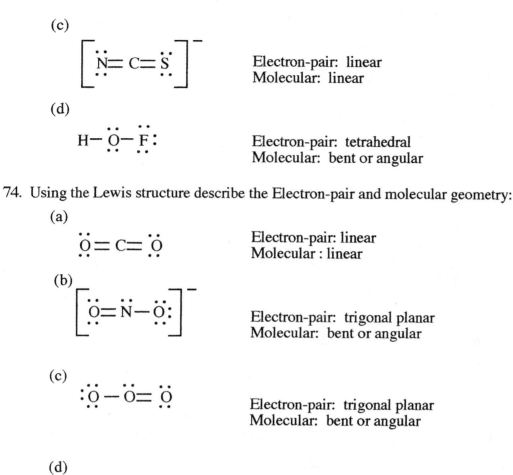

Electron-pair: linear
Molecular: linear

(d)

Electron-pair: tetrahedral
Molecular: bent or angular

74. Using the Lewis structure describe the Electron-pair and molecular geometry:

(a)

Electron-pair: linear
Molecular : linear

(b)

Electron-pair: trigonal planar
Molecular: bent or angular

(c)

Electron-pair: trigonal planar
Molecular: bent or angular

(d)

Electron-pair: tetrahedral
Molecular: bent or angular

For the species shown above, having at least one lone pair on the central atom **changes the molecular geometry from linear to bent.**

76. Using the Lewis structure describe the electron-pair and molecular geometry.

[Lone pairs on F have been omitted for clarity.]

(a)

Electron-pair: trigonal bipyramidal
Molecular : linear

(b)

Electron-pair: trigonal bipyramidal

Molecular: T-shaped

(c)

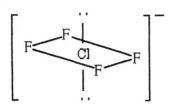

Electron-pair: octahedral
Molecular: square planar

(d)

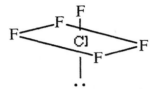

Electron-pair: octahedral
Molecular: square pyramidal

78.   (a) O-S-O angle in SO$_2$:

Slightly less than 120°; The lone pair of S should reduce the predicted 120° angle slightly.

(b) F-B-F angle in BF$_3$:

120°

(c) Cl-C-Cl in Cl$_2$CO

Slightly less than 120°; The two lone pairs of electrons on O will reduce the predicted 120° angle slightly.

(d) (1) H-C-H angle in CH$_3$CN: 109°
     (2) C-C ≡N angle in CH$_3$CN: 180°

80.   Estimate the values of the angles indicated in the model of phenylalanine below:

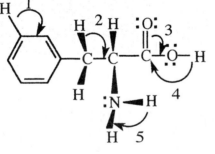

Angle 1:   H-C-C 120°   three groups around the C atom
Angle 2:   H-C-C 109°   four groups around the C atom
Angle 3:   O-C-O 120°   three groups around the C atom
Angle 4:   C-O-H 109°   four groups around the O atom
Angle 5:   H-N-H 109°   four groups around the N atom

The CH_2-$CH(NH_2)$-CO_2H can not be a straight line, since the first two carbons will have bond angles of 109 degrees(with their connecting atoms) and the third C (the C of the CO_2H group) has a 120 bond angle with the C and O on either side.

Molecular Polarity

82. For the molecules:
$$H_2O \qquad NH_3 \qquad CO_2 \qquad ClF \qquad CCl_4$$

(i) Using the electronegativities to determine bond polarity:

$\dfrac{\Delta\chi}{\Sigma\chi}$	$\dfrac{1.4}{5.6}$	$\dfrac{0.9}{5.1}$	$\dfrac{1.0}{6.0}$	$\dfrac{10}{7.0}$	$\dfrac{0.5}{5.5}$

Reducing these fractions to a decimal form indicates that the H-O bonds in water are the most polar of these bonds. [Note: $\Delta\chi$ represents the **difference** in electronegativities between the elements while $\Sigma\chi$ is the **sum** of the electronegativities of the elements.]

(ii) The nonpolar compounds are:

CO_2 The O-C-O bond angle is 180°, thereby canceling the C-O dipoles.

CCl_4 The Cl-C-Cl bond angles are approximately 109°, with the Cl atoms directed at the corners of a tetrahedron. Such an arrangement results in a net dipole moment of zero.

(iii) The F atom in ClF is more negatively charged.(Electronegativity of F = 4.0, Cl = 3.0)

84. Molecular polarity of the following: (a) $BeCl_2$, (b) HBF_2, (c) CH_3Cl, (d) SO_3

$BeCl_2$ and SO_3 are nonpolar—since the linear geometry (of $BeCl_2$) and the trigonal planar geometry (of SO_3) would give a net dipole moment of zero. For HBF_2, the hydrogen and fluorine atoms are arranged at the corners of a triangle. The "negative end" of the molecule lies on the plane between the fluorine atoms, and the H atom is the "positive end." For CH_3Cl, with the H and Cl atoms arranged at the corners of a tetrahedron, the chlorine atom is the negative end and the H atoms form the positive end.

General Questions on Bonding and Molecular Structure

86. Lewis structure(s) for the following: What are similarities and differences ?

(a) CO_2

$$\left[\ddot{O}=C=\ddot{O}\right] \longleftrightarrow \left[:\ddot{O}-C\equiv O:\right] \longleftrightarrow \left[:O\equiv C-\ddot{O}:\right]$$

(b) N_3^-

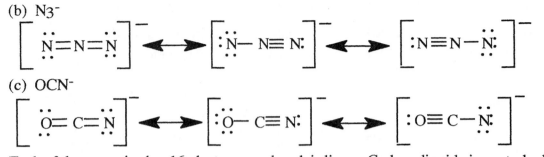

(c) OCN^-

Each of these species has 16 electrons, and each is linear. Carbon dioxide is neutral while the azide ion and isocyanate ion are charged. The isocyanate ion is also polar, while carbon dioxide and azide are not polar.

88. NO_2^+ has two structural pairs around the N atom. We predict that the O-N-O bond angle would be approximately 180°. NO_2^- has three structural pairs (one lone pair). The geometry around this central atom would be trigonal planar with a bond angle of approximately 120°.

90. The electron dot structure for the cyanide ion may be represented :
     The ion is symmetric with respect to the electron distribution—
     that is each atom has **one** lone pair and participates in a **triple bond**. N
     has a formal charge of 0, while C has a formal charge of –1, so the positive H ion would be
     attracted more to the C "end " of the ion.

92. Enthalpy change for the reaction : $2 N_2O$ (g) $\rightarrow$ $2 N_2$ (g) + 1 O_2 (g)

Energy input:	2 mol N≡N	=	2 mol • 945 kJ/mol	=	1890 kJ
	2 mol N-O	=	2 mol • 201 kJ/mol	=	402 kJ
			Total input	=	2292 kJ
Energy release:	2 mol N≡N	=	2 mol • 945 kJ/mol	=	1890 kJ
	1 mol O=O	=	1 mol • 498 kJ/mol	=	498 kJ
			Total released	=	2388 kJ
Energy change:	2292 kJ - 2388 kJ	=	- 96 kJ		

94. The much more negative lattice energy of NaF is attributable to the very small ionic radius of the fluoride ion. The result is a greater attraction between the Na and F ions in the solid lattice.

96.   (a) Angle 1 = 120°; Angle 2 = 180°; Angle 3 = 120°

(b) The C=C double bond is shorter than the C-C single bond.

(c) The C=C double bond is stronger than the C-C single bond.

(d) The C≡N bond is the most polar bond, with the N atom being the negative end of the bond dipole.

98.   (a) In XeF$_2$ the bonding pairs occupy the axial positions (of a trigonal bipyramid) with the lone pairs located in the equatorial plane. (See question 44(b) for an isoelectronic specie, (I$_3^-$).

(b) In ClF$_3$ two of the three equatorial positions are occupied by the lone pairs of electrons on the Cl atom. (See question 76b)

100. Hydroxyproline has the structure:

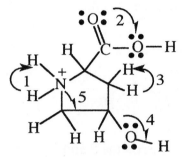

(a) Values for the selected angles:

Angle 1: 109° since the N has four groups of electrons around the atom.

Angle 2: 120° since the C atom has three groups of electrons around it.

Angle 3: 109° since the C atom has four groups of electrons around it.

Angle 4: 109° since the O atom has four groups of electrons around it
(2 bonding and 2 non-bonding pairs)

Angle 5: 109° since the C atom has four groups of bonding pairs of electrons around it.

(b) The most polar bonds in the molecule are the O-H bonds. The electronegativity of O is 3.5, while that of H is 2.1 (a Δ of 1.4). This difference represents the **greatest** differences in electronegativity in the hydroxyproline molecule.

102. Enthalpy change for decomposition of urea:

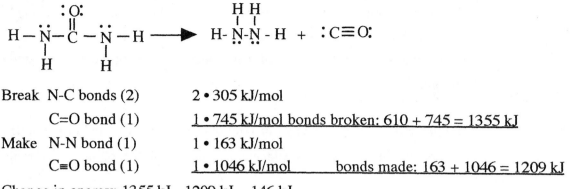

Break	N-C bonds (2)	2 • 305 kJ/mol
	C=O bond (1)	1 • 745 kJ/mol bonds broken: 610 + 745 = 1355 kJ
Make	N-N bond (1)	1 • 163 kJ/mol
	C≡O bond (1)	1 • 1046 kJ/mol bonds made: 163 + 1046 = 1209 kJ

Change in energy: 1355 kJ - 1209 kJ = 146 kJ

Since there was no change in the number of N-H bonds (in urea) or hydrazine, those bond energies were omitted in the calculation.

104. For the molecule 2-furylmethanethiol:

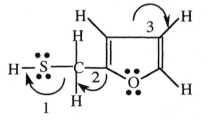

(a) The formal charges on the S and O atoms:

Formal charge = Group # - nonbonding electrons – 1/2(bonding electrons)

$S = 6 – 4 - 1/2(4) = 0$ and for $O = 6 – 4 - 1/2(4) = 0$

(b) Angles 1,2, and 3:

Angle 1: 109 ° since there are 4 groups of electrons around the S atom

Angle 2: 109 ° since there are 4 groups of electrons around the C atom.

Angle 3: 120 ° since there are 3 groups of electrons around the C atom.

(c) The shorter carbon-carbon bonds are the C=C bonds in the five-member ring—since C=C double bonds are shorter than C-C single bonds.

(d) The most polar bond in the molecule:

The electronegativity of C is 2.5 while that of O is 3.5 with a Δ of 1.0. All other electronegativity differences are smaller than this, so the C-O bonds are the most polar bonds.

(e) The molecule as a whole is **polar** since the S-H sidechain will be polar. No other off-setting dipole reduces that polarity.

(f) The O atom in the furan ring assumes an sp^2 hybridization (and a 120° bond angle) to reside within the same plane as the other atoms in the furan ring.

106. Resonance structures for nitric acid:

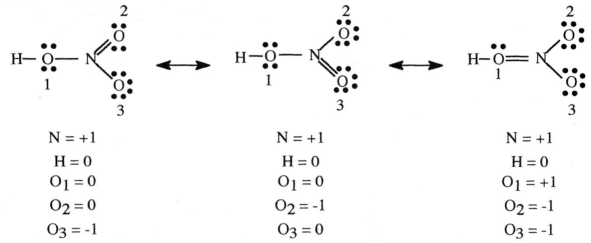

N = +1	N = +1	N = +1
H = 0	H = 0	H = 0
$O_1 = 0$	$O_1 = 0$	$O_1 = +1$
$O_2 = 0$	$O_2 = -1$	$O_2 = -1$
$O_3 = -1$	$O_3 = 0$	$O_3 = -1$

The least likely structure is the rightmost one, since the formal charge on O1 is positive, an unlikely charge for a very electronegative atom.

108. For the synthesis of acrolein from ethylene:

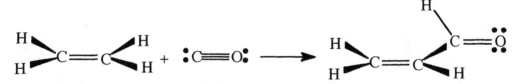

(a) The C=C bond in acrolein is stronger than the C-C bond.

(b) The C-C bond is longer than a C=C bond.

(c) Polarity of ethylene and acrolein:

.For the bonds in acrolein the polarities are as follows:

	H-C	C-C	C=O
$\dfrac{\Delta\chi}{\Sigma\chi}$	0.09	0	0.16

The C-C and C=C bonds are nonpolar, the C-H bonds are very slightly polar, and the C=O bond is polar. [$\Delta\chi$ represents the **difference** in electronegativities between the two bonded atoms while $\Sigma\chi$ represents the **sum** of the electronegativities of the two bonded atoms.] The structure of ethylene (coupled with the relative lack of polarity of the C-H and C=C bonds) results in a molecule that is nonpolar (net dipole moment of 0). Acrolein however substitutes the polar C=O group (essentially inserted into a C-H bond, and gives a **polar** molecule.

(d) Reaction endothermic or exothermic:

Careful examination of the structures reveals that to form acrolein from ethylene, we must break a C-H bond (which reforms in the product), and a $C \equiv O$ bond (which forms C=O bond in the product). Using bond energies, the energy change is:

Break $C \equiv O$ bond (1)   1 • 1046 kJ/mol

C-H bond (1) 1 • 413 kJ/mol bonds broken: 413 + 1046 = 1459 kJ

Make   C-C bond (1) 1 • 346 kJ/mol

C-H bond (1) 1 • 413 kJ/mol

C=O bond (1) 1 • 745 kJ/mol bonds made: 346 + 413 + 745 = 1504 kJ

Change in energy: 1459 kJ - 1504kJ = -45 kJ , so the reaction is **exothermic**.

110. For the molecule epinephrine:

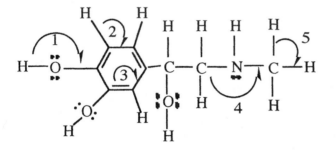

(a) The indicated bond angles are:

Angle 1: 109°   Angle 2: 120 °

Angle 3: 120 °   Angle 4: 109 °

Angle 5: 109 °

(b) The most polar bonds in the molecule:

The O-H bonds are most polar (with a difference in electronegativity of 3.5-2.1)

Using Electronic Resources:

111. Using data from the CD, the following are obtained:

Formula	Measured Bond Distance (A)		Bond Order	
Ethane, C_2H_6	C-C = 1.540 A	C-H = 1.117 A	C-C:1	C-H = 1
Butane. C_4H_{10}	C-C = 1.540 A	C-H = 1.116 A	C-C:1	C-H = 1
Ethylene, C_2H_4	C-C= 1.352 A	C-H = 1.098 A	C-C:2	C-H = 1
Acetylene,C_2H_2	C-C = 1.226 A	C-H = 1.061 A	C-C:3	C-H = 1
Benzene. C_6H_6	C-C = 1.397 A	C-H = 1.386 A	C-C:2*	C-H = 1

Note that Bond order for C-C bonds and C-H is deduced, since greater bond orders should have a shorter measured bond distance. Ethane and Butane have similar C-C bond lengths. Ethylene has a shorter C-C bond, and Acetylene an even shorter one. Note that the C-H bond distance is approximately the same in all five molecules, with benzene showing a slightly longer C-H bond length.

*Using the bond distances, the C-C bond in beneze appears to have a bond order that is approximately that of the double bond in ethylene and considerably shorter than the C-C single bonds in ethane or butane. Using a Lewis dot structure for the benzene molecule, one calculates that alternating C-C bonds have a bond order of 1 and 2, with an average bond order of 1.5.

Chapter 10
Bonding and Molecular Structure: Orbital Hybridization and Molecular Orbitals

Reviewing Important Concepts

2. The maximum number of hybrid orbitals that a carbon atom can form is 4, involving the s and p orbitals. The minimum number that can be formed is two. Mixing more than 4 orbitals would involve orbitals other than those which carbon atoms normally occupy. It's impossible to form a hybrid with only one orbital—hence the two orbital minimum

4. For atoms with sp-hybridization, **two** pure p orbitals remain. The remaining p orbitals are available to form π bonds—so two π bonds can form.

6. For the species BF_4^-, SiF_4, and SF_4:

   (a) A molecule isoelectronic with BF_4^- would have to have the same number of valence electrons and the same Lewis structures. While there could be many molecules that fit this description, **CF_4** is one example.

   (b) SiF_4, and SF_4 are **not** isoelectronic . Sulfur has two more valence electrons than silicon.

   (c) The hybridization of the central atom:

species	hybridization
BF_4^-	sp^3
SiF_4	sp^3
SF_4	sp^3d

8. The maximum number of hybrid orbitals available to third-period elements is increased, owing to the availability of d orbitals. Usually two d orbitals mix with s and p orbitals providing a maximum of **six** hybrid orbitals.

10. Four principles of Molecular Orbital Theory:

   (a) Orbital conservation— The number of molecular orbitals in a molecule **is equal** to the number of atomic orbitals brought by all the atoms in the molecule.

   (b) Energy released in bonding MOs— **Bonding MOs are lower in energy** that their parent orbitals while **antibonding MOs are higher in energy** than the parent orbitals.

(c) Pauli principle & Hund's rule obeyed—When electrons are placed in MOs, the placement is done so that **electrons are placed in the lowest energy orbitals** available, and when placed **in the same orbital the spins are paired.**

(d) Similar atomic orbitals combine most effectively to form MOs—e.g. 1s orbitals combine with 1s orbitals--not 2s orbitals.

12. Connection between bond order, bond length, and bond energy:

molecule	bond order
C_2H_6	C-C bond = 1
C_2H_4	C-C bond = 2
C_2H_2	C-C bond = 3

In the three compounds noted above, the bond order (of the C-C bonds) increases from 1 to 2 to 3. This change is accompanied by a **shortening of the bond**—that is the C-C bond in ethylene is shorter than the C-C bond in ethane and the C-C bond in acetylene is **shorter than** the C-C bond in ethylene. The shorter bonds are also more difficult to rupture, so the bond energy **increases** as the **length decreases.**

14. The term **localized** as it pertains to bonding theories refers to the idea that electron pairs are contained (or localized) in orbitals between two --and only two--atoms, while **delocalized** refers to the orbitals (in MO theory) which are thought of as being spread out (or delocalized) over the entire molecule.

16. The band gap in insulators is large, in contrast to that of metals. This gap can range from a few kilojoules to several hundred kJ/mol. Metals have only partially filled occupied valence bands while insulators have completely filled valence bands. With the next available empty level for an insulator at a much higher energy, the electrons aren't promoted.

Practicing Skills

Valence Bond Theory

18. NF_3 has **tetrahedral** electron-pair geometry and **pyramidal** molecular geometries. With four electron-pairs attached, the N is sp^3 hybridized. The lone pair occupies one of the four sp^3 hybrid orbitals with the remaining three orbitals overlapping the **p** orbitals on F, forming the N-F sigma bonds. [Lone pairs on the fluorine atoms have been omitted for the sake of clarity.]

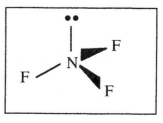

20. The Lewis electron dot structure of $CHCl_3$:

    Electron pair geometry = tetrahedral

    Molecular geometry = tetrahedral

    The H-C bonds are a result of the overlap of the

    hydrogen **s** orbital with **sp³** hybrid orbitals on carbon.

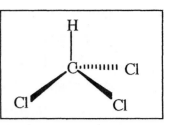

    The Cl-C bonds are formed by the overlap of the **sp³** hybrid orbitals on carbon with the **p**
    orbitals on chlorine. [Lone pairs on the chlorine atoms have been omitted for the sake of
    clarity.]

22. Orbital sets used by the underlined atoms:

    (a) $\underline{B}Br_3$: sp²   3 groups to be attached to the B
    (b) $\underline{C}O_2$: sp   2 groups to be attached to the C
    (c) $\underline{C}H_2Cl_2$: sp³   4 groups to be attached to the C
    (d) $\underline{C}O_3{}^{2-}$: sp²   3 groups to be attached to the C

24. Hybrid orbital sets used by the underlined atoms:
    (a) the C atoms and the O atom in dimethylether: $\underline{C}$: sp³ ; $\underline{O}$: sp³

        In the case of either the carbon OR oxygen atoms in dimethylether, each atom is bound
        to *four other groups*. This would require **four orbitals**
    (b) The carbon atoms in propene: $\underline{C}H_3$: sp³ ; $\underline{C}H$ and $\underline{C}H_2$: sp²

        The methyl carbon is attached to four groups (three H and 1 C), and needs then four
        orbitals. The methylene and methine have bonds to four groups (2H and 2C in the case
        of CH_2 and four groups (1H and 3 C) in the case of CH-.
    (c) The C atoms and the N atom in glycine: $\underline{N}$: sp³; $\underline{C}H_2$: sp³, $\underline{C}=O$: sp²

        The N atom is attached to 4 groups (3 atoms and 1 lone pair), the CH_2 carbon has 4
        groups attached (2 H atoms and 1N and 1 C atom). The carbonyl carbon is attached to
        only 3 groups (1C and 2 O atoms)

26. Hybrid orbital sets used by the underlined atoms:

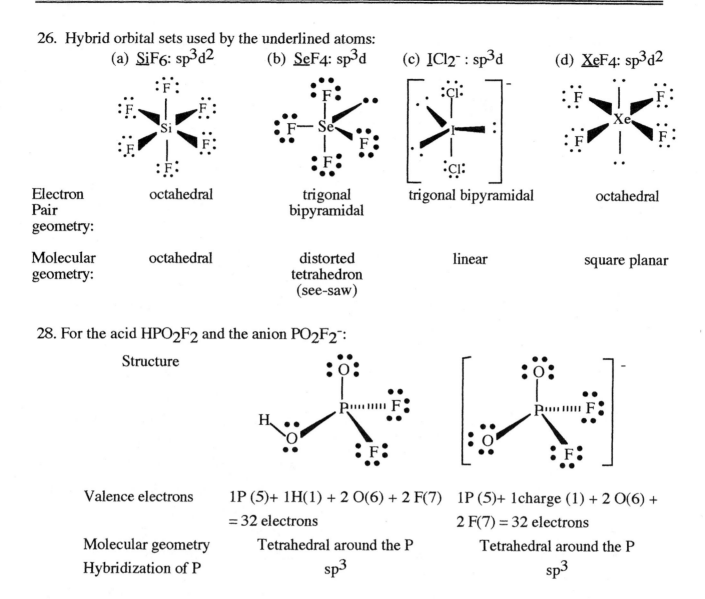

(a) $\underline{Si}F_6$: sp^3d^2 (b) $\underline{Se}F_4$: sp^3d (c) $\underline{I}Cl_2^-$: sp^3d (d) $\underline{Xe}F_4$: sp^3d^2

Electron Pair geometry:	octahedral	trigonal bipyramidal	trigonal bipyramidal	octahedral
Molecular geometry:	octahedral	distorted tetrahedron (see-saw)	linear	square planar

28. For the acid HPO_2F_2 and the anion $PO_2F_2^-$:

Structure

Valence electrons	1P (5)+ 1H(1) + 2 O(6) + 2 F(7) = 32 electrons	1P (5)+ 1charge (1) + 2 O(6) + 2 F(7) = 32 electrons
Molecular geometry	Tetrahedral around the P	Tetrahedral around the P
Hybridization of P	sp^3	sp^3

30. For the molecule $COCl_2$: Hybridization of C = sp^2

Bonding:1 sigma bond between each chlorine and carbon

(**sp** hybrid orbitals)

1 sigma bond between carbon and oxygen

(**sp** hybrid orbitals)

1 pi bond between carbon and oxygen

(**p** orbital)

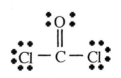

32. For the following compounds, the other isomer is:

(a)	H_3C \ $C=C$ / H ; H / $C=C$ \ CH_3	H_3C \ $C=C$ / CH_3 ; H / $C=C$ \ H
(b)	Cl \ $C=C$ / CH_3 ; H / $C=C$ \ H	Cl \ $C=C$ / H ; H / $C=C$ \ CH_3

Molecular Orbital Theory

34.   Configuration for H_2^+: $(\sigma 1s)^1$

Bond order for H_2^+: 1/2 (no. bonding e^- - no. antibonding e^-) = 1/2

The bond order for molecular hydrogen is <u>one</u> (1), and the H-H bond is stronger in the H_2 molecule than in the H_2^+ ion.

36.   The molecular orbital diagram for C_2^{2-} , the acetylide ion:

σ^*2p	_____	There are 2 net pi bonds and 1 net
π^*2p	____ ____	sigma bond in the ion, giving a bond
$\sigma 2p$	↑↓	order of 3. On adding two electrons
$\pi 2p$	↑↓ ↑↓	to C_2 (added to $\sigma 2p$)to obtain C_2^{2-},
σ^*2s	↑↓	the bond order increases by one. The
$\sigma 2s$	↑↓	ion is **diamagnetic**.

38.   (a) The electron configuration (showing only the outer level electrons) for CO is:

σ^*2p	_____	(b) The HOMO is the $\sigma 2p$
π^*2p	____ ____	(c) There are no unpaired electrons,
$\sigma 2p$	↑↓	hence CO is diamagnetic.
$\pi 2p$	↑↓ ↑↓	(d) There is one net sigma bond, and
σ^*2s	↑↓	two net pi bonds for an overall bond
$\sigma 2s$	↑↓	order of 3.

Metals and Semiconductors

40. See Figure 10.25 for a similar example with elemental lithium. Magnesium has a 2s and three 2p orbitals which form MO's. So from 4 atomic orbitals, 4 MO arise. With 100 atoms, there are then 4 • 100 or 400 MO. Each Mg contributes 2 valence electrons, for a total of 200 electrons (2 •100), With each MO capable of holding 2 electrons, only **100** of the MO are filled.

General Questions on Valence Bond and Molecular Orbital Theory

42. The O-S-O bond angle and the hybrid orbitals used by sulfur:

(a)	SO_2	120° angle	3 electron-pair groups (1 lone pair on S)	sp^2
(b)	SO_3	120° angle	3 electron-pair groups	sp^2
(c)	$SO_3{}^{2-}$	109° angle	4 electron-pair groups (1 lone pair on S)	sp^3
(d)	$SO_4{}^{2-}$	109° angle	4 electron-pair groups	sp^3

44. Resonance structures for the nitrite ion:

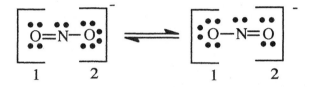

   1   2   1   2

The electron-pair geometry of the ion is *trigonal planar*, and the molecular geometry is *bent* (or angular). With three electron-pairs around the N, the O-N-O bond angle is 120°. The average bond order is 3/2 (three bonds divided connecting the two O atoms). The hybridization associated with 3 electron-pair groups is sp^2.

46. Resonance structures for N_2O:

$$N\equiv N-\overset{..}{\underset{..}{O}}: \rightleftharpoons \overset{..}{\underset{..}{N}}=N=\overset{..}{\underset{..}{O}} \rightleftharpoons :\overset{..}{\underset{..}{N}}-N\equiv O:$$

  1 2   1 2   1 2

Hybridization	N1:sp **N2:sp**		N1:sp^2 **N2: sp**		N1: sp^3 **N2:sp**
Central N orbitals	The s & p orbitals		The s & p orbitals		The s & p orbitals

48. Regarding acrolein:

(a) Both carbon 1 and 2 are **sp^2** hybridized (3 electron-pair groups).

(b) Since angles A, B, and C are associated with sp^2 hybridized carbons, each of the three angles is approximately **120°**.

(c) cis-trans isomerization **isn't possible** since C1 has two identical groups (H).

50. Regarding ethylene oxide, acetaldehyde, and vinyl alcohol:

(a)Formula:	C_2H_4O	C_2H_4O	C_2H_4O — these **are** isomers of one another
(b)Hybridization	C1 and C2 - sp^3	C1 sp^3 C2-sp^2	C1 and C2 - sp^2
(c) Bond angles	109° (anticipated) 60° due to geometry	C1 109° C2-120°	120°
(d) Polarity	Polar	Polar	Polar
(e) Strongest bonds		Strongest C-O bond	Strongest C-C bond

52. (a) There is 1 π bond in lactic acid (the C=O). There are 12 atoms in the molecule, so there must be 11 σ bonds.

   (b) Hybridizations: C1 (4 groups attached)-sp^3

                         C2 (3 groups attached)-sp^2

                         O3 (3 groups attached)-sp^3

   (c) The *shortest* and *strongest* CO bond is the C=O bond.

   (d) Bond Angle: A—109° (sp^3 geometry); B--109° (sp^3 geometry); C—120°(sp^2 geometry)

54. For BF_3 and NH_3BF_3 :

   (a) With 3 groups attached to the B in BF_3 , the geometry is trigonal planar. With 4 groups attached to the B in NH_3BF_3, the geometry is tetrahedral.

   (b) In BF_3 the B is sp^2 hybridized, in NH_3BF_3 the B is sp^3 hybridized.

   (c) The hybridization of B changes.

56.   (a) Hybridizations: In SbF_5, the Sb atom has five pairs of electrons around it, and the hybridization would be sp^3d. In the SbF_6^- anion, with six pairs of electrons around it, the hybridization would be sp^3d^2

   (b) The Lewis structure for H_2F^+ :

The electron-pair geometry is *tetrahedral* and the molecular geometry is *bent (or angular)*. With 4 electron pairs around the F, the hybridization would be sp^3.

58. The molecule cinnamaldehyde:

    (a) The C=O is the most polar bond.

    (b) 2π bonds (outside the ring) and 3π bonds (in the aromatic ring) There are 18σ bonds.

    (c) cis-trans isomerism is possible. The trans-isomer is shown below.

    (d) Since each C is attached to 3 electron pairs, all C are sp^2 hybridized.

    (e) Since the angles indicated have an sp^2 hybridized C at the center, all the angles are 120°.

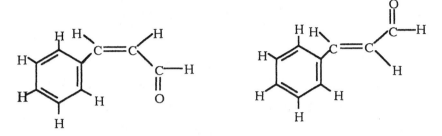

60. (a) A Lewis dot structure of the peroxide ion indicates that the bond order is 1.

    (b) The molecular orbital electron configuration

    of the peroxide ion is:
    (Showing only valence electrons) $(\sigma 2s)^2(\sigma^* 2s)^2(\pi 2p)^4(\sigma 2p)^2(\pi^* 2p)^4$ Using the MO

    configuration, we have 8 bonding electrons and 6 antibonding electrons or 2 bonding electrons (Bond order =1)

    (c) Both theories show *no unpaired electrons*, and a bond order of 1. The magnetic character of both is the same: diamagnetic.

62. Consider the diatomic molecules of Li_2 through Ne_2

molecule	electron configuration	magnetic property	bond order
Li_2	$(\sigma 2s)^2$	diamagnetic	1
Be_2	$(\sigma 2s)^2 (\sigma^* 2s)^2$	diamagnetic	0
B_2	$(\sigma 2s)^2(\sigma^* 2s)^2(\pi 2p)^2$	paramagnetic	1
C_2	$(\sigma 2s)^2(\sigma^* 2s)^2(\pi 2p)^4$	diamagnetic	2
N_2	$(\sigma 2s)^2(\sigma^* 2s)^2(\pi 2p)^4(\sigma 2p)^2$	diamagnetic	3
O_2	$(\sigma 2s)^2(\sigma^* 2s)^2(\pi 2p)^4(\sigma 2p)^2(\pi^* 2p)^2$	paramagnetic	2
F_2	$(\sigma 2s)^2(\sigma^* 2s)^2(\pi 2p)^4(\sigma 2p)^2(\pi^* 2p)^4$	diamagnetic	1
Ne_2	$(\sigma 2s)^2(\sigma^* 2s)^2(\pi 2p)^4(\sigma 2p)^2(\pi^* 2p)^4(\sigma 2p)^2$	diamagnetic	0

64. Using the orbital diagram we predict the electron configuration to be:
(with 4 electrons from C and 5 from N) $(\sigma2s)^2(\sigma^*2s)^2(\pi2p)^4(\sigma2p)^1$

    (a) The highest energy MO to which an electron is assigned is the $\sigma2p$.

    (b) Bond order = 1/2 (# bonding electrons - # non-bonding electrons)

$$= 1/2 \ (7 - 2) \ \text{or} \ 5/2 \ \text{or} \ 2.5.$$

    (c) There is a **net of 1/2** sigma bond $-(\sigma2p)^1$, and 2 net π bonds— $(\pi2p)^4$.

    (d) The molecule has an unpaired electron so it is paramagnetic

66. For the molecule menthol:

    (a) Every carbon is surrounded by 4 electron pairs, so the hybridization is sp^3.

    (b) The O atom has 4 electron pairs (two lone pairs and two bonded pairs), so the C-O-H bond angle is predicted to be the tetrahedral angle of 109°.

    (c) With the pendant OH group, the molecule will be slightly polar.

    (d) The ring is *not planar*. Each of the carbons is bonded to two other carbons with a 109° bond angle, and *unlike the benzene ring—which is planar*, this ring will be puckered.

68. The primary difference in the band structure of conductors and insulators is the band gap. For conductors (e.g. metals) the highest occupied band of electrons is only *partially occupied*. By contrast, insulators have the highest occupied band filled, with the next available empty levels at a much higher energy (a gap). Such electrons aren't promoted, and the material acts as an insulator.

70. See Study Questions 68 and 80 for additional information on this concept. Since Ge has a conduction band that is *higher* than that of a metal (e.g. Li), and *lower* than the conduction band for diamond or silicon, we would anticipate that Ge is a poorer conductor of electrons than lithium, but a better conductor than diamond or silicon.

<u>72</u>. The essential part of the Lewis structure for the peptide linkage is shown here:

    (a) The carbonyl carbon (C=O) is attached to 3 groups and so has a hybridization of sp^2. The N atom is connected to 4 groups (3 atoms, 1 lone pair) and has a hybridization of sp^3.

    (b) Another structure is feasible:

       This structure would have a formal charge on: carbonyl carbon of 0, on the oxygen of -1 , and on the N a formal charge +1.

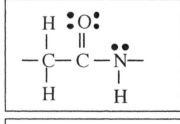

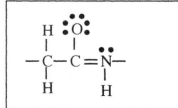

Since the "preferred" structure has 0 formal charges on **all three atoms**, the first structure is the more favorable.

(c) If one views the "less preferred" resonance structure as a contributor, one can see that both the carbonyl carbon and the N are sp^2 hybridized. This leaves one "p" orbital on O,C, and N unhybridized, and capable of side-to-side overlap (also known as π type overlap) between the C and the O and between the C and the N, forming a planar region in the molecule.

74.   The structure for ethylene oxide is: (a and b) Since the 3 atoms in the ring all use sp^3 hybridization, the angles are anticipated to be approximately 109°.

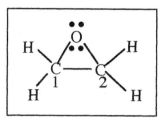

(c) The hybridization would lead us to believe that the angles will be approximately 109°, while the geometry would suggest that the sum of the angles is 180°, so each angle would be approximately 60°. This strain accounts for the reactivity (and instability) of this 3-member ring.

Using Electronic Resources

76. Regarding hybrid orbitals:
   (a) If n atomic orbitals are combined, n hybrid orbitals will be formed—in other words orbitals are conserved.
   (b) Hybrid orbitals do not form between p orbitals without involvement of an s orbital.
   (c) Hybrid orbitals are intermediate in energy to the atomic orbitals from which they were formed. An sp hybrid orbital would be lower in energy than the p orbital and higher in energy than the s orbital from which the hybrid orbital was formed.
   (d) The shapes of a sp$_x$ hybrid and an sp$_z$ hybrid are identical, and differ only in their orientation. The sp$_x$ hybrid lies along the p$_x$ axis while the p$_z$ hybrid would lie along the z axis.
   (e) The shapes of the orbital formed from the s,p$_x$ and p$_y$ orbitals are identical to the orbital formed from the s, p$_x$, p$_z$. As in (d) above, the only difference is in the orientation of the orbital in space.

78.   (a) The vast difference in energies between the rotations of trans-2-butene and butane is a reflection of the low energy required for rotation around C-C single bonds, and the *greatly restricted* rotation around C=C double bonds.

(b) In the molecule propene, the methyl group (CH$_3$-) can rotate freely about the C-C single bond. The methylene group (CH$_2$-) is hindered from free rotation by the C=C double bond—much as trans-2-butene was hindered in its rotation.

(c) Given that the terminal methylene groups in allene are connected to the central C atom with C=C double bonds, rotation of those methylene groups with respect to one another is not expected to occur easily.

80.   (a) See Study Question 68 for a discussion of conductors and insulators. Semiconductors have a band gap between the valence band and the conduction band, *but* it is very much smaller than the gap found in insulators.

(b) To provide vacant orbitals a Group 3A metal (like Al) would provide a band whose energies are only slightly higher than the semiconductor, and which can therefore accept electrons.

Chapter 11
Carbon: More Than Just Another Element

Reviewing Important Concepts

2. The hindered rotation about C=C double bonds is due to the overlap of orbitals forming the π bond. Shown here is a schematic indicate the σ and π type overlaps.

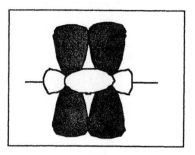

4. The structure of 2-butanol and its mirror image are shown below.

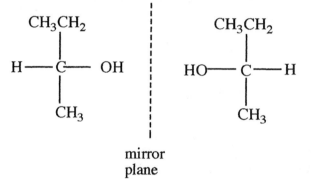

mirror
plane

The drawing is a Fischer projection of the molecule, with the chiral C at the "intersection". The mirror image of the molecule is on the right. These two are non-superimposable mirror images, also known as enantiomers.

6. The structures and names of the alkenes with formula C_4H_8

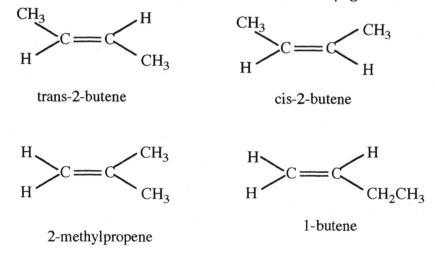

trans-2-butene

cis-2-butene

2-methylpropene

1-butene

8. Structure and names of products formed when the following alcohols are oxidized:

   (a) $CH_3CH_2CH_2CH_2OH$ - a primary alcohol should form a carboxylic acid

$$CH_3CH_2CH_2CH_2OH \longrightarrow CH_3CH_2CH_2\overset{\overset{O}{\|}}{C}\text{-OH}$$

   butanoic acid

   (b) 2-Butanol- a secondary alcohol should form a ketone:

   2-butanone

   (c) 2-Methyl-2-propanol- a tertiary alcohol should *not react*.

   (d) 2-Methyl-1-propanol- a primary alcohol should form a carboxylic acid

$$\overset{\overset{CH_3}{|}}{CH_3CHCH_2OH} \longrightarrow \overset{\overset{CH_3}{|}}{CH_3CH\overset{\overset{}{\underset{\underset{O}{\|}}{C}}}{}OH}$$

   2-methylpropanoic acid

10. Reactions:

   (a) Formation of ethyl acetate from acetic acid and ethanol:

$$CH_3\overset{\overset{O}{\|}}{C}\text{-OH} + HOCH_2CH_3 \longrightarrow CH_3\overset{\overset{O}{\|}}{C}\text{-OCH}_2CH_3 + H_2O$$

   (b) Hydrolysis of glyceryl tristearate:

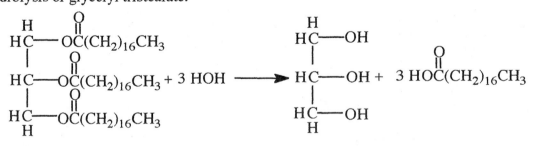

12. Equations for

   (a) the hydrolysis of the amide:

$$C_6H_5\overset{\overset{O}{\|}}{C}NHCH_3 \longrightarrow C_6H_5\overset{\overset{O}{\|}}{C}OH + H_2NCH_3$$

(b) the hydrolysis of nylon-66:

14. Properties imparted by the following characteristics:

(a) Cross-linking in polyethylene: Brings structural integrity to the polymeric chain. This also increases the rigidity of the chain. CLPE (the material of soda bottle caps) is a good example of a cross-linked polymer.

(b) OH groups in polyvinyl alcohol: These OH groups increase water solubility as hydrogen-bonding between polyvinyl alcohol and water is possible. Additionally these OH group provide a locus for cross-linking agents (e.g. borax is a good cross-linking agent in the commercial polymer called Slime™.

(c) Hydrogen bonding in a peptide: Causes peptides to form coils and sheets.

Practicing Skills
Alkanes and Cycloalkanes

16. The straight chain alkane with the formula C_7H_{16} is heptane. The prefix "hept" tells us that there are seven carbons in the chain.

18. Of the formulas given, which represents an alkane? Which a cycloalkane?

Alkanes have saturated carbon atoms—with 4 bonds. Since bridging C atoms have two bonds to C, then the other 2 bonds are to H atoms. Terminal C atoms on the end of the chains have only 1 bond to the chain, leaving 3 bonds to H. The net result is that alkanes have the general formula: C_nH_{2n+2} (c) has this general formula: $C_{14}H_{30}$. Cycloalkanes "lose 2H" atoms as the ends of the chains are "joined", so cycloalkanes have the general formula: C_nH_{2n} so (b) C_5H_{10} has this general formula. While (a) fits the general formula, it is impossible to make a ring of fewer than 3 C atoms.

20. Systematic name for the alkane: Numbering the longest C chain from one end to the other indicates that the longest C chain has 4 carbons, the root name is therefore *butane*. On the 2nd and 3rd C there is a 1-C chain. Since we truncate the –ane ending and change it to –yl, these are *methyl groups*. Indicating that the methyl groups are on the 2nd and 3rd carbon atoms, we get 2,3-dimethylbutane.

22. Structures for the compounds:

(a) 2,3-dimethylhexane

$$CH_3$$
$$|$$
$$CH_3 \ CHCHCH_2CH_2CH_3$$
$$|$$
$$CH_3$$

(b) 2,3-dimethyloctane

$$CH_3$$
$$|$$
$$CH_3 \ CHCHCH_2CH_2CH_2CH_2CH_3$$
$$|$$
$$CH_3$$

(c) 3-ethylheptane

$$CH_3 \ CH_2CHCH_2CH_2CH_2CH_3$$
$$|$$
$$CH_2CH_3$$

(d) 2-methyl-3-ethylhexane

$$CH_3$$
$$|$$
$$CH_3 \ CHCHCH_2CH_2CH_3$$
$$|$$
$$CH_2CH_3$$

24. Structures for all compounds with a seven-carbon chain and one methyl substituent.

$$CH_3$$
$$|$$
$$CH_3CHCH_2CH_2CH_2CH_2CH_3$$

2-methylheptane

$$CH_3$$
$$|$$
$$CH_3CH_2CHCH_2CH_2CH_2CH_3$$

3-methylheptane

$$CH_3$$
$$|$$
$$CH_3CH_2CH_2CHCH_2CH_2CH_3$$

4-methylheptane

Assuming that we deal *only* with alkanes, there are 3 structures. *3-methylheptane has a chiral carbon*, since on C-3, there is a methyl group (CH_3) an ethyl group (CH_3CH_2), a H, and a butyl group ($CH_2CH_2CH_2CH_3$).

26. Structures, and names of the two ethylheptanes:

$$CH_2CH_3$$
$$|$$
$$CH_3CH_2CHCH_2CH_2CH_2CH_3$$

3-ethylheptane
(has chiral carbon)

$$CH_2CH_3$$
$$|$$
$$CH_3CH_2CH_2CHCH_2CH_2CH_3$$

4-ethylheptane

28. Physical properties of C_4H_{10}: Assuming that we're talking about butane, the following properties are pertinent: mp = -138.4 °C, bp = –0.5°C (it's a colorless gas at room T). According to the CRC Handbook of Chemistry and Physics, it is soluble in water, ether, chloroform, and alcohol—slightly polar substances.
Predicted properties for $C_{12}H_{26}$: One would guess that the material is colorless (no features that would invoke color). Given the much longer chain, one could guess that it might be a liquid (It is! mp = -9.6°C and bp = 216°C. We would also guess that it would *not* be water soluble—much less polar than butane--(and it is reported *not* to be water soluble), but soluble in alcohol and ether.

Alkenes and Alkynes

30. Cis- and trans- isomers of 4-methyl-2-hexene:

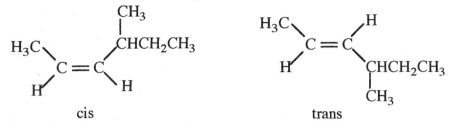

cis   trans

32. (a) Structures and names for alkenes with formula C_5H_{10}.

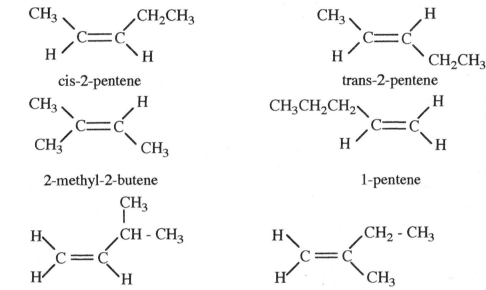

cis-2-pentene   trans-2-pentene

2-methyl-2-butene   1-pentene

3-methyl-1-butene   2-methyl-1-butene

(b) The cycloalkane with the formula C_5H_{10}

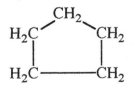

34. Structure and names for products of:

(a) $CH_3CH=CH_2$ + Br_2 → $CH_3CHBrCH_2Br$

1,2-dibromopropane

(b) $CH_3CH_2CH=CHCH_3 + H_2$ → $CH_3CH_2CH_2CH_2CH_3$

pentane

36. The alkene which upon addition of HBr yields: CH3CH2CH2CH2Br

The addition of HBr is accomplished by adding (H) on one "side" of the double bond and (Br) on the other. If we can mentally "subtract" an H and Br from the formula above we get:

CH3CH2CH2CH2**Br**   →   CH3CH2CH=CH2

38. Alkenes with the formula C3H5Cl:

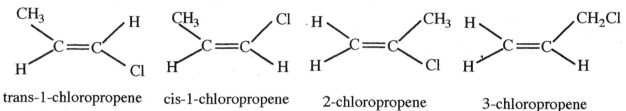

trans-1-chloropropene   cis-1-chloropropene   2-chloropropene   3-chloropropene

Aromatic Compounds

40. Structural formulas for :

(a) m-dichlorobenzene   (b) p-bromotoluene

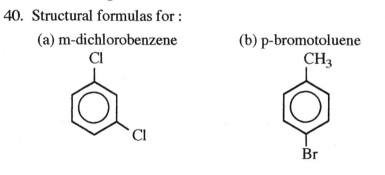

42. Ethylbenzene may be prepared from benzene in the following manner:

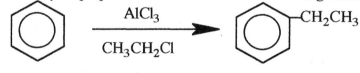

44. The alkylated product of p-xylene:

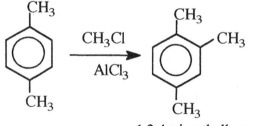

1,2,4-trimethylbenzene

Alcohols, Ethers, and Amines

46. Systematic names of the alcohols:

    (a) 1-propanol   primary alcohol

    (b) 1-butanol   primary alcohol

    (c) 2-methyl-2-propanol   tertiary alcohol

    (d) 2-methyl-2-butanol   tertiary alcohol

48. Formulas and structures

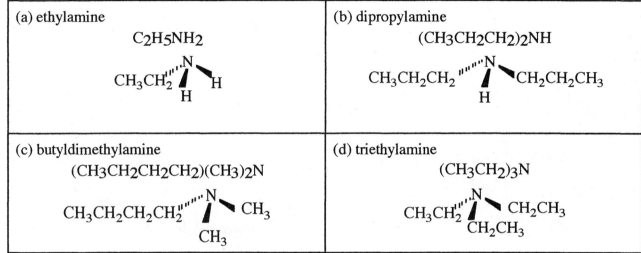

50. Structural formulas for alcohols with the formula $C_4H_{10}O$:

    1-butanol $CH_3CH_2CH_2CH_2OH$

    2-butanol $CH_3CH_2CHCH_3$
                                    |
                                 OH

    2-methyl-1-propanol CH_3CHCH_2OH
                                |
                               CH_3

                           OH
                           |
    2-methyl-2-propanol CH_3CCH_3
                                |
                             CH_3

52. Amines treated with acid:

    (a) $C_6H_5NH_2 + HCl \longrightarrow C_6H_5NH_3^+Cl^-$

    (b) $(CH_3)_3N + H_2SO_4 \longrightarrow (CH_3)_3NH^+ HSO_4^-$

141

Compounds with a Carbonyl Group

54. Structural formulas for :

(a) 2-pentanone

$$CH_3-CH_2-CH_2-\underset{\underset{O}{||}}{C}-CH_3$$

(b) hexanal

$$CH_3-CH_2-CH_2-CH_2-CH_2-\underset{\underset{O}{||}}{C}-H$$

(c) pentanoic acid

$$CH_3 CH_2CH_2CH_2CO_2H \ \text{ or } \ CH_3(CH_2)_3CO_2H$$

56. Name the following compounds:

(a) $CH_3CH_2\underset{\underset{CH_3}{|}}{C}HCH_2CO_2H$ 3-methylpentanoic acid
                                                a carboxylic acid

(b) $CH_3CH_2\overset{\overset{O}{||}}{C}OCH_3$ methyl propanoate
                                                  an ester

(c) $CH_3\overset{\overset{O}{||}}{C}OCH_2CH_2CH_2CH_3$ butyl ethanoate
                                                  an ester

(d) 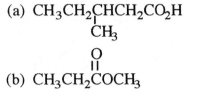   p-bromobenzoic acid
                                                  an aromatic carboxylic acid

58.   (a) The product of oxidation is pentanoic acid:

$$CH_3CH_2CH_2CH_2\overset{\overset{O}{||}}{C}-OH$$

(b) The reduction product is 1-pentanol:

$$CH_3CH_2CH_2CH_2CH_2 OH$$

(c) The reduction yields 2-octanol:

$$CH_3CH_2CH_2CH_2CH_2\underset{\underset{|}{|}}{\overset{\overset{OH}{|}}{C}}H CH_3$$

(d) The oxidation of 2-octanone with potassium permanganate gives <u>no reaction</u>.

60. The following equations show the preparation of propyl propanoate:

$$CH_3CH_2CH_2OH \xrightarrow{\ KMnO_4\ } CH_3CH_2CO_2H$$

$$CH_3CH_2CO_2H + HOCH_2CH_2CH_3 \xrightarrow{\ H^+\ } CH_3CH_2COCH_2CH_2CH_3$$

62. The products of the hydrolysis of the ester are 1-butanol and sodium acetate:

$$CH_3CH_2CH_2CH_2OH \qquad CH_2CO_2Na$$

64. Regarding phenylalanine:

  (a) Carbon 3 has three groups attached, and a trigonal planar geometry.

  (b) The O-C-O bond angle would be 120°.

  (c) Carbon 2 has four different groups attached, so the molecule is chiral.

  (d) The H attached to the carboxylic acid group (Carbon 3) is acidic.

Funtional Groups

66. Functional groups present:

  (a) the –OH group makes this molecule an *alcohol*.

  (b) the carbonyl group (C=O) adjacent to the N-H makes this molecule an *amide*.

  (c) the carbonyl group (C=O) adjacent to the O-H makes this molecule a *carboxylic acid*.

  (d) the carbonyl group (C=O) with an attached O-C makes this molecule an *ester*.

Polymers

68.　　(a) An equation for the formation of polyvinyl acetate from vinyl acetate:

  n CH2CHOCOCH3 $\longrightarrow$ [-CH2CH(OCOCH3)-] n

  (b) The structure for polyvinylacetate:

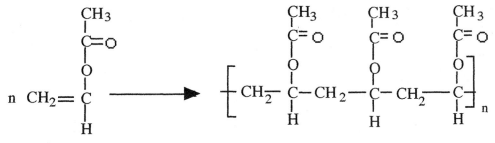

  (c) Prepare polyvinyl alcohol from polyvinyl acetate:

  Polyvinyl acetate is an ester. Hydrolysis of the ester (structure shown above) with NaOH will produce the sodium salt, NaC2H3O2, and polyvinyl alcohol. Acidification

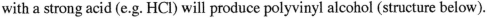

  with a strong acid (e.g. HCl) will produce polyvinyl alcohol (structure below).

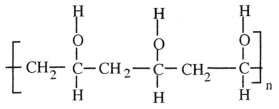

70.

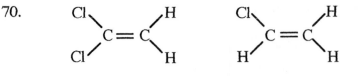

1,1-dichloroethene   chloroethene

The reaction proceeds with the free radical addition of the copolymers:

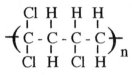

72. A tripeptide that can form from glycine, alanine , and histidine.

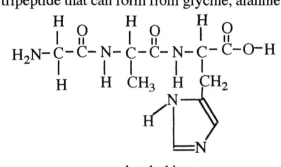

gly-ala-his

General Questions on Organic Chemistry

74. Structures for:

(a) 2,2-dimethylpentane

$$H—\underset{\underset{H}{|}}{\overset{\overset{H}{|}}{C}}—\underset{\underset{CH_3}{|}}{\overset{\overset{CH_3}{|}}{C}}—\underset{\underset{H}{|}}{\overset{\overset{H}{|}}{C}}—\underset{\underset{H}{|}}{\overset{\overset{H}{|}}{C}}—\underset{\underset{H}{|}}{\overset{\overset{H}{|}}{C}}—H$$

(b) 3,3-diethylpentane

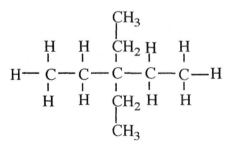

144

(c) 2-methyl-3-ethylpentane

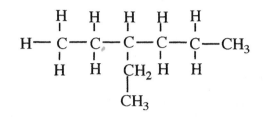

(d) 3-ethylhexane

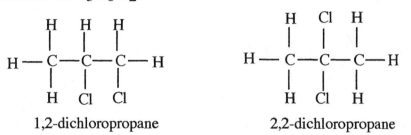

<u>76</u>. Structural isomers for $C_3H_6Cl_2$:

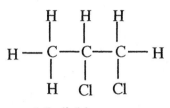

1,2-dichloropropane

2,2-dichloropropane

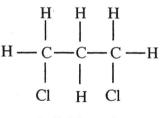

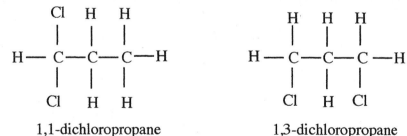

1,1-dichloropropane

1,3-dichloropropane

78. Structural isomers for trimethylbenzene:

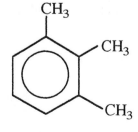

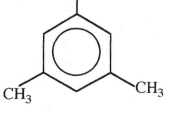

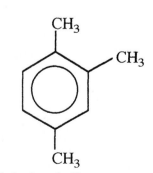

1,2,3-trimethylbenzene   1,3,5-trimethylbenzene   1,2,4-trimethylbenzene

80. The decarboxylase enzyme would remove the COOH functionality-- releasing CO_2 and appending the H in the former location of the COOH group.

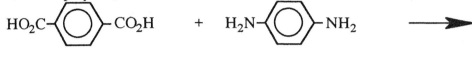

lysine   cadaverine

82. Repeating unit of the polymer, Kevlar

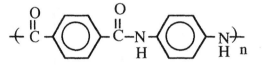

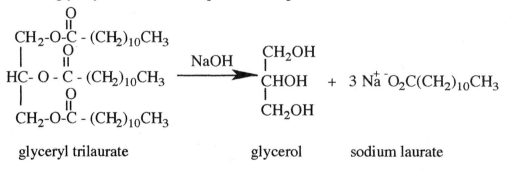

84. Structure for glyceryl trilaurate and saponification products:

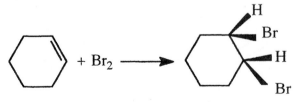

glyceryl trilaurate   glycerol   sodium laurate

86. The reaction between bromine and cyclohexene:

The double bond in cyclohexene absorbs elemental bromine—adding across the double bond so that the adjacent carbon atoms (originally participating in the double bond) each have one Br atom bond to them.

88. (a) Two structures with formula C_3H_6O:

$$\underset{\text{propanone (acetone)}}{CH_3\overset{\overset{\displaystyle O}{\|}}{C}CH_3} \qquad \underset{\text{propanal}}{CH_3CH_2\overset{\overset{\displaystyle O}{\|}}{C}H}$$

   (b) Oxidation of the compound gives an acidic solution (a carboxylic acid, perhaps?)

   (c) Oxidation of propanal gives propanoic acid (structure below):

$$CH_3CH_2\overset{\overset{\displaystyle O}{\|}}{C}OH$$

90. Addition of water to an alkene , X, gives an alcohol, Y. Oxidation of Y

   gives 3,3-dimethyl-2-pentanone. The structure for X and Y:

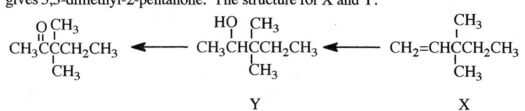

                          Y   X

92. Show products of the following reactions of CH_2CHCH_2OH:

   (a) Hydrogenation of the compound
   $$CH_2=CHCH_2OH \longrightarrow CH_3CH_2CH_2OH$$

   (b) Oxidation of the compound:
   $$CH_2=CHCH_2OH \qquad CH_2=CHCO_2H$$

   (c) Addition polymerization:
   $$n\ CH_2=CHCH_2OH \longrightarrow [-CH_2-\underset{\underset{\displaystyle CH_2OH}{|}}{C}H-]_n$$

   (d) Ester formation with acetic acid:
   $$CH_2=CHCH_2OH\ +\ CH_3CO_2H \longrightarrow CH_3CO_2CH_2CH=CH_2\ +\ H_2O$$

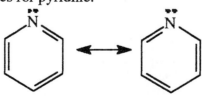

94. Resonance structures for pyridine:

   Both pyridine and benzene have the same number of electrons (30 valence). The molecular

   formulas also differ in that for pyridine one C-H unit has been replaced by a N atom.

96. To discriminate between the two isomers, react the two with elemental bromine. Cyclopentane will not react with bromine, while 1-pentene will react with bromine.

98. Regarding polymers:

Symbol	(a) Polymer	(b) Use
1 PETE	Polyethylene terephthalate	Soft-drink bottles
2 HDPE	High-Density Polyethylene	Bottles, Food & drink containers (gallon milk container)
3 V	Poly(vinyl chloride)	Floor tiles, plumbing pipes, raincoats
4 LDPE	Low-Density Polyethylene	Grocery bags
5 PP	Polypropylene	Carpet, Bottles(aspirin container), Film, Furniture

(c) To separate items of different polymers, one can float them in water. PP and HDPE, with densities less than 1, will float. PETE will sink to the bottom of the container. Careful heating of a mixture of PP and HDPE (to say approximately 150°C) should cause HDPE to melt, while the PP samples remain as a solid.

Using Electronic Resources

100. Comparison of benzene and cyclohexane:

(a) C hybridization: In cyclohexane, each C is attached to 4 groups (2 H and 2C). This requires sp^3 hybridization. For benzene, each C is attached to 3 groups (1 H and 2C). This arrangement calls for sp^2 hybridization.

(b) π electron delocalization is possible in benzene owing to adjacent C atoms with an unhybridized p orbital which is perpendicular to the ring. Overlap of these p orbitals, above and below the plane results in a π bond—and electron delocalization.

(c) Cyclohexane is not planar since adjacent sp^3 hybridized Carbon atoms would be connected at angles of approximately 109°. With six such C atoms attached in a ring, the geometry of 6-109° angles would *not* give a planar arrangement.(a 120° angle would provide for a planar arrangement).

102. Hybrid orbitals used by the following atoms:

(a) O in an alcohol; sp^{3}hybridization accommodates the two atoms (and the two lone pairs of electrons on the O atom.

(b) (C=O) C in an aldehyde; sp^2 hybridization accommodates the C,O, and H atoms, and leaves a p orbital to participate in π bond formation with the O.

(c) (C=O) C in a carboxylic acid; sp^2 hybridization accommodates the C,O, and a second O atom, and leaves a p orbital to participate in π bond formation with the carbonyl O atom.

(d) (C-O-C) O atom in an ester; sp^{3}hybridization accommodates the two C atoms (and the two lone pairs of electrons on the O atom.

(e) N atom in an amine; sp^{3}hybridization accommodates the C atom, and two H atoms (and the lone pair of electrons on the N atom.

104. Regarding Fats and Oils:

(a) Substitution; Removal of an H atom from glycerol and an OH from the acid (forms H_2O) and the ester linkage. (This is sometimes referred to as an *esterification* reaction.)

(b) The primary structural difference between fats and oils is in the presence of unsaturated C-C bonds. Oils tend to have unsaturated (C=C) bonds while fats tend to have saturated C-C bonds.

(c) The presence of C=C double bonds reduces the flexibility of the carbon chains and makes it impossible for the chains to bend to the extent possible when only C-C single bonds (a saturated bond) exists.

106. Addition polymerization:

(a) The primary structural feature to form addition polymers is a C=C double bond.

(b) The chain can be shorter OR longer than 14-Carbons. The necessary feature is a C=C double bond to begin the addition process.

(c) Reaction conditions control the length of the polymer chain formed. If the monomer concentration is too great, for example, a frequent termination of the chain-growth sequence results when two chains—each containing a free radical—combines.

(d) Addition polymerization would most reasonably be classified as an *Addition*.

Chapter 12
Gases and Their Properties

Reviewing Important Concepts

4. Avogadro's hypothesis states equal volumes of gases under the same conditions of T and P
   have equal numbers of molecules. Consider the formation of water from gaseous hydrogen
   and oxygen: $2H_2 + O_2 \rightarrow 2\,H_2O$.

   Using the fact that at STP, 1 mol of a gas occupies 22.4L, 3 mol of H_2 will occupy 3 •22.4L

   or 67.2 L. The balanced equation indicates that for each 2 moles of hydrogen gas, 1 mol of

   oxygen is required and 2 moles of H_2O will be produced, so 67.2 L of H_2 will require

   (67.2/2) or 33.6 L of O_2, and 67.2 L of gaseous water will be produced.

Practicing Skills
Pressure

10.   (a) 440 mm Hg • $\dfrac{1\ \text{atm}}{760\ \text{mm Hg}}$ = 0.58 atm (2 significant figures in 440)

   (b) 440 mm Hg • $\dfrac{1.013\ \text{bar}}{760\ \text{mm Hg}}$ = 0.59 bar

   (c) 440 mm Hg • $\dfrac{101.325\ \text{kPa}}{760\ \text{mm Hg}}$ = 59 kPa

12. The higher pressure in each of the pairs: (using appropriate conversion factors)
   (a) 534 mm Hg or 0.754 bar

   534 mm Hg • $\dfrac{1.013\ \text{bar}}{760\ \text{mm Hg}}$ = 0.712 bar so 0.754 bar is higher

   (b) 534 mm Hg or 650 kPa

   534 mm Hg • $\dfrac{101.325\ \text{kPa}}{760\ \text{mm Hg}}$ = 71.2 kPa so 650 kPa is higher

   (c) 1.34 bar or 934 kPa

   1.34 bar • $\dfrac{101.325\ \text{kPa}}{1.01325\ \text{bar}}$ = 134 kPa so 934 kPa is higher

Boyle's Law and Charles's Law

14. **Boyle's law** states that the pressure a gas exerts is inversely proportional to the volume it
   occupies, or for a given amount of gas--PV = constant . We can write this as:
   $$P_1V_1 = P_2V_2$$

   So   (67.5 mm Hg)(500. mL) = (P_2)(125 mL)

and $\dfrac{(67.5 \text{ mm Hg})(500. \text{ mL})}{125 \text{ mL}}$ = 270. mm Hg

Note that the **volume decreased** by a factor of 4, and the **pressure increased** by a factor of 4.

16. **Charles' law** states that V α T (in Kelvin) or $\dfrac{V_1}{T_1}$ = $\dfrac{V_2}{T_2}$

$\dfrac{3.5 \text{ L}}{295 \text{ K}}$ = $\dfrac{V_2}{310 \text{ K}}$ and rearranging to solve for V_2 yields

V_2 = $\dfrac{(3.5 \text{ L})(310 \text{ K})}{(295 \text{ K})}$ = 3.7 L

Note that with the **increase in T** there has been an **increase in volume**.

The General Gas Law

18. The reaction can be written: $2 H_2 + O_2 \rightarrow 2 H_2O$. Avogadro's Law states that, since the volume of a gas is **proportional to** the number of moles of gas, 3.6 L of H_2 would require 1.8 L of O_2, if both volumes are measured at the same temperature and pressure.

20. Using the general gas law we can write: $\dfrac{P_1V_1}{T_1}$ = $\dfrac{P_2V_2}{T_2}$

and for a fixed volume: $\dfrac{P_1}{T_1}$ = $\dfrac{P_2}{T_2}$ or if we rearrange, we obtain P_2 = $P_1 \cdot \dfrac{T_2}{T_1}$

P_2 = 360 mm Hg $\cdot \dfrac{268.2 \text{ K}}{298.7 \text{ K}}$ = 320 mm Hg (2 significant figures)

22. Using the general gas law we can write: $\dfrac{P_1V_1}{T_1}$ = $\dfrac{P_2V_2}{T_2}$

The volume is changing from 400. cm^3 to 50.0 cm^3, and the temperature from 15°C to 77 °C. We can rearrange the equation to solve for the new pressure:

$P_2 = P_1 \cdot \dfrac{T_2}{T_1} \cdot \dfrac{V_1}{V_2}$ = 1.00 atm $\cdot \dfrac{350. \text{ K}}{288 \text{ K}} \cdot \dfrac{400. \text{ cm}^3}{50.0 \text{ cm}^3}$ = 9.72 atm

Avogadro's Hypothesis

24.   (a) The balanced equation indicates that 1 O_2 is needed for 2 NO. At the same conditions of T and P, the amount of O_2 is 1/2 the amount of NO, so 75.0 mL of O_2 is required.

(b). The amount of NO_2 produced will have the same volume as the amount of NO (since their coefficients are equal in the balanced equation—150 mL of NO_2 .

Ideal Gas Law

26. The pressure of 1.25 g of gaseous carbon dioxide may be calculated with the ideal gas law:

$$1.25 \text{ g CO}_2 \cdot \frac{1 \text{ mol CO}_2}{44.01 \text{ g CO}_2} = 0.0284 \text{ mol CO}_2$$

Rearranging PV = nRT to solve for P, we obtain:

$$P = \frac{nRT}{V} = \frac{(0.0284 \text{ mol})(0.082057 \frac{L \cdot atm}{K \cdot mol})(295.7 \text{ K})}{0.750 \text{ L}} = 0.919 \text{ atm}$$

28. The volume of the flask may be calculated by realizing that the gas will expand to fill the flask.

$$2.2 \text{ g CO}_2 \cdot \frac{1 \text{ mol CO}_2}{44.0 \text{ g CO}_2} = 0.050 \text{ mol CO}_2$$

$$P = 318 \text{ mm Hg} \cdot \frac{1 \text{ atm}}{760 \text{ mm Hg}} = 0.418 \text{ atm}$$

$$V = \frac{(0.050 \text{ mol})(0.082057 \frac{L \cdot atm}{K \cdot mol})(295 \text{ K})}{0.418 \text{ atm}} = 2.9 \text{ L}$$

30. Rearranging PV = nRT to solve for n, we obtain: $n = \dfrac{PV}{RT}$

Converting 737 mm Hg to atmospheres, we obtain

$$737 \text{ mm Hg} \cdot \frac{1 \text{ atm}}{760 \text{ mm Hg}} = 0.970 \text{ atm}$$

$$n = \frac{(0.970 \text{ atm})(1.2 \times 10^7 \text{ L})}{(0.082057 \frac{L \cdot atm}{K \cdot mol})(298.2 \text{ K})} = 4.8 \times 10^5 \text{ moles of He}$$

and since each mole of He has a mass of 4.00 g,

$$4.8 \times 10^5 \text{ mol He} \cdot \frac{4.00 \text{ g He}}{1 \text{ mol He}} = 1.9 \times 10^6 \text{ g He}$$

Gas Density

32. Write the ideal gas law as: Molar Mass $= \dfrac{dRT}{P}$ where d = density in grams per liter.

Solving for d, we obtain: $\dfrac{(\text{Molar Mass}) \cdot P}{R \cdot T} = d$

The average molar mass for air is approximately 29 g/mol.

$$\frac{(29 \text{ g/mol})(0.20 \text{ mm Hg} \cdot 1 \text{ atm}/760 \text{ mm Hg})}{(0.082057 \frac{L \cdot atm}{K \cdot mol})(250 \text{ K})} = 3.7 \times 10^{-4} \text{ g/L} = d$$

34. Molar mass $= \dfrac{(0.355 \text{ g/L})(0.082057 \frac{L \cdot atm}{K \cdot mol})(290. \text{K})}{(189 \text{mmHg} 1 \text{atm}/760 \text{mmHg})} = 34.0 \text{ g/mol}$

Ideal Gas Laws and Determining Molar Mass

36. Rearranging $PV = nRT$ to solve for n, we obtain: $n = \dfrac{PV}{RT}$

Converting 715 mm Hg to atmospheres, we obtain

$$715\text{mm Hg} \cdot \frac{1 \text{ atm}}{760 \text{ mm Hg}} = 0.941 \text{ atm}$$

$$n = \frac{(0.941 \text{ atm})(0.452 \text{ L})}{(0.082057\frac{\text{L} \cdot \text{atm}}{\text{K} \cdot \text{mol}})(296.2 \text{ K})} = 0.0175 \text{ moles of unknown gas.}$$

Since this number of moles of the gas has a mass of 1.007 g, we can calculate the molar mass

$$\frac{1.007 \text{ g of unknown gas}}{0.0175 \text{ moles of unknown gas}} = 57.5 \text{ g/mol}$$

38. To calculate the molar mass:

$$\text{Molar mass} = \frac{(0.0125 \text{ g}/0.125\text{L})(0.082057\frac{\text{L} \cdot \text{atm}}{\text{K} \cdot \text{mol}})(298.2 \text{ K})}{(24.8 \text{ mmHg} \cdot 1\text{atm}/760\text{mmHg})} = 74.9 \text{ g/mol}$$

B_6H_{10} has a molar mass of 74.94 grams.

See SQ 34 for a similar problem.

Gas Laws and Stoichiometry

40. Determine the amount of H_2 generated when 2.2 g Fe reacts:

$$2.2 \text{ g Fe} \cdot \frac{1 \text{ mol Fe}}{55.85 \text{ g Fe}} \cdot \frac{1 \text{ mol H}_2}{1 \text{ mol Fe}} = 0.0394 \text{ mol H}_2 \text{ (0.039 to 2 sf)}$$

Note that the latter factor is achieved by examining the balanced equation!

The pressure of this amount of H_2 is :

$$P = \frac{nRT}{V} = \frac{(0.039 \text{ mol H}_2)(62.4\frac{\text{L} \cdot \text{torr}}{\text{K} \cdot \text{mol}})(298 \text{ K})}{10.0 \text{ L}}$$

$$= 73 \text{ torr or 73 mm Hg}$$

42. Calculate the moles of N_2 needed:

$$n = \frac{PV}{RT} = \frac{(1.3 \text{ atm})(75.0 \text{ L})}{(0.082057\frac{L \cdot atm}{K \cdot mol})(298 \text{ K})} = 3.99 \text{ mol } N_2$$

The mass of NaN_3 needed is obtained from the stoichiometry of the equation:

$$3.99 \text{ mol } N_2 \cdot \frac{2 \text{ mol } NaN_3}{3 \text{ mol } N_2} \cdot \frac{65.0 \text{ g } NaN_3}{1 \text{ mol } NaN_3} = 170 \text{ g } NaN_3 \text{ (to 2 sf)}$$

44. $N_2H_4 (g) + O_2 (g) \rightarrow N_2 (g) + 2 H_2O (g)$

$$1.00 \text{ kg } N_2H_4 \cdot \frac{1.0 \times 10^3 \text{ g } N_2H_4}{1.0 \text{ kg } N_2H_4} \cdot \frac{1 \text{ mol } N_2H_4}{32.0 \text{ g } N_2H_4} \cdot \frac{1 \text{ mol } O_2}{1 \text{ mol } N_2H_4}$$

$$= 3.13 \times 10^1 \text{ mole } O_2$$

$$P(O_2) = \frac{n(O_2) \cdot R \cdot T}{V} = \frac{(3.13 \times 10^1 \text{mol})(0.082057 \text{ L} \cdot atm/K \cdot mol)(296 \text{ K})}{450 \text{ L}}$$

$$P(O_2) = 1.69 \text{ atm or } 1.7 \text{ atm to 2 sf}$$

Gas Mixtures and Dalton's Law

46. We know that the total pressure will be equal to the sum of the pressure of each gas (also called the *partial pressure* of each gas).

$$1.0 \text{ g } H_2 \cdot \frac{1 \text{ mol } H_2}{2.02 \text{ g } H_2} = 0.50 \text{ mol } H_2 \text{ and } 8.0 \text{ g Ar} \cdot \frac{1 \text{ mol Ar}}{39.9 \text{ g Ar}} = 0.20 \text{ mol Ar}$$

We can calculate the **total pressure** using the ideal gas law:

$$P = \frac{n \cdot R \cdot T}{V} = \frac{(0.70 \text{ mol})(0.082057 \text{ L} \cdot atm/K \cdot mol)(300 \text{ K})}{3.0 \text{ L}} = 5.7 \text{ atm}$$

The pressure of **each gas** can be calculated by multiplying the total pressure (5.7 atm) by the mole fraction of the gas.

$$\text{Pressure of } H_2 = \frac{0.50 \text{ mol } H_2}{0.70 \text{ mol } H_2 + Ar} \cdot 5.7 = 4.1 \text{ atm and the}$$

$$\text{Pressure of Ar} = \frac{0.20 \text{ mol Ar}}{0.70 \text{ mol } H_2 + Ar} \cdot 5.7 = 1.6 \text{ atm}$$

Note that the total pressure is indeed (4.1+1.6) or 5.7 atm

48. $P_{total} = P_{halothane} + P_{oxygen} = 170 \text{ mm Hg} + 570 \text{ mm Hg} = 740 \text{ mm Hg}$

(a) Since we know that the pressure a gas exerts is **proportional** to the # of moles of gas present we can calculate the ratio of moles by using their partial pressures:

$$\frac{\text{moles of halothane}}{\text{moles of oxygen}} = \frac{170 \text{ mm Hg}}{570 \text{ mm Hg}} = 0.30$$

(b) $160 \text{ g oxygen} \cdot \dfrac{1 \text{ mol oxygen}}{32.0 \text{ g oxygen}} \cdot \dfrac{0.30 \text{ mol halothane}}{1 \text{ mol oxygen}} \cdot \dfrac{197.38 \text{ g halothane}}{1 \text{ mol halothane}} =$

3.0×10^2 g halothane (2 significant figures)

Kinetic-Molecular Theory

50. (a) Kinetic energy depends only on the temperature so the average kinetic *energies of these two gases are equal.*

(b) Since the kinetic energies are equal, we can state:

$$KE(H_2) = KE(CO_2)$$

$$1/2 \ m(H_2) \cdot \overline{V}^2(H_2) = 1/2 \ m(CO_2) \cdot \overline{V}^2(CO_2)$$

Where m = mass of a molecule and $\overline{V}$ = average velocity of a molecule

So $m(H_2) \cdot \overline{V}^2(H_2) = m(CO_2) \cdot \overline{V}^2(CO_2)$

and $\dfrac{\overline{V}^2(H_2)}{\overline{V}^2(CO_2)} = \dfrac{m(CO_2)}{m(H_2)}$

Now the molar mass of $H_2 = 2.0$ g and the molar mass of $CO_2 = 44$ g

$$\dfrac{\overline{V}_{H_2}}{\overline{V}_{CO_2}} = \sqrt{\dfrac{m_{CO_2}}{m_{H_2}}} = \sqrt{\dfrac{44}{2.0}} = 4.7$$

The hydrogen molecules have an average velocity which is 4.7 times the average velocity of the CO_2 molecules.

(c) Since the volumes are equal for these two gas samples, the pressure is proportional to the amount of gas present.

$$V_A = \dfrac{n_A R T_A}{P_A} \qquad \text{and} \quad V_B = \dfrac{n_B R T_B}{P_B} \quad \text{and } V_A = V_B \text{ so}$$

$\dfrac{n_A R T_A}{P_A} = \dfrac{n_B R T_B}{P_B}$ and rearranging to solve for the ratio of molecules present

$\dfrac{P_B T_A}{T_B P_A} = \dfrac{n_B}{n_A}$ substituting yields: $\dfrac{2 \text{ atm} \cdot 273 \text{ K}}{298 \text{ K} \cdot 1 \text{ atm}} = \dfrac{n_B}{n_A} = 1.8$

There are 1.8 times as many moles (and molecules) of gas in Flask B (CO_2) as there are in Flask A (H_2).

(d) Since Flask B contains 1.8 times as many moles of CO_2 as Flask A contains of H_2, the *ratio* of masses of gas present are:

$$\frac{\text{Mass (Flask B)}}{\text{Mass (Flask A)}} = \frac{(1.8 \text{ mole } CO_2)(44 \text{ g } CO_2/\text{mol } CO_2)}{(1 \text{ mol } H_2)(2 \text{ g } H_2/\text{mol } H_2)} = \frac{40}{1}$$

Note that any number of moles of CO_2 and H_2 (in the ratio of 1.8:1) would provide the same answer.

52. Since two gases at the same temperature have the same kinetic energy

$$KE_{O_2} = KE_{CO_2}$$

and since the average $KE = 1/2 \, m\bar{u}^2$

where $\bar{u}$ is the average speed of a molecule, we can write.

$$1/2 \, M_{O_2}\bar{U}_{O_2}^2 = 1/2 \, M_{CO_2}\bar{U}_{CO_2}^2 \qquad \text{or} \qquad M_{O_2}\bar{U}_{O_2}^2 = M_{CO_2}\bar{U}_{CO_2}^2$$

and $$\frac{M_{O_2}}{M_{CO_2}} = \frac{\bar{U}_{CO_2}^2}{\bar{U}_{O_2}^2}$$

and solving for the average velocity of CO_2 :

$$\bar{U}_{CO_2}^2 = \frac{M_{O_2}}{M_{CO_2}} \cdot \bar{U}_{O_2}^2$$

Taking the square root of both sides

$$\bar{U}_{CO_2} = \sqrt{\frac{M_{O_2}}{M_{CO_2}}} \cdot \bar{U}_{O_2} = \sqrt{\frac{32.0 \text{ g } O_2 / \text{mol } O_2}{44.0 \text{ g } CO_2 / \text{mol } CO_2}} \cdot 4.28 \times 10^4 \text{cm/s}$$

$$= 3.65 \times 10^4 \text{ cm/s}$$

54. The species will have average molecular speeds which are inversely proportional to their molar masses.

Slowest				Fastest
CH_2F_2	<	Ar	<	N_2	<	CH_4
54		40		28		16	(to integral values)

Diffusion and Effusion

56. Relative rates of effusion for the following pairs of gases:

(a) CO_2 or F_2 : *Fluorine* effuses faster, since the molar mass of F_2 is 38 g/mol and that of CO_2 is 44 g/mol

(b) O_2 or N_2: *Nitrogen* effuses faster. (MM N_2 = 28 g/mol; for O_2 = 32 g/mol)

(c) C_2H_4 or C_2H_6 : *Ethylene* effuses faster. (MM C_2H_4 = 28 g/mol; for C_2H_6=30 g/mol)

(d) $CFCl_3$ or $C_2Cl_2F_4$: *CFCl3* effuses faster.(MM CFCl3 = 137 g/mol; for $C_2Cl_2F_4$ = 171 g/mol)

58. Determine the molar mass of a gas which effuses at a rate 1/3 that of He:

$$\frac{\text{Rate of effusion of He}}{\text{Rate of effusion of unknown}} = \sqrt{\frac{\text{M of unknown}}{\text{M of He}}}$$

$$\frac{3}{1} = \sqrt{\frac{\text{M of unknown}}{4.0 \text{ g/mol}}}$$

Squaring both sides gives: $9 = \frac{M}{4.0}$ or M = 36 g/mol

Nonideal Gases

60. According to the Ideal Gas Law, the pressure would be:

$$P = \frac{n \cdot R \cdot T}{V} = \frac{(8.00 \text{ mol})(0.082057 \frac{L \cdot atm}{K \cdot mol})(300 \text{ K})}{4.00 \text{ L}} = 49.3 \text{ atm}$$

The van der Waal's equation is: $\left[P + a\left(\frac{n}{V}\right)^2\right]\left[V - bn\right] = nRT$.

Substituting we get

$$\left[P + 6.49\frac{atm \cdot L^2}{mol^2}\left(\frac{8.00 \text{ mol}}{4.00 \text{ L}}\right)^2\right]\left[4.00L - 0.0562\frac{L}{mol} \cdot 8.00mol\right] = 8.00mol \cdot 0.082057\frac{atm \cdot L}{K \cdot mol} \cdot 300K$$

Simplifying : $[P + 25.96atm][4.00L - 0.45L] = 196.94atm \cdot L$ and

$$P = \frac{196.94 \text{ atm} \cdot L}{3.55L} - 25.96 \text{ atm} = 29.5 \text{ atm}$$

General Questions on Gas Behavior

62.	atm	mm Hg	kPa	bar
Standard atmosphere:	1	1 atm • $\frac{760. \text{ mm Hg}}{1 \text{ atm}}$ = 760. mm Hg	1 atm • $\frac{101.325 \text{ kPa}}{1 \text{ atm}}$ = 101.325 kPa	1 atm • $\frac{1.013 \text{ bar}}{1 \text{ atm}}$ = 1.013 bar

Partial pressure of N_2 in the atmosphere	593 mm Hg • $\dfrac{1 \text{ atm}}{760 \text{ mm Hg}}$ $= 0.780$ atm	**593 mm Hg**	0.780 atm • $\dfrac{101.3 \text{ kPa}}{1 \text{ atm}}$ $= 79.1$ kPa	0.780 atm • $\dfrac{1.013 \text{ bar}}{1 \text{ atm}}$ $=0.791$ bar
Tank of compressed H_2	133 bar • $\dfrac{1 \text{ atm}}{1.013 \text{ bar}}$ $= 131$ atm	131 atm • $\dfrac{760. \text{ mm Hg}}{1 \text{ atm}}$ $= 99800$ mm Hg	131 atm • $\dfrac{101.3 \text{ kPa}}{1 \text{ atm}}$ $= 13300$ kPa	**133 bar**
Atmospheric pressure at top of Mt. Everest	33.7 kPa • $\dfrac{1 \text{ atm}}{101.3 \text{ kPa}}$ $= 0.333$ atm	0.333 atm • $\dfrac{760 \text{ mm Hg}}{1 \text{ atm}}$ $= 253$ mm Hg	**33.7 kPa**	0.333 atm • $\dfrac{1.013 \text{ bar}}{1 \text{ atm}}$ $= 0.337$ bar

64. 1.0 L of a compound gives 2.0 L CO_2, 2.82 g H_2O, and 0.50 L of N_2 at STP.

At STP, 1 mol of a gas occupies 22.4 L so we can calculate moles of CO_2 and N_2 .

$$2.0 \text{ L } CO_2 \cdot \frac{1 \text{ mol } CO_2}{22.4 \text{ L}} \cdot \frac{1 \text{ mol C}}{1 \text{ mol } CO_2} = 0.089 \text{ mol C}$$

$$0.50 \text{ L } N_2 \cdot \frac{1 \text{ mol } N_2}{22.4 \text{ L}} \cdot \frac{2 \text{ mol N}}{1 \text{ mol } N_2} = 0.045 \text{ mol N}$$

Given the mass of water formed, we can calculate the moles of H in the compound:

$$2.82 \text{ g } H_2O \cdot \frac{1 \text{ mol } H_2O}{18.02 \text{ g } H_2O} \cdot \frac{2 \text{ mol H}}{1 \text{ mol } H_2O} = 0.31 \text{ mol H}$$

Establish the ratio of C:N:H by dividing the amount of each element by the *smallest amount present* (in this case, 0.045 mol N).

$$\frac{0.089 \text{ mol C}}{0.045 \text{ mol N}} = 2 \qquad \frac{0.31 \text{ mol H}}{0.045 \text{ mol N}} = 7 \quad \text{The empirical formula is then } C_2H_7N.$$

66. To increase the average speed of helium atoms by 10.0%, we must know the average speed initially.

$$\sqrt{\overline{u^2}} = \sqrt{\frac{3RT}{M}} \quad \text{and substituting for He: } = \sqrt{\frac{3 \cdot (8.314 \text{ J/K} \cdot \text{mol}) \cdot 240K}{4.00 \times 10^{-3} \text{ kg/mol}}} = 1220 \text{ m/s}$$

NOTE: A **J**oule is a kg•m/s, so it's necessary to express the molar mass of helium in units of kg/mol.

The new average speed = 110%(1220 m/s) = 1350 m/s. Substituting this value as u:

$$\frac{M u^2}{3R} = T; \quad \frac{4.00 \times 10^{-3} \text{ kg/mol} \cdot (1350 \text{ m/s})^2}{3 \cdot 8.314 \text{ J/K} \cdot \text{mol}} = 290 \text{ K} \quad \text{or } (290\text{-}273)°C \text{ or } 17°C.$$

159

68. (a) To calculate the balloon containing the greater number of molecules we need only to calculate the ratio of the amounts of gas present in the balloons. Using the Ideal Gas Law allows us to solve for the number of moles of each gas:

Pick a volume of gas—say 5L for He. Since the hydrogen balloon is twice the size of the He balloon,

$$n = \frac{P \cdot V}{R \cdot T}$$

we'll use 10 L for the volume of H. We can then establish a ratio of this expression for the two gases.

$$\frac{n_{He}}{n_{H_2}} = \frac{\dfrac{P_{He} \cdot V_{He}}{R \cdot T_{He}}}{\dfrac{P_{H_2} \cdot V_{H_2}}{R \cdot T_{H_2}}} = \frac{\dfrac{2\,atm \cdot 5L}{296K}}{\dfrac{1atm \cdot 10\,L}{268K}} = \frac{268\ K}{296\ K} = \frac{0.9\ mol\ He}{1\ mol\ H_2}$$

Note that **R** cancels in the numerator and denominator, simplifying the calculation.

(b) Two calculate which balloon contains the greater mass of gas, we can use the ratio found: 0.9 mol He • 4.0 g/mol He = 3.6 g He; 1 mol H_2 • 2.0 g/mol H_2 = 2 g H_2

The balloon containing the *HELIUM* has the greater mass. Note that the *ratio* of moles of helium and hydrogen are **independent** of the sizes of the balloons.

70. Using the Ideal Gas Law,

$$n = \frac{P \cdot V}{R \cdot T} = \frac{\left(8\ mm\ Hg \cdot \dfrac{1\ atm}{760\ mm\ Hg}\right) \cdot \left(10.\ m^3 \cdot \dfrac{1000L}{1\ m^3}\right)}{0.082057\ \dfrac{L \cdot atm}{K \cdot mol} \cdot 300.\ K} = 4\ mol\ (1sf)$$

72. This problems has two parts: 1) How many moles of Ni are present?

2) How many moles of CO are present?

1) # moles of Ni present:

$$0.450\ g\ Ni \cdot \frac{1\ mol\ Ni}{58.693\,g\ Ni} = 7.67\ x\ 10^{-3}\ mol\ Ni$$

and since 1 mol Ni(CO)$_4$ is formed for **each mol of Ni**, one can form

$$7.67\ x\ 10^{-3}\ mol\ Ni(CO)_4$$

2) # moles of CO present:

Using the Ideal Gas Law:

$$n = \frac{\dfrac{418}{760}\,atm \cdot (1.50L)}{(0.082057\ \dfrac{L \cdot atom}{K \cdot mol})(298K)} = 0.0337\ mol\ CO$$

which would be capable of forming

$$0.0337\ mol\ CO \cdot \frac{1\ mol\ Ni(CO)_4}{4\ mol\ CO} = 8.43\ x\ 10^{-3}\ mol\ Ni(CO)_4$$

Since the amount of *nickel limits the maximum amount of Ni(CO)4 that can be formed*, the maximum mass of $Ni(CO)_4$ is then:

$$7.67 \times 10^{-3} \text{ mol Ni(CO)}_4 \cdot \frac{170.7 \text{ g Ni(CO)}_4}{1 \text{ mol Ni(CO)}_4} = 1.31 \text{ g Ni(CO)}_4$$

74. Calculate the empirical formula:

$$\% \text{ F} = 100.0\% - (11.79 \% \text{ C} + 69.57 \% \text{ Cl}) = 18.64 \% \text{ F}$$

The moles of each element:

$$18.64 \text{ g F} \cdot \frac{1 \text{ mol F}}{18.998 \text{ g F}} = 0.9812 \text{ mol F}$$

$$11.79 \text{ g C} \cdot \frac{1 \text{ mol C}}{12.011 \text{ g C}} = 0.9816 \text{ mol C}$$

$$69.57 \text{ g Cl} \cdot \frac{1 \text{ mol Cl}}{35.453 \text{ g Cl}} = 1.962 \text{ mol Cl}$$ and if we express these # of moles in a ratio

(by dividing each of the moles by the smallest # — 0.9812) we get an empirical formula of $FCCl_2$.

Calculate the molar mass:

$$21.3 \text{ mm Hg} \cdot \frac{1 \text{ atm}}{760. \text{ mm Hg}} = 0.0280 \text{ atm}$$

$$n = \frac{PV}{RT} = \frac{(0.0280 \text{ atm})(0.458 \text{ L})}{(0.082057 \frac{\text{L} \cdot \text{atm}}{\text{K} \cdot \text{mol}})(298 \text{ K})} = 5.25 \times 10^{-4} \text{ mol}$$

Since 0.107 g corresponds to 5.25×10^{-4} mol , the molar mass is:

$$\frac{0.107 \text{ g}}{5.25 \times 10^{-4} \text{ mol}} = 204 \text{ g/mol}$$

With an empirical formula of CCl_2F (Empirical formula weight = 102), the molecular formula must be $C_2Cl_4F_2$.

76. Let's calculate the # of moles of ammonium dichromate (and from that the # of moles of gaseous products expected:

$$0.950 \text{ g (NH}_4)_2\text{Cr}_2\text{O}_7 \cdot \frac{1 \text{ mol (NH}_4)_2\text{Cr}_2\text{O}_7}{252 \text{ g (NH}_4)_2\text{Cr}_2\text{O}_7} = 3.77 \times 10^{-3} \text{mol (NH}_4)_2\text{Cr}_2\text{O}_7$$

(3 significant figures)

The balanced equation tells us that 1 mol of $(NH_4)_2Cr_2O_7$ produces 1 mol of N_2 and 4 mol of H_2O so we anticipate $(1 \cdot 3.77 \times 10^{-3})$ mol N_2 and $(4 \cdot 3.77 \times 10^{-3})$mol H_2O.

[Summing the # of mole of gaseous products, one gets $(5 \cdot 3.77 \times 10^{-3})$mol of gas (total)]

The **total** pressure would then be

$$P = \frac{n \cdot R \cdot T}{V} = \frac{(1.88 \times 10^{-2} \text{ mol})(0.082057\frac{L \cdot atm}{K \cdot mol})(296K)}{15.0 \text{ L}} = 0.0305 \text{ atm}$$

Partial pressures are easily obtained by noting that the amount of water composes 4 out of 5 moles of gaseous products, so the partial pressure of water would be $4/5 \cdot 0.0305$ atm $=$ 0.0244 atm , and the partial pressure of nitrogen would be $1/5 \cdot 0.0305$ atm or 0.0061 atm

78. The amount of N_2 can be calculated.

$$P(N_2) = 713 \text{ mm Hg} = 0.938 \text{ atm}$$

$$n(N_2) = \frac{(0.938 \text{ atm})(0.295 \text{ L})}{(0.082057\frac{L \cdot atm}{K \cdot mol})(294.2 \text{ K})} = 1.15 \times 10^{-2} \text{ mol } N_2$$

According to the equation in which sodium nitrite reacts with sulfamic acid, one mole of $NaNO_2$ produces one mole of N_2.

$$1.15 \times 10^{-2} \text{ mol } N_2 \cdot \frac{1 \text{ mol } NaNO_2}{1 \text{ mol } N_2} \cdot \frac{69.00 \text{ g } NaNO_2}{1 \text{ mol } NaNO_2} = 0.791 \text{ g } NaNO_2$$

$$\text{Weight percentage of } NaNO_2 = \frac{0.791 \text{ g } NaNO_2}{1.232 \text{ g sample}} \times 100 = 64.2\% \text{ } NaNO_2$$

80. The partial pressure of each gas can be calculated from

$$P_2 = P_1 \cdot \frac{V_1}{V_2} \qquad \text{where } V_2 \text{ in each case is 5.0 L}$$

Partial pressure of He $= 145 \text{ mm Hg} \cdot \frac{3.0 \text{ L}}{5.0 \text{ L}} = 87 \text{ mm Hg}$

Partial pressure of Ar $= 355 \text{ mm Hg} \cdot \frac{2.0 \text{ L}}{5.0 \text{ L}} = 140 \text{ mm Hg}$

$P_{total} = P_{He} + P_{Ar} = 87 \text{ mm Hg} + 140 \text{ mm Hg} = 227 \text{ mm Hg}$ (or 230 mm Hg to 2 sf)

82. Recall that unless gases react, they behave (i.e. exert pressure) as if they were alone in a container. We can then calculate the pressure of fluorine in a container that *already* contains xenon by subtracting from the **total pressure** the partial pressure of xenon—or vice versa. If (before reaction), the total pressure was 0.72atm and the pressure of xenon was 0.12 atm, then the pressure of fluorine was (0.72-0.12)atm. Following the reaction, the pressure of xenon was zero (i.e. consumed completely) . Since the pressure of fluorine remaining in the flask was 0.36 atm, we can calculate the amount of fluorine that reacted:

$$P(\text{fluorine}) = (0.72\text{-}0.12) \text{ atm} - 0.36 \text{ atm} = 0.24 \text{ atm.}$$

From the Ideal Gas Law, we can calculate the *moles of each substance* that reacted:

$$n = \frac{PV}{RT} = \frac{0.24 \text{ atm} \cdot 0.25 \text{ L}}{0.082057 \frac{\text{L} \cdot \text{atm}}{\text{K} \cdot \text{mol}} \cdot 273.2 \text{ K}} = 0.0027 \text{ mol F}_2 \text{ (and 0.0054 mol F)}$$

For xenon:

$$n = \frac{PV}{RT} = \frac{0.12 \text{ atm} \cdot 0.25 \text{ L}}{0.082057 \frac{\text{L} \cdot \text{atm}}{\text{K} \cdot \text{mol}} \cdot 273.2 \text{ K}} = 0.0013 \text{ mol Xe}$$

The ratio of Xe:F that reacts is 0.0013 mol: 0.0054 mol or a 1:4 ratio. The empirical formula of the compound must be XeF_4

84. The number of moles of He in the balloon can be calculated with the Ideal Gas Law.

First calculate the P of He in the balloon:

gauge = total pressure - barometric   or

gauge + barometric = total pressure = 22 mm Hg + 755 mm Hg = 777 mm Hg

and since 1 mm Hg = 1 torr,

$$n = \frac{PV}{RT} = \frac{(777 \text{ torr})(0.305 \text{ L})}{(62.4 \frac{\text{L} \cdot \text{torr}}{\text{K} \cdot \text{mol}})(298 \text{ K})} = 0.0127 \text{ mol He}$$

86. We need to know the number of molecules/cm^3. Rearrangement of the Ideal Gas Law can

help: $\dfrac{n}{V} = \dfrac{P}{RT} = \dfrac{23.8 \text{ torr}}{62.4 \frac{\text{L} \cdot \text{torr}}{\text{K} \cdot \text{mol}} \cdot 298 \text{ K}} = 1.28 \times 10^{-3} \text{ mol/L}$

Recalling that 1 L = 1000 cm^3, and that 1 mol = 6.022 $\times$ 10^{23} molecules:

$$\frac{1.28 \times 10^{-3} \text{ mol}}{1 \text{ L}} \cdot \frac{1 \text{ L}}{1000 \text{ cm}^3} \cdot \frac{6.022 \times 10^{23} \text{ molecules}}{1 \text{ mol}} = 7.71 \times 10^{17} \text{ molecules/cm}^3$$

88. According to Dalton's Law of Partial Pressures, the total P is the sum of the pressures of the gases present (assuming no reaction occurs). Hence

$P_{total} = P_{H2O} + P_{O2} + P_{CO2} + P_{N2}$

253 mm Hg = 47.1 mm Hg + 35 mm Hg + 7.5 mm Hg + P_{N2}

The pressure of nitrogen is then 253 mm Hg – (47.1 + 35 + 7.5) mm Hg = 163 mm Hg

90. In a 1.0-L flask containing 10.0 g each of O_2 and CO_2 at 25 °C.

(a) The gas with the greater partial pressure:

Partial pressure is a relative measure of the number of moles of gas present. The molar mass of oxygen is approximately 32 g/mol while that of CO_2 is approximately

44 g/mol. There will be a greater number of moles of oxygen in the flask—hence the **partial pressure of O$_2$ will be greater**.

(b) The gas with the greater average speed:

The kinetic energy of each gas is given as: KE = 1/2 mu^2, where u is the average speed of the molecules. Since the average KE of both gases are the same (They are at the same T), The lighter of the gases (O$_2$)**will have a greater average speed**.

(c) The gas with the greater average kinetic energy:

The KE depends upon temperature, and since both gases are at the same T, the **average KE of the two gases is the same**.

92. Two containers with 1.0 kg of CO and 1.0 kg C$_2$H$_2$

(a) Cylinder with the greater pressure:

Since P is proportional to the amount of substance present, let's calculate the # of moles of each gas.

Molar Mass of CO = 28 g/mol   Molar Mass of C$_2$H$_2$= 26 g/mol

While we could calculate the # of moles, note that since acetylene has a small molar mass, 1.0 kg of acetylene would have more moles of gas (and *a greater pressure*) than 1.0 kg of CO.

(b) Cylinder with the greater number of molecules:

Since we know that the cylinder with acetylene has more moles of gas, the cylinder *with the acetylene* will have the greater number of molecules (since to convert between moles of gas and molecules of gas, we multiply by the constant, Avogadro's number.

94. Which of the following samples is a gas?

(a) Material expands 10% when a sample of gas originally at 100 atm is suddenly allowed to exist at one atmosphere pressure: This sample is **not a gas**, since a gas would expand in volume by 100-fold (to exist at 1 atm).

(b) A 1.0-ml sample of material weighs 8.2 g: This sample is **not a gas** since the density of the sample (at 8.2 g/mL) is too great.

(c) Material is transparent and pale green in color: **Insufficient information** to tell. Liquids could also be pale green and transparent.

(d) One cubic meter of material contains as many molecules as an equal volume of air at the same temperature and pressure: This material **is a gas**, since one cubic meter of a liquid or solid would contain a greater number of molecules than a cubic meter of air.

96. Four tires contain four different gases:

    (a) Since the four tires have the same volume and pressure at the same temperature, **each contains the same number of molecules** as the other three.

    (b) The relative mass of an atom of the unknown gas compared to an atom of helium. Since the tires have the same number of molecules, and the unknown gas has a mass of 160. g while that of He is 16.0 g, the **atoms of unknown gas are (160/16.0) or 10.0 times heavier than atoms of helium.**

    (c) The molecules of each of the four gases **all have the same kinetic energy** since kinetic energy depends on temperature.   However since kinetic energy is a function of the mass of the gas (= 1/2 mu^2), the **molecules of the lightest gas—helium—will have the greatest average speed.**

98. Examine the balanced equation for the combustion of methane:
$$CH_4(g) + 2\ O_2\ (g) \rightarrow CO_2\ (g) + 2\ H_2O\ (g)$$

The stoichiometry tells us that we'll need 2 moles of oxygen/1 mole of methane.

How much methane is present?   Substitute into the Ideal Gas Law—and use 5.0 L/min as the Volume of methane (even though it represents the *rate* of methane flow).

$$n_{CH_4} = \frac{PV}{RT} = \frac{773\ torr\ \bullet\ 5.0\ L/min}{62.4\ \dfrac{L \bullet torr}{K \bullet mol} \bullet 301K} = 0.21 mol/min$$

Since we need 2 mol of oxygen/mol methane, we'll need (2 • 0.21 mol) O$_2$/min

Substituting this rate (*as the number of moles*) into the Ideal Gas Law we obtain:

$$V = \frac{nRT}{P} = \frac{0.42\ \dfrac{mol\ O_2}{min} \bullet 62.4\ \dfrac{L \bullet torr}{K \bullet mol} \bullet 299\ K}{742\ torr} = 10.\ L/min \ \ (to\ 2sf)$$

100. What amount of carbon dioxide is evolved?
$$n_{CO_2} = \frac{PV}{RT} = \frac{69.8\ torr\ \bullet\ 0.285\ L}{62.4\ \dfrac{L \bullet torr}{K \bullet mol} \bullet 298\ K} = 1.07\ x\ 10^{-3} mol\ CO_2$$

The balanced equation indicates that we get 1 mol of CO$_2$ for each mole of the carbonate.

So the 0.158g of the metal carbonate represents 1.07 x 10^{-3} mol of the carbonate.

The molar mass is then $\dfrac{0.158\ g\ carbonate}{1.07 x 10^{-3}\ mol\ carbonate} = 148\ g/mol$

Removing the mass associated with "CO$_3$ ": 148-60= 88 g/mol. Strontium (with an atomic weight of 87.6) seems a likely candidate.

102. (a) Mass of NiO that reacts with ClF_3 gas (250 torr at 20°C in a 2.5 L flask)

$$n_{ClF_3} = \frac{PV}{RT} = \frac{250\ torr \bullet 2.5\ L}{62.4\ \dfrac{L \bullet torr}{K \bullet mol} \bullet 293\ K} = 3.4 \times 10^{-2} mol\ ClF_3$$

The balanced equation indicates that 6/4 or 1.5 mol of NiO react / mol of ClF_3. The amount of NiO is then:

3.4×10^{-2} mol $ClF_3 \bullet$ 1.5 mol NiO/1 mol $ClF_3 \bullet$ 74.69 g NiO/1 mol NiO = 3.8 g NiO

(b) Partial pressures and total pressure after the reaction:

The balanced equation indicates that 2 moles of Cl_2 are formed when 4 mol of ClF_3 react.

The amount of Cl_2 formed would be: $(0.5 \bullet 3.4 \times 10^{-2}$ mol $ClF_3) = 1.7 \times 10^{-2}$ mol Cl_2

Likewise there are 3 mol of O_2 formed when 4 mol of ClF_3 react.

The amount of O_2 formed would be $(0.75 \bullet 3.4 \times 10^{-2}$ mol $ClF_3) = 2.6 \times 10^{-2}$ mol O_2

The pressure of Cl_2 formed would

$$be: P = \frac{nRT}{V} = \frac{1.7 \times 10^{-2}\ mol\ Cl_2 \bullet 62.4 \dfrac{L \bullet torr}{K \bullet mol} \bullet 293\ K}{2.5\ L} = 124\ torr\ (or\ 120\ torr\ to\ 2sf)$$

The pressure of O_2 formed would

$$P = \frac{nRT}{V} = \frac{2.6 \times 10^{-2}\ mol\ Cl_2 \bullet 62.4 \dfrac{L \bullet torr}{K \bullet mol} \bullet 293\ K}{2.5\ L} = 190\ torr$$

The *total pressure would be*: (190 + 120) or 310 torr or 310 mm Hg.

104. (a) Theoretical yield of sodium azide:

65.0 g Na $\bullet$ 1 mol Na/23.00 g Na = 2.83 mol Na

$$n_{ClF_3} = \frac{PV}{RT} = \frac{2.12\ atm \bullet 35.0\ L}{0.082057 \dfrac{L \bullet atm}{K \bullet mol} \bullet 296\ K} = 3.05\ mol\ N_2O$$

The ratio of required substances is: $\dfrac{4\ mol\ Na}{3\ mol\ N_2O} = 1.33$

The ratio that is available is: $\dfrac{2.83\ mol\ Na}{3.05\ mol\ N_2O} = 0.928$, so Na is the Limiting Reagent.

2.83 mol Na $\bullet \dfrac{1\ mol\ NaN_3}{4\ mol\ Na} \bullet \dfrac{65.01g\ NaN_3}{1\ mol\ NaN_3} = 46.0\ g\ NaN_3$

(b) Lewis structure for the azide ion:

$$\left[\overset{\bullet\bullet}{N} \equiv N - \overset{\bullet\bullet}{\underset{\bullet\bullet}{N}} \overset{\bullet\bullet}{} \right]^{-} \quad \rightleftharpoons \quad \left[\overset{\bullet\bullet}{\underset{\bullet\bullet}{N}} = N = \overset{\bullet\bullet}{\underset{\bullet\bullet}{N}} \right]^{-} \quad \rightleftharpoons \quad \left[\overset{\bullet\bullet}{\underset{\bullet\bullet}{N}} - N \equiv \overset{\bullet\bullet}{N} \right]^{-}$$

 0   +1   -2 -1   +1   -1 -2   +1   0

The central structure (having the smallest set of formal charges on each atom) is the most likely

(c) The central N has two groups of electron pairs around it, making the *linear* geometry the preferred geometry for the azide ion.

Using Electronic Resources

106. (a) The experiment is impossible since the gas will assume the *solid state* before reaching absolute zero.

(b) With 12 H_2 molecules and 4 N_2 molecules, 8 molecules of NH_3 will form.

The stoichiometry for the process is: $N_2 + 3H_2 \rightarrow 2 NH_3$

Note that the ratio of nitrogen and hydrogen molecules is in the proper ratio for complete reaction to form the ammonia. The volume of the ammonia formed would be *two times* the volume of N_2 and 2/3 the volume of H_2. [This assumes that the P and T remain unchanged!]

108. (a) The characteristic odors of fish, coffee, oranges, etc, are brought to our olfactory nerves by the process of *diffusion.*

(b) The change of average speed of a gaseous molecules when T doubles:
Since the speed of a gaseous molecule is related to the square root of the *Absolute* T (Equation 12.9 on page 492 of your text.), a change by a factor of 2 in the Absolute T will lead to a change of $\sqrt{2}$ in the speed,

(c) The Kinetic Molecular Theory does **not** take into account the shape, size, or chemical properties of gas molecules.

110. Regarding CD-ROM Screen 12.11:

(a) In the reaction of NH_3 with HCl, the ammonia molecule (with molar mass less than that of HCl) covers a greater distance than the HCl—indicating that *molecules with smaller molar masses move with a greater average speed*

(b) We know that gases move more rapidly at higher T than they move at lower T. Were the NH_3 and HCl reaction carried out at a higher T than the one presented, the reaction would occur faster, with the less massive ammonia molecules moving faster than the HCl molecules.

Chapter 13
Intermolecular Forces, Liquids, and Solids

Practicing Skills
Intermolecular Forces

16. The <u>intermolecular</u> forces one must overcome to:

<u>change</u>	<u>intermolecular force</u>
(a) melt ice	hydrogen bonds and dipole-dipole
	(molecule with OH bonds)
(b) melt solid I_2	induced dipole-induced dipole (nonpolar molecule)
(c) convert $NH_3(l)$ to $NH_3(g)$	hydrogen bonds and dipole-dipole
	(molecule with NH bonds)

18. To convert <u>species</u> from a liquid to a gas the <u>intermolecular</u> forces one must overcome:

<u>species</u>	<u>intermolecular force</u>
(a) liquid O_2	induced dipole-induced dipole (nonpolar molecule)
(b) mercury	induced dipole-induced dipole (nonpolar atom)
(c) methyl iodide	dipole-dipole (polar molecule)
(d) ethanol	hydrogen bonding and dipole-dipole (polar molecule with OH bonds)

20. Increasing strength of intermolecular forces:

$$Ne \; < \; CH_4 \; < \; CO \; < \; CCl_4$$

Neon and methane are nonpolar species and possess only induced dipole-induced dipole interactions. Neon has a smaller molar mass than CH_4, and therefore weaker London (dispersion) forces. Carbon monoxide is a polar molecule. Molecules of CO would be attracted to each other by dipole-dipole interactions, but the CO molecule is not a very strong dipole. The CCl_4 molecule is a non-polar molecule, but very heavy (when compared to the other three). Hence the greater London forces that accompany larger molecules would result in the strongest attractions of this set of molecules.

The lower molecular weight molecules with weaker interparticle forces should be gases at 25 °C and 1 atmosphere: Ne, CH_4, CO.

22. Compounds which are capable of forming hydrogen bonds with water are those containing polar O-H bonds and lone pairs of electrons on N,O, or F.

(a) CH_3-O-CH_3 no; no "polar H's" and the C-O bond is not very polar

    (b) CH_4 no

    (c) HF   yes: lone pairs of electrons on F and a "polar hydrogen".

    (d) CH_3COOH yes: lone pairs of electrons on O atoms, and a "polar hydrogen"

                                attached to one of the oxygen atoms

    (e) Br_2 no; nonpolar molecules

    (f) CH_3OH yes: "polar H" and lone pairs of electrons on O

Liquids

24. Heat required: $125 \text{ mL} \cdot \dfrac{0.7849 \text{g}}{1 \text{ mL}} \cdot \dfrac{1 \text{ mol}}{46.07 \text{ g}} \cdot \dfrac{42.32 \text{ kJ}}{1 \text{mol}} = 90.1 \text{ kJ}$

26. Using Figure 13.16:

    (a) The equilibrium vapor pressure of water at 60 °C is approximately 150 mm Hg.
        Appendix G lists this value as 149.4 mm Hg

    (b) Water has a vapor pressure of 600 mm Hg at 93 °C.

    (c) At 70 °C the vapor pressure of water is approximately 225 mm Hg while that of
        ethanol is approximately 520 mm Hg.

28. The vapor pressure of $(C_2H_5)_2O$ at 30. °C is **590 mm Hg**.

    Calculate the amount of $(C_2H_5)_2O$ to furnish this vapor pressure at 30 °C (303K).

$$n = \frac{PV}{RT} = \frac{600 \text{ mm} \cdot \dfrac{1 \text{ atm}}{760 \text{ mm}} \cdot 0.100 \text{ L}}{0.082057 \dfrac{L \cdot atm}{K \cdot mol} \cdot 303 \text{ K}} = 3.2 \times 10^{-3} \text{ mol}$$

    The total mass of $(C_2H_5)_2O$ [FW = 74.1 g] needed to create this pressure is about 0.24 g.

    Since there is adequate ether to provide this pressure, we anticipate the pressure in the flask

    to be approximately 600 mm Hg.

    As the flask is cooled from 30.°C to 0 °C, **some of the gaseous ether will condense** to

    form liquid ether.

30. Member of each pair with the higher boiling point:

    (a) O_2 would have a higher boiling point than N_2 owing to its greater molar mass.

    (b) SO_2 would boil higher since SO_2 is a polar molecule while CO_2 is non-polar.

    (c) HF would boil higher since strong hydrogen bonds exist in HF but not in HI

    (d) GeH_4 would boil higher. While both molecules are non-polar, germane has the

        greater molar mass and therefore stronger London forces.

32.   (a) From the figure, we can read the vapor pressure of CS_2 as approximately 620 mmHg

and for nitromethane as approximately 80 mm Hg.

(b) The principle intermolecular forces for CS_2 (a non-polar molecule) are **induced**

dipole-induced dipole; for nitromethane (polar molecule)--**dipole-dipole**.

(c) The normal boiling point from the figure for CS_2 is 46 °C and for CH_3NO_2, 100 °C.

(d) The temperature at which the vapor pressure of CS_2 is 600 mm Hg is about 39 °C.

(e) The vapor pressure of CH_3NO_2 is 60 mm Hg at approximately 34 °C.

34. CO can be liquefied at or below its critical temperature. The Tc for CO is 132.9 K (or

approximately -140 °C), so CO **cannot** be liquefied **at or above room temperature**.

Metallic and Ionic Solids

36. This compound would have the formula AB_8 since each black square (A) has eight

corresponding white squares (B).

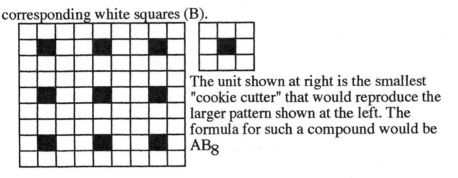

The unit shown at right is the smallest
"cookie cutter" that would reproduce the
larger pattern shown at the left. The
formula for such a compound would be
AB_8

38. To determine the perovskite formula, determine the number of each atom belonging **uniquely**

to the unit cell shown. The Ca atom is wholly contained within the unit cell. There are Ti

atoms at each of the eight corners. Since each of these atoms belong to eight unit cells, the

portion of each Ti atom belonging to the pictured unit cell is 1/8 so 8 Ti atoms x 1/8 = 1 Ti

atom. The O atoms on an edge belong to 4 unit cells, so the fraction contained within the

pictured cell is 1/4. There are twelve such O atoms, leading to 12 x 1/4 = 3 O atoms— and

a formula of $CaTiO_3$.

40. For cuprite:

(a) Formula for cuprite: There are 8 oxygen atoms at the corner of the cell (each 1/8 within

the cell) and 1 oxygen atom internal to the cell (wholly within the cell). The number of

oxygen atoms is then [(8 • 1/8) +(1)] 2. There are 4 Cu atoms wholly within the cell, so

the ratio was Cu_2O.

(b) With a formula of Cu_2O, and the oxidation state of O = -2, the oxidation state of copper

must be +1.

Other Types of Solids

42. For the unit cell of diamond:

(a) The unit cell has 8 corner atoms (1/8 in the cell), 6 face atoms (1/2 in the cell), and 4 atoms wholly within the cell, for a total of **8 carbon atoms**.

(b) Diamond uses a fcc unit cell (The structure shown also has 4 atoms occupying holes in the lattice. The holes are **tetrahedral**.

Physical Properties of Solids

44. The heat evolved when 15.5 g of benzene freezes at 5.5 °C :

$$15.5 \text{ g benzene} \cdot \frac{1 \text{ mol benzene}}{78.1 \text{ g benzene}} \cdot \frac{9.95 \text{ kJ}}{1 \text{ mol benzene}} = -1.97 \text{ kJ}$$

Note once again the negative sign indicates that heat is evolved.

The quantity of heat needed to remelt this 15.5 g sample of benzene would be +1.97 kJ.

Phase Diagrams and Phase Changes

46. (a) The positive slope of the solid/liquid equilibrium line means the liquid CO_2 is **less dense** than solid CO_2.

(b) At 5 atm and 0 °C, CO_2 is in the **gaseous phase**.

(c) The phase diagram for CO_2 shows the critical pressure for CO_2 to be 73 atm, and the critical temperature to be +31 °C.

48. The heat required is a summation of three "steps":

1. heat the liquid(at -50.0 °C) to its boiling point (-33.3 °C)

2. "boil" the liquid—converting it to a gas and

3. warm the gas from -33.3 °C to 0.0 °C

Let's do them one at a time:

1. To heat the liquid(at -50.0 °C) to its boiling point (-33.3 °C):

$$q_{liquid} = 1.2 \times 10^4 \text{ g} \cdot 4.7 \frac{J}{g \cdot K} \cdot (239.9 \text{ K} - 223.2 \text{ K}) = 9.4 \times 10^5 \text{ J}$$

2. To boil the liquid:

$$23.3 \times 10^3 \frac{J}{mol} \cdot \frac{1 \text{ mol NH}_3}{17.03 \text{ g NH}_3} \cdot 1.2 \times 10^4 \text{ g} = 1.6 \times 10^7 \text{ J}$$

3. To heat the gas from -33.3 °C to 0.0 °C:

$$q_{gas} = 1.2 \times 10^4 \text{ g} \cdot 2.2 \frac{J}{g \cdot K} \cdot (273.2 \text{ K} - 239.9 \text{ K}) = 8.8 \times 10^5 \text{ J}$$

The total heat required is then:

$$9.4 \times 10^5 \text{ J} + 1.6 \times 10^7 \text{ J} + 8.8 \times 10^5 \text{ J} = 1.8 \times 10^7 \text{ J or } 1.8 \times 10^4 \text{ kJ}$$

General Questions

50. Increasing strength of intermolecular forces: $Ar < CO_2 < CH_3OH$

Argon and CO_2 are nonpolar species and possess only induced dipole-induced dipole (London) interactions. Ar has a smaller mass than CO_2, so London forces are expected to be weaker for Ar than for CO_2. The polar molecular CH_3OH is capable of forming the stronger hydrogen-bonds(with an O-H bond in the molecule)

52.

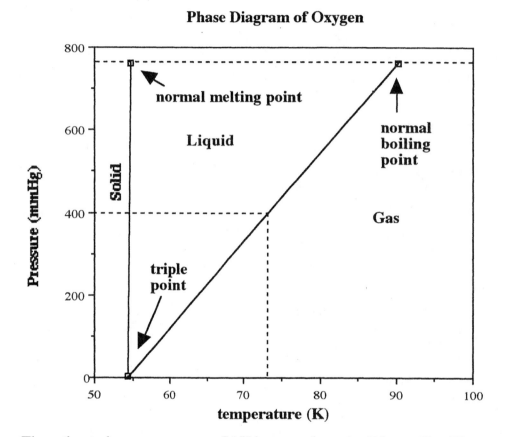

Phase Diagram of Oxygen

The estimated vapor pressure at 74 K is approximately 400 mm Hg. The very slight positive slope of the solid/liquid equilibrium line indicates the **solid is more dense than the liquid**

54. Acetone readily absorbs water owing to *hydrogen bonding* between the C = O oxygen atom and the O—H bonds of water.

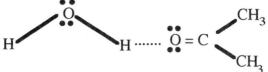

56. The **viscosity of ethylene glycol would be predicted to be greater** than that of ethanol since the glycol possesses two O-H groups per molecule while ethanol possesses one. Two OH groups/molecule would provide more hydrogen bonding!

58. Liquid methanol is, in many ways, equivalent to water—since both molecules are polar and can hydrogen bond. We would expect the meniscus of methanol therefore to be like that for water—concave.

60. Explain the fact that

(a) Ethanol has a lower boiling point than water. Both molecules are polar, and both can form hydrogen bonds. Water has **two** polar H atoms, and two lone pairs of electrons on the O atom. Ethanol has only **one** polar H atom, with the accompanying two lone pairs on the O atom. Given the increased ability of water to form hydrogen bonds with other water molecules, one would expect that water would have a higher boiling point.

(b) Mixing 50 mL of water with 50 mL of ethanol results in less than 100 mL of solution. The H-bonding that occurs not only between water molecules and other water molecules and ethanol molecules and other molecules also occurs between ethanol and water molecules. The attraction results in the molecules occupying less space than one would anticipate—and a non-additive volume.

<u>62</u>. Silver crystallizes in the face-centered cubic cell, with a side of 409 pm. The radius of a silver atom can be found by examining the geometry of a face of the cell. The diagram shows such a face with an edge of 409 pm. A look at the diagram will reveal that the diagonal is

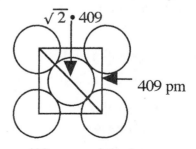

$\sqrt{2} \cdot 409$

409 pm

the hypotenuse of a right triangle—two sides of which are 409 pm, and the hypotenuse^2 is then $(409pm)^2 + (409pm)^2$. Let d be the hypotenuse.

Then we have: $d^2 = 2 \cdot (409pm)^2$.

The interest in d lies in the fact that d equals 1 diameter (of the center atom) and two-halves of two other atoms (or 4 radii).

$d = \sqrt{2} \cdot 409pm$ or $(1.414 \cdot 409pm) = 578$ pm (to 3sf) So 4 radii = 578pm and 1 radius= 145 pm (to 3sf). This radius compares favorably with the reported radius of Ag (See the CD-ROM Periodic Table).

64. The noncubic unit cell for calcium carbide shows 8 Ca atoms at the corners, and 1 Ca atom wholly within the cell for a total of [(8*1/8) +1] 2 Ca atoms. There are 8 C atoms along edges (each 1/4 within a cell) and 2 C atoms wholly within the cell for a total of [(8 • 1/4) +2] 4 C atoms. The ratio of atoms indicates a formula for calcium carbide of CaC_2.

66. $CaCl_2$ cannot have the NaCl structure. As shown in Figure 13.24 of your text (page 534), the cubic structure possesses 4 net lattice ions (occupied by anions) per face-centered lattice and 4 octahedral holes (occupied by cations). This is suitable for salts of a 1:1 composition.

68. Evidence that water molecules in the liquid state exert attractive forces on one another:

1) Water has a relatively large specific heat capacity (4.18 J/g •K). This large value is a reflection of the strong forces that hold water molecules together and require a large amount of energy to overcome.

2) Water is a liquid at room temperature—even though it has a molar mass of approximately 18 g/mol. With this molar mass, one would anticipate a boiling point well below 0°C (and a *gaseous* physical state).

70. The can collapses as a result of the condensation of the gas in the can—which has filled the heated can—to a liquid. The distances between the particles of liquid are **much less** than the distances between the particles of gas. The resulting decrease in pressure inside the can causes the greater pressure outside the can to crush the can.

72. Regarding dichlorodifluoromethane:

(a) The normal boiling point for this substance is the temperature *at which the vapor pressure of the substance is equal to atmospheric pressure.* From the diagram we see that this is approximately –27 °C.

(b) The pressure will be equal to the vapor pressure of the substance at 25 °C. From the diagram, this pressure is approximately 6.5 atmospheres.

174

(c) The flow is rapid at first, since the liquid has evaporated to form the gas to the maximum extent possible at that temperature. As the gas leaves the cylinder, the lower energy molecules remain behind, and the gas absorbs heat—resulting in the formation of ice on the exterior of the cylinder.

(d) Knocking the valve off the top is a **big safety risk**. Gas cylinders in laboratories are required to be anchored to a fixed surface so that they do not overturn and possibly knock the valve off in the process. The safest alternative given is ©. Cooling the container to −78 °C should cause most of the vapor to condense to a liquid. The valve can safely be opened.

74. Regarding benzene:

(a) Using the data provided, note that the vapor pressure of benzene is 760 mm Hg at 80.1°C (the definition of the normal boiling point).

(b) The graph of the data:

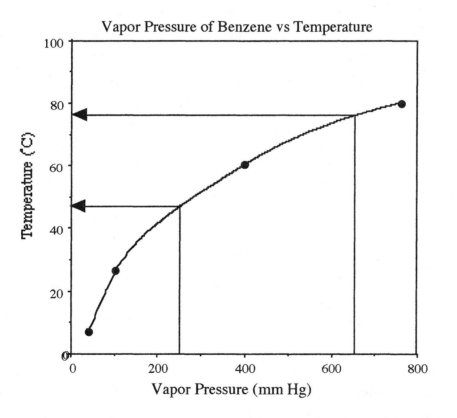

Vapor Pressure of Benzene vs Temperature

The temperature at which the liquid has a vapor pressure of 250 mm Hg is about 47°C.
The temperature at which the vapor pressure is 650 mm Hg is about 77 °C.

(c) The molar enthalpy of vaporization from the Clausius-Clapeyron equation:

The Clausius-Clapeyron equation is: $\ln\left(\dfrac{P_2}{P_1}\right) = \dfrac{\Delta H}{R}\left(\dfrac{1}{T_1} - \dfrac{1}{T_2}\right)$

From the graph using (arbitrarily) two of the data points:
$T_1 = 47\,°C$ and $P_1 = 250.$ mm Hg and $T_2 = 77\,°C$ and $P_2 = 650.$ mm Hg. Recall however that the units of R include temperature units of K, So convert the two temperatures to the K scale and substitute.

$$\ln\left(\dfrac{650.}{250.}\right) = \dfrac{\Delta H}{8.3145 \dfrac{J}{K \bullet mol}}\left(\dfrac{1}{320\ K} - \dfrac{1}{350\ K}\right)$$

Simplifying the equation:

$8.3145\ \dfrac{J}{K \bullet mol} \bullet \ln(2.60) = \Delta H\left(2.180 \text{ x } 10^{-4}\right)$ and $\Delta H = 29660$ J/mol or 30.0 kJ/mol

<u>76.</u> D for Iridium = 22.56 g/cm^3 The radius of an iridium atom is:

Ir crystallizes in the fcc lattice. There are 4 atoms in the unit cell (8 •1/8 corner atoms + 6•1/2 face atoms). See SQ13.62 for a diagram of a face that may be useful to you.

The mass/unit cell is: $\dfrac{4\text{ Ir atoms}}{1\text{ unit cell}} \bullet \dfrac{192.22\text{ g Ir}}{6.0221 \text{ x } 10^{23}\text{ atoms}} = 1.277 \text{ x } 10^{-21}$ g/unit cell

The volume of the unit cell is:

$$\dfrac{1.277 \text{ x } 10^{-21}\text{g}}{1\text{ unit cell}} \bullet \dfrac{1\text{ cm}^3}{22.56\text{ g}} = \dfrac{5.659 \text{ x } 10^{-23}\text{ cm}^3}{\text{unit cell}}$$

Since the cell is a cube, we can determine an edge by taking the cube root of this volume:

Edge = 3.839 x 10^{-8} cm

The diagonal of the unit cell (which corresponds to 4•radii) = 1.414 • 3.839 x 10^{-8} cm

So 4•radii = 1.414 • 3.839 x 10^{-8} cm and the radius =

$\dfrac{1.414 \bullet 3.839 \text{ x } 10^{-8}\text{ cm}}{4} = 1.356 \text{ x } 10^{-8}$ cm = or 135.6 x 10^{-12} m or 135.6 pm.

This value compares favorably with the literature value of 136pm (See the CD-ROM Periodic Table).

<u>78.</u> Let's assume that copper has a face-centered cubic unit cell, and calculate the density from this assumption. Comparing it with the stated density (8.95 g/cm^3) should give us an idea about the correctness of our assumption.

The fcc unit cell has 4 Cu atoms/cell. The mass of **one** Cu atom is:

$$\dfrac{63.546\text{ g Cu}}{1\text{ mol Cu}} \bullet \dfrac{1\text{ mol Cu}}{6.0221 \text{ x } 10^{23}\text{ Cu atoms}} = \dfrac{1.0552 \text{ x } 10^{-22}\text{ g}}{\text{Cu atom}}$$

$$\frac{4 \text{ Cu atoms}}{1 \text{ unit cell}} \cdot \frac{1.0552 \times 10^{-22} \text{ g}}{\text{Cu atom}} = \frac{4.2209 \times 10^{-22} \text{ g}}{\text{unit cell}}$$

The diagonal of the face of the unit cell = 4 atomic radii = $\sqrt{2}$ x edge

Solving for the edge gives:

$$\text{edge} = \frac{4 \text{ atomic radii}}{\sqrt{2}} = \frac{4 \cdot 127.8 \text{ pm}}{1.414} = 361.5 \text{ pm} = 3.615 \times 10^{-8} \text{ cm}$$

The volume of this cell is then the edge^3 = $(3.615 \times 10^{-8} \text{ cm})^3$ = $4.723 \times 10^{-23} \text{ cm}^3$

$$\text{Density} = \frac{M}{V} = \frac{4.2209 \times 10^{-22} \text{ g}}{4.723 \times 10^{-23} \text{ cm}^3} = 8.937 \text{ g/cm}^3$$

This is in good agreement with the measured density, hence **Cu is a face-centered cubic unit cell.**

80. For iron, whose density is 7.874g/cm^3, and whose body-centered cubic cell has a cell dimension (edge) of 286.65pm, the value of Avogadro's number may be calculated if we:

(1) calculate the volume of a unit cell

(2) determine the number of atoms in the unit cell

(3) use the molar mass to determine the number of atoms/mol (Avogadro's number).

(1) The volume of the cell can be determined by determining the length of one edge of the cell. The edge of the cell is 286.65 x 10^{-12}m

The volume of the unit cell is that length cubed. Since the density is given in terms of cm^3, first convert the length to units of centimeters then calculate the volume = (286.65 x 10^{-10} cm)^3

(2) BCC cells have

2 atoms per unit cell:1 in the center of the cube + 8 corner atoms • 1/8 = 2

(3) The molar mass of Iron = 55.847 g/mol. Combining these data:

$$\frac{2.355 \times 10^{-23} \text{ cm}^3}{1 \text{ unit cell}} \cdot \frac{7.845 \text{ g}}{1 \text{ cm}^3} \cdot \frac{1 \text{ mol}}{55.847 \text{ g}} \cdot \frac{1 \text{ unit cell}}{2 \text{ atoms}} = \frac{1.654 \times 10^{-24} \text{ mol}}{\text{atom}}$$

and the reciprocal gives: $\dfrac{1 \text{ atom}}{1.654 \times 10^{-24} \text{ mol}} = 6.0448 \times 10^{23} \text{atom/mol}$

82. Calculate the number of atoms represented by the vapor:

The pressure of Hg is 0.00169 mm Hg • $\dfrac{1 \text{atm}}{760 \text{ mm Hg}}$ = 2.22 x 10^{-6} atm

The number of moles/L is: $\dfrac{P}{RT} = \dfrac{n}{V} = \dfrac{2.22 \times 10^{-6} \text{ atm}}{(0.082057 \frac{\text{L•atm}}{\text{K•mol}})(297 \text{ K})} = 9.12 \times 10^{-8} \text{ mol/L}$

Converting this to atoms/m^3:

$$9.12 \times 10^{-8} \frac{mol}{L} \cdot \frac{1000L}{1m^3} \cdot \frac{6.022 \times 10^{23} \, atoms}{1 mol} = 5.49 \times 10^{19} \frac{atoms}{m^3}$$

Note that the information that the air was saturated with mercury vapor obviates the need to calculate the volume of the room.

<u>84.</u> For a simple cubic unit cell, each corner is occupied by an atom or ion. Each of these is contained within EIGHT unit cells contributing 1/8 to each. Within one unit cell, therefore, there is ($8 \times \dfrac{1}{8}$) 1 atom or ion. The volume occupied by that one net atom would be equal to 4/3 πr^3 with r representing the radius of the spherical atom or ion. The volume of the unit cell may be calculated by noting that the length of one side of the cell (an edge) corresponds to two radii (2r)-- since the spheres touch. The volume of this cube is therefore $(2r)^3$. The empty space within the cell is therefore:

$(2r)^3$ - 4/3 πr^3 and the fraction of space unoccupied is:

$$\frac{(8 - 4/3 \, \pi)r^3}{8r^3} = \frac{8 - 4/3 \, \pi}{8} = 0.476 \text{ or approximately 48 \%}$$

Using Electronic Resources

86. Regarding intermolecular forces:

   (a) The connection between the strength of a compound's intermolecular forces and its boiling point is that *the stronger the intermolecular forces between molecules of a substance, the higher the boiling point of the substance.*

   (b) Two factors controlling the strength of intermolecular forces:

   (1) Between nonpolar molecules, induced dipole-induced dipole forces (sometimes called dispersion forces) increase with increasing molecular volume (and mass). Compare boiling points for SiH_4 and GeH_4

   (2) Between polar molecules, dipole-dipole forces are important, with increased polarity resulting in higher boiling points for polar molecules compared to nonpolar molecules of approximately the same molar mass (compare the non polar molecule Br_2 and the polar ICl).

88. Solid Structures:

(a) Crystalline solids have a regular repeating pattern of atoms in an arrangement called a crystal lattice. Amorphous solids, on the other hand, have no regular repeating pattern.

(b) From the screen we see 12 unit cells with alternating chloride ions, and sodium ions—filling the octahedral holes.

(c) Predict crystalline or amorphous solids from the following:

    1. Table salt—the regularly shaped appearance of the smallest particles of table salt dumped from any salt shaker indicates a *crystalline solid*.

    2. Ice- (snow flakes) An examination of snow flakes indicates a regularity—with great diversity, indicating a solid that is organized due to H-bonding (a regular *crystalline* solid)

    3. Glass-An amorphous solid; One only has to break some glass to observe the irregular breakage—indicating an *irregular or amorphous* solid.

    4. Wood- An *amorphous* solid; Same observations as with glass.

Chapter 14
Solutions and Their Behavior

Practicing Skills

Concentration

16. For 2.56 g of succinic acid in 500. mL of water:

The molarity of the solution:

We need the # of moles of solute (succinic acid).

$$2.56 \text{ g } C_4H_6O_4 \bullet \frac{1 \text{mol } C_4H_6O_4}{118.09 \text{ g } C_4H_6O_4} = 0.0217 \text{ mol } C_4H_6O_4$$

$$M = \frac{0.0217 \text{ mol}}{0.500 \text{ L}} = 0.0434 \text{ M}$$

Note the *assumption* that the addition of 2.56 g of the acid will not change the volume of solution from that of the water.

The molality of the solution:

Molality = #mole solute/kg solvent:

With a density of water of 1.00 g/cm3, 500. mL= 0.500 kg

$$\text{Molality} = \frac{0.0217 \text{ mol}}{0.500 \text{ kg}} = 0.0434 \text{ molal}$$

The mole fraction of succinic acid in the solution:

For mole fraction we need *both* the #moles of solute *and* #moles of solvent.

$$\text{Moles of water= 500. g } H_2O \bullet \frac{1 \text{ mol } H_2O}{18.02 \text{ g } H_2O} = 27.7 \text{ mol } H_2O$$

$$\text{The mf of acid} = \frac{0.0217 \text{ mol}}{(0.0217 \text{ mol } + 27.7 \text{ mol})} = 7.81 \times 10^{-4}$$

The weight percentage of succinic acid in the solution:

The fraction of *total* mass of solute +solvent which is solute:

$$\text{Weight percentage=} \frac{2.56 \text{ g succinic acid}}{502.56 \text{ g acid } + \text{ water}} \bullet 100 = 0.509\% \text{ succinic acid}$$

18. Complete the following transformations for

NaI:

Weight percent:

$$\frac{0.15 \text{ mol NaI}}{1 \text{ kg solvent}} \bullet \frac{149.9 \text{ g NaI}}{1 \text{ mol NaI}} = \frac{22.5 \text{ g NaI}}{1 \text{ kg solvent}}$$

$$\frac{22.5 \text{ g NaI}}{1000 \text{ g solvent} + 22.5 \text{ g NaI}} \bullet 100 = 2.2 \% \text{ NaI}$$

Mole fraction:

$$1000 \text{ g H}_2\text{O} = 55.51 \text{ mol H}_2\text{O}$$

$$X_{NaI} = \frac{0.15 \text{ mol NaI}}{55.51 \text{ mol H}_2\text{O} + 0.15 \text{ mol NaI}} = 2.7 \times 10^{-3}$$

C_2H_5OH:

Molality:

$$\frac{5.0 \text{ g C}_2\text{H}_5\text{OH}}{100 \text{ g solution}} \cdot \frac{1 \text{ mol C}_2\text{H}_5\text{OH}}{46.07 \text{ g C}_2\text{H}_5\text{OH}} \cdot \frac{100 \text{ g solution}}{95 \text{ g solvent}} \cdot \frac{1000 \text{ g solvent}}{1 \text{ kg solvent}}$$

$$= 1.1 \text{ molal}$$

Mole fraction:

$$\frac{5.0 \text{ g C}_2\text{H}_5\text{OH}}{1} \cdot \frac{1 \text{ mol C}_2\text{H}_5\text{OH}}{46.07 \text{ g C}_2\text{H}_5\text{OH}} = 0.11 \text{ mol C}_2\text{H}_5\text{OH}$$

and for water : $\dfrac{95 \text{ g H}_2\text{O}}{1} \cdot \dfrac{1 \text{ mol H}_2\text{O}}{18.02 \text{ g H}_2\text{O}} = 5.27 \text{ mol H}_2\text{O}$

$$X_{C_2H_5OH} = \frac{0.11 \text{ mol C}_2\text{H}_5\text{OH}}{5.27 \text{ mol H}_2\text{O} + 0.11 \text{ mol C}_2\text{H}_5\text{OH}} = 0.020$$

$C_{12}H_{22}O_{11}$:

Weight percent:

$$\frac{0.15 \text{ mol C}_{12}\text{H}_{22}\text{O}_{11}}{1 \text{ kg solvent}} \cdot \frac{342.3 \text{ g C}_{12}\text{H}_{22}\text{O}_{11}}{1 \text{ mol C}_{12}\text{H}_{22}\text{O}_{11}} = \frac{51.3 \text{ g C}_{12}\text{H}_{22}\text{O}_{11}}{1 \text{ kg solvent}}$$

$$\frac{51.3 \text{ g C}_{12}\text{H}_{22}\text{O}_{11}}{1000 \text{ g H}_2\text{O} + 51.3 \text{ g C}_{12}\text{H}_{22}\text{O}_{11}} \times 100 = 4.9 \% \text{ C}_{12}\text{H}_{22}\text{O}_{11}$$

Mole fraction:

$$X_{C_{12}H_{22}O_{11}} = \frac{0.15 \text{ mol C}_{12}\text{H}_{22}\text{O}_{11}}{55.51 \text{ mol H}_2\text{O} + 0.15 \text{ mol C}_{12}\text{H}_{22}\text{O}_{11}} = 2.7 \times 10^{-3}$$

20. To prepare a solution that is 0.200 m Na_2CO_3:

$$\frac{0.200 \text{ mol Na}_2\text{CO}_3}{1 \text{ kg H}_2\text{O}} \cdot \frac{0.125 \text{ kg H}_2\text{O}}{1} \cdot \frac{106.0 \text{ g Na}_2\text{CO}_3}{1 \text{ mol Na}_2\text{CO}_3} = 2.65 \text{ g Na}_2\text{CO}_3$$

$$\text{mol Na}_2\text{CO}_3 = \frac{0.200 \text{ mol Na}_2\text{CO}_3}{1 \text{ kg H}_2\text{O}} \cdot \frac{0.125 \text{ kg H}_2\text{O}}{1} = 0.025 \text{ mol}$$

The mole fraction of Na_2CO_3 in the resulting solution:

$$\frac{125.\ g\ H_2O}{1} \cdot \frac{1\ mol\ H_2O}{18.02\ g\ H_2O} = 6.94\ mol\ H_2O$$

$$X_{Na_2CO_3} = \frac{0.025\ mol\ Na_2CO_3}{0.025\ mol\ Na_2CO_3 + 6.94\ mol\ H_2O} = 3.59 \times 10^{-3}$$

22. To calculate the number of mol of $C_3H_5(OH)_3$:

$$0.093 = \frac{x\ mol\ C_3H_5(OH)_3}{x\ mol\ C_3H_5(OH)_3 + (425\ g\ H_2O \cdot \frac{1\ mol\ H_2O}{18.02\ g\ H_2O})}$$

$$0.093 = \frac{x\ mol\ C_3H_5(OH)_3}{x\ mol\ C_3H_5(OH)_3 + 23.58\ mol\ H_2O}$$

$0.093(x + 23.58) = x$ and solving for x we get 2.4 mol $C_3H_5(OH)_3$

Grams of glycerol needed: 2.4 mol $C_3H_5(OH)_3 \cdot \frac{92.1\ g}{1\ mol} = 220\ g\ C_3H_5(OH)_3$

The molality of the solution is (2.4 mol $C_3H_5(OH)_3$, 0.425 kg H_2O)= 5.7 m

24. Concentrated HCl is 12.0M and has a density of 1.18 g/cm^3.

  (a) The molality of the solution:

   Molality is defined as moles HCl/kg solvent, so begin by deciding the mass of 1 L, and the mass of water in that 1L. Since the density = 1.18g/mL, then 1 L (1000 mL) will have a mass of 1180g.

   The mass of HCl present in 12.0 mol HCl =

$$12.0\ mol\ HCl \cdot \frac{36.46\ g\ HCl}{1\ mol\ HCl} = 437.52\ g\ HCl$$

   Since 1 L has a mass of 1180 g and 437.52 g is HCl, the difference (1180-437.52) is solvent. So 1 L has 742.98 g water.

$$\frac{12.0\ mol\ HCl}{1\ L} \cdot \frac{1\ L}{742.98\ g\ H_2O} \cdot \frac{1000\ g\ H_2O}{1\ kg\ H_2O} = 16.2\ m$$

  (b) Weight percentage of HCl:

   12.0 mol HCl has a mass of 437.52 g, and the 1 L of solution has a mass of 1180 g.

$$\%HCl = \frac{437.52\ g\ HCl}{1180\ g\ solution} \cdot 100 = 37.1\ \%$$

26. The concentration of ppm expressed in grams is:

$$0.18\ ppm = \frac{0.18\ g\ solute}{1.0 \times 10^6\ g\ solvent} = \frac{0.18\ g\ solute}{1.0 \times 10^3\ kg\ solvent}\ \ or\ \ \frac{0.00018\ g\ solute}{1\ kg\ water}$$

$$\frac{0.00018 \text{ g Li}^+}{1 \text{ kg water}} \bullet \frac{1 \text{ mol Li}^+}{6.939 \text{ g Li}^+} = 2.6 \times 10^{-5} \text{ molal Li}^+$$

The Solution Process

28. Pairs of liquids that will be miscible:

    (a) $H_2O/CH_3CH_2CH_2CH_3$

        Will **not** be miscible. Water is a polar substance, while butane is nonpolar.

    (b) C_6H_6/CCl_4

        Will **be** miscible. Both liquids are nonpolar and are expected to be miscible.

    (c) H_2O/CH_3CO_2H

        Will **be** miscible. Both substances can hydrogen bond, and we know that they mix—since a 5% aqueous solution of acetic acid is sold as "vinegar"

30. The enthalpy of solution for LiCl:

    The process can be represented as LiCl (s) $\rightarrow$ LiCl (aq)

    The $\Delta H_{reaction} = \Sigma \Delta H_f$ (product) - $\Sigma \Delta H_f$ (reactant)

$$= (-445.6 \text{ kJ/mol})(1\text{mol}) - (-408.6 \text{ kJ/mol})(1\text{mol}) = -37.0 \text{ kJ}$$

    The similar calculation for NaCl is + 3.9 kJ. Note that the enthalpy of solution for NaCl is endothermic while that for LiCl is exothermic.

32. Raising the temperature of the solution will increase the solubility of NaCl in water. Hence to increase the amount of dissolved NaCl in solution one must **(c) raise the temperature of the solution and add some NaCl.**

34. More likely to have a more negative heat of hydration:

    (a) **LiCl** or CsCl—See SQ14.30 for similar data. The aquation of the larger Cs ion requires more energy (ΔH_f is less negative) and the resulting $\Delta H_{hydration}$ is less negative than for the smaller lithium cation.

    (b) NaNO$_3$ or **Mg(NO$_3$)$_2$**: The greater charge of the Mg^{2+} compared to the Na^+ ion will have a stronger attraction to water, and a *more negative* heat of hydration.

    (c) RbCl or **NiCl$_2$**: The smaller dipositive Ni ion will cause NiCl$_2$ to have the more negative heat of hydration. Here two factors are in play: (1)the smaller size of the metal cation, and (2) the greater charge of the nickel cation over that of the rubidium cation.

Henry's Law

36. Solubility of O_2 = $k \cdot P_{O_2}$

$$= (1.66 \times 10^{-6} \frac{M}{mm\ Hg}) \cdot 40\ mm\ Hg = 6.6 \times 10^{-5}\ M\ O_2$$

and $6.6 \times 10^{-5} \frac{mol}{L} \cdot \frac{32.0\ g\ O_2}{1\ mol\ O_2} = 2 \times 10^{-3} \frac{g\ O_2}{L}$

38. Solubility = $k \cdot P_{CO_2}$

$$0.0506\ M = (4.48 \times 10^{-5} \frac{M}{mm\ Hg}) \cdot P_{CO_2}$$

$1130\ mm\ Hg = P_{CO_2}$ or expressed in units of atmospheres:

$$1130\ mm\ Hg \cdot \frac{1\ atm}{760\ mm\ Hg} = 1.49\ atm$$

Raoult's Law

40. Since $P_{water} = X_{water}P°_{water}$, to determine the vapor pressure of the solution (P_{water}), we need the mf of water.

$$35.0\ g\ glycol \cdot \frac{1\ mol\ glycol}{62.07\ g\ glycol} = 0.564\ mol\ glycol\ and$$

$$500.0\ g\ H_2O \cdot \frac{1\ mol\ H_2O}{18.02\ g\ H_2O} = 27.75\ mol\ H_2O.\ The\ mf\ of\ water\ is\ then:$$

$$\frac{27.75\ mol\ H_2O}{(27.75\ mol\ +\ 0.564\ mol)} = 0.9801\ and$$

$P_{water} = X_{water}P°_{water} = 0.9801 \cdot 35.7\ mm\ Hg = 35.0\ mm\ Hg$

42. Using Raoult's Law, we know that the vapor pressure of pure water ($P°$) multiplied by the mole fraction(X) of the solute gives the vapor pressure of the solvent above the solution (P).

$$P_{water} = X_{water}P°_{water}$$

The vapor pressure of pure water at 90 °C is 525.8 mmHg (from Appendix G).

Since the P_{water} is given as 457 mmHg, the mole fraction of the water is:

$$\frac{457\ mmHg}{525.8\ mmHg} = 0.869$$

The 2.00 kg of water correspond to a mf of 0.869. This mass of water corresponds to:

$$2.00 \times 10^3\ g\ H_2O \cdot \frac{1 mol H_2O}{18.02 g H_2O} = 111\ mol\ water.$$

Representing moles of ethylene glycol as x we can write:

$$\frac{mol\ H_2O}{mol\ H_2O\ +\ mol\ C_2H_4(OH)_2} = \frac{111}{111 + x} = 0.869$$

$$\frac{111}{0.869} = 111 + x\ ;\ 16.7 = x\ (mol\ of\ ethylene\ glycol)$$

$$16.7\ mol\ C_2H_4(OH)_2 \cdot \frac{62.07\ g\ C_2H_4(OH)_2}{1\ mol\ C_2H_4(OH)_2} = 1.04 \times 10^3\ g\ C_2H_4(OH)_2$$

Boiling Point Elevation

44. From Table 14.4 we see that benzene normally boils at a temperature of 80.10 °C. If the solution boils at a temperature of 84.2 °C, the change in temperature is (84.2 - 80.10 °C) or 4.1 °C.

Let's calculate the Δt, using the equation $\Delta t = K_{bp} \cdot m_{solute}$:

The molality of the solution is $\dfrac{0.200 \text{ mol}}{0.125 \text{ kg solvent}}$ or 1.60 m

The K_{bp} for benzene (from Table 14.4) is +2.53 °C/m

So $\Delta t = K_{bp} \cdot m_{solute} = +2.53 \text{ °C/m} \cdot 1.60 \text{ m} = +4.1 \text{ °C}$.

46. Calculate the molality of $C_{12}H_{10}$ in the solution.

$$0.515 \text{ g } C_{12}H_{10} \cdot \frac{1 \text{ mol } C_{12}H_{10}}{154.2 \text{ g } C_{12}H_{10}} = 3.34 \times 10^{-3} \text{ mol } C_{12}H_{10}$$

and the molality is : $\dfrac{3.34 \times 10^{-3} \text{ mol acenaphthalene}}{0.0150 \text{ kg } CHCl_3} = 0.223 \text{ molal}$

the boiling point elevation is:

$$\Delta t = m \cdot K_{bp} = 0.223 \cdot \frac{+3.63 \text{ °C}}{\text{molal}} = 0.808 \text{ °C}$$

and the boiling point will be $61.70 + 0.808 = 62.51 \text{ °C}$

48. The change in the temperature of the boiling point is (80.51 - 80.10)°C or 0.41 °C.

Using the equation $\Delta t = m \cdot K_{bp}$; 0.41 °C = m • +2.53 °C/m, and the molality is:

$$\frac{0.41 \text{ °C}}{+2.53 \text{ °C/m}} = m = 0.16 \text{ molal}$$

The solution contains 50.0 g of solvent (or 0.0500 kg solvent). We can calculate the # of moles of phenanthrene:

$$0.16 \text{ molal} = \frac{x \text{ mol } C_{14}H_{10}}{0.0500 \text{ kg}} \text{ or } 8.0 \times 10^{-3} \text{ mol } C_{14}H_{10}, \text{ and since 1 mol of}$$

$C_{14}H_{10}$ has a mass of 178 g, $8.0 \times 10^{-3} \text{ mol } C_{14}H_{10} \cdot \dfrac{178 \text{ g } C_{14}H_{10}}{1 \text{ mol } C_{14}H_{10}} = 1.4 \text{ g } C_{14}H_{10}$

Freezing Point Depression

50. The solution freezes 16.0 °C lower than pure water.

(a) We can calculate the molality of the ethanol:

$$\Delta t = mK_{fp}$$

$$-16.0 \text{ °C} = m(-1.86 \text{ °C/molal})$$

$$8.60 = \text{ molality of the alcohol}$$

(b) If the molality is 8.60 then there are 8.60 moles of C_2H_5OH (8.60 x 46.07 g/mol= 396 g) in 1000 g of H_2O.

The weight percent of alcohol is $\frac{396 \text{ g}}{1396 \text{ g}}$ x 100 = 28.4 % ethanol

52. Freezing point of a solution containing 15.0 g sucrose in 225 g water:

(1) Calculate the molality of sucrose in the solution:

$$15.0 \text{ g } C_{12}H_{22}O_{11} \bullet \frac{1 \text{ mol } C_{12}H_{22}O_{11}}{342.30 \text{ g } C_{12}H_{22}O_{11}} = 0.0438 \text{ mol}$$

$$\frac{0.0438 \text{ mol } C_{12}H_{22}O_{11}}{0.225 \text{ kg } H_2O} = 0.195 \text{ molal}$$

(2) Use the Δt equation to calculate the freezing point change:

$\Delta t = mK_{fp} = 0.195 \text{ molal} \bullet (-1.86\,°C/molal) = -0.362\,°C$

The solution is expected to begin freezing at $-0.362\,°C$.

Colligative Properties and Molar Mass Determination

54. The change in the temperature of the boiling point is $(80.26 - 80.10)°C$ or $0.16\,°C$.

Using the equation $\Delta t = m \bullet K_{bp}$; $0.16\,°C = m \bullet +2.53\,°C/m$, and the molality is:

$$\frac{0.16\,°C}{+2.53\,°C/m} = m = 0.063 \text{ molal}$$

The solution contains 11.12 g of solvent (or 0.01112 kg solvent). We can calculate the # of moles of the orange compound, since we know the molality:

$$0.063 \text{ molal} = \frac{x \text{ mol compound}}{0.01112 \text{ kg solvent}} \quad \text{or } 7.0 \text{ x } 10^{-4} \text{ mol compound.}$$

This number of moles of compound has a mass of 0.255 g, so 1 mol of compound is:

$$\frac{0.255 \text{ g compound}}{7.0 \text{ x } 10^{-4} \text{ mol}} = 360 \text{ g/mol.}$$

The empirical formula , $C_{10}H_8Fe$, has a mass of 184 g, so the # of "empirical formula units" in one molecular formula is : $\frac{360 \text{g/mol}}{184 \text{ g/empirical formula}} = 2$ mol/empirical formulas or a molecular formula of $C_{20}H_{16}Fe_2$.

56. The change in the temperature of the boiling point is $(61.82 - 61.70)°C$ or $0.12\,°C$.

Using the equation $\Delta t = m \bullet K_{bp}$; $0.12\,°C = m \bullet +3.63\,°C/m$, and the molality is:

$$\frac{0.12\,°C}{+3.63\,°C/m} = m = 0.033 \text{ molal}$$

The solution contains 25.0 g of solvent (or 0.0250 kg solvent). We can calculate the # of moles of benzyl acetate:

$$0.033 \text{ molal} = \frac{x \text{ mol compound}}{0.0250 \text{ kg solvent}} \quad \text{or } 8.3 \text{ x } 10^{-4} \text{ mol compound.}$$

This number of moles of benzyl acetate has a mass of 0.125 g, so 1 mol of benzyl acetate is:

$$\frac{0.125 \text{ g compound}}{8.3 \times 10^{-4} \text{ mol}} = 150 \text{ g/mol. (to 2 significant figures)}$$

58. To determine the molar mass, first determine the molality of the solution

$$-0.040\,°C = m \cdot -1.86\,°C/molal = 0.0215 \text{ molal (or 0.022 to 2 sf)}$$

and $0.022 \text{ molal} = \dfrac{\dfrac{0.180 \text{ g solute}}{MM}}{0.0500 \text{ kg water}}$

$$MM = 167 \text{ or } 170 \text{ (to 2 sf)}$$

60. The change in the temperature of the freezing point is (69.40 - 70.03)°C or -0.63 °C.

Using the equation $\Delta t = m \cdot K_{fp}$; $-0.63\,°C = m \cdot -8.00\,°C/m$, and the molality is:

$$\frac{-0.63\,°C}{-8.00\,°C/m} = m = 0.079 \text{ molal (to 2 significant figures)}$$

The solution contains 10.0 g of biphenyl (or 0.0100 kg solvent).

We can calculate the # of moles of naphthalene:

$$0.079 \text{ molal} = \frac{x \text{ mol naphthalene}}{0.0100 \text{ kg solvent}} \text{ or } 7.9 \times 10^{-4} \text{ mol compound.}$$

This number of moles of naphthalene has a mass of 0.100 g, so 1 mol of naphthalene is:

$$\frac{0.100 \text{ g naphthalene}}{7.9 \times 10^{-4} \text{ mol}} = 130 \text{ g/mol (to 2 significant figures)}$$

Colligative Properties of Ionic Compounds

62. The number of moles of LiF is : $52.5 \text{ g LiF} \cdot \dfrac{1 \text{ mol LiF}}{25.94 \text{ g LiF}} = 2.02 \text{ mol LiF}$

So $\Delta t_{fp} = \dfrac{2.02 \text{ mol LiF}}{0.306 \text{ kg H}_2\text{O}} \cdot -1.86\,°C/molal \cdot 2 = -24.6\,°C$

The anticipated freezing point is then 24.6 °C lower than pure water (0.0°C) or -24.6 °C

64. Solutions given in order of decreasing freezing point (lowest freezing point listed last):

The solution with the greatest **number** of particles will have the lowest freezing point.
The total molality of solutions is then:

(a) 0.1 m sugar   x 1 particle/formula unit = 0.1 m   [covalently bonded molecules]

(b) 0.1 m NaCl   x 2 particles/formula unit = 0.2 m   [Na^+ , Cl^-]

(c) 0.08 m $CaCl_2$ x 3 particles/formula unit = 0.24 m   [Ca^{2+}, $2\,Cl^-$]

(d) 0.04 m Na_2SO_4 x 3 particles/formula unit = 0.12 m   [$2\,Na^+$, SO_4^{2-}]

The freezing points would decrease in the order: sugar> Na_2SO_4> NaCl > $CaCl_2$

Osmosis

66. Assume we have 100 g of this solution, the number of moles of phenylalanine is

$$3.00 \text{ g phenylalanine} \bullet \frac{1 \text{ mol phenylalanine}}{165.2 \text{ g phenylalanine}} = 0.0182 \text{ mol phenylalanine}$$

The molality of the solution is $\dfrac{0.0182 \text{ mol phenylalanine}}{0.09700 \text{ kg water}} = 0.187 \text{ molal}$

(a) The freezing point :

$$\Delta t = 0.187 \text{ molal} \bullet -1.86 \text{ °C/molal} = -0.348 \text{ °C}$$

The new freezing point is $0.0 - 0.348 \text{ °C} = -0.348 \text{ °C}$.

(b) The boiling point of the solution

$$\Delta t = m \, K_{bp} = 0.187 \text{ molal} \bullet + 0.5121 \text{°C/molal} = +0.0959 \text{ °C}$$

The new boiling point is then $100.00 + 0.0959 = +100.10 \text{ °C}$

(c) The osmotic pressure of the solution:

If we assume that the **Molarity** of the solution is equal to the **molality**, then the osmotic pressure should be

$$\Pi = (0.187 \text{ mol/L})(0.0821 \frac{L \bullet atm}{K \bullet mol})(298 \text{ K}) = 4.58 \text{ atm}$$

The osmotic pressure will be most easily measured, since the magnitudes of osmotic pressures (large values) result in decreased experimental error.

68. The molar mass of bovine insulin with a solution having an osmotic pressure of 3.1 mm Hg:

$$3.1 \text{ mm Hg} \bullet \frac{1 \text{ atm}}{760 \text{ mm Hg}} = (M)(0.08205 \frac{L \bullet atm}{K \bullet mol})(298 \text{ K})$$

$$1.67 \times 10^{-4} = \text{Molarity or } 1.7 \times 10^{-4} \text{ (to 2 sf)}$$

The definition of molarity is #mol/L. Substituting into the definition we obtain:

$$1.7 \times 10^{-4} \frac{\text{mol bovine insulin}}{L} = \frac{\frac{1.00 \text{ g bovine insulin}}{MM}}{1 \text{ L}}$$

Solving for $MM = 6.0 \times 10^{3} \text{ g/mol}$

Colloids

70.   (a) $BaCl_2(aq) + Na_2SO_4(aq) \rightarrow BaSO_4(s) + 2 \, NaCl(aq)$

(b) The $BaSO_4$ initially formed is of a colloidal size — not large enough to precipitate fully.

(c) The particles of BaSO4 grow with time, owing to a gradual loss of charge and become large enough to have gravity affect them — and settle to the bottom.

General Questions

72. Li$_2$SO$_4$ is expected to have the more exothermic (negative) heat of solution. See SQ14.34 and 14.92 for additional information on this concept.

74. Arranged the solutions in order of (i) increasing vapor pressure of water and (ii) increasing boiling points:

   (i) The solution with the highest water vapor pressure would have the **lowest particle concentration**, since according to Raoult's Law, the vapor pressure of the water in the solution is directly proportional to the mole fraction of the water. The lower the number of particles, the greater the mf of water, and the greater the vapor pressure. Hence the order of *increasing* vapor pressure is:

$$Na_2SO_4 < sugar < KBr < glycol$$

   (See part (ii) for particle concentrations — (m•i)

   (ii). Recall that $\Delta t = m•Kfp • i$. The difference in these four solutions will be in the product (m • i). The products for these solutions are:

   glycol $= 0.35 • 1 = 0.35$
   sugar $\quad= 0.50 • 1 = 0.50$
   KBr $\quad\;= 0.20 • 2 = 0.40$
   Na$_2$SO$_4 = 0.20 • 3 = 0.60$

   Arranged in *increasing* boiling points: glycol < KBr < sugar < Na$_2$SO$_4$

76. For DMG, (CH$_3$CNOH)$_2$, the MM is 116.1 g/mol

   So 53.0 g is: $53.0 \text{ g} • \dfrac{1 \text{ mol DMG}}{116.1 \text{ g DMG}} = 0.456$ mol DMG

   525. g of C$_2$H$_5$OH is : $525. \text{ g} • \dfrac{1 \text{ mol C}_2\text{H}_5\text{OH}}{46.07 \text{ g C}_2\text{H}_5\text{OH}} = 11.4$ mol C$_2$H$_5$OH

   (a) the mole fraction of DMG: $\dfrac{0.456 \text{ mol}}{(11.4 + 0.456) \text{ mol}} = 0.0385$ mf DMG

   (b) The molality of the solution: $\dfrac{0.456 \text{ mol DMG}}{0.525 \text{ kg}} = 0.869$ molal DMG

   (c) $P_{alcohol} = P°_{alcohol} • X_{alcohol}$

$$= (760. \text{ mm Hg})(1 - 0.0385) = 730.7 \text{ mm Hg}$$

(d) The boiling point of the solution:

$$\Delta t = m \cdot K_{bp} \cdot i = (0.870)(+1.22 \,°C/\text{molal})(1)$$

$$= 1.06 \,°C$$

The new boiling point is $78.4 \,°C + 1.06 \,°C = 79.46 \,°C$ or $79.5 \,°C$

78. Concentrated NH_3 is 14.8 M and has a density of 0.90 g/cm^3.

(1) The molality of the solution:

Molality is defined as moles NH_3/kg solvent, so begin by deciding the mass of 1 L, and

the mass of water in that 1L. Since the density = 0.90g/mL, then 1 L (1000 mL) will

have a mass of 900g.

The mass of NH_3 present in 14.8 mol NH_3=

$$14.8 \text{ mol NH}_3 \cdot \frac{17.03 \text{ g NH}_3}{1 \text{ mol NH}_3} = 252 \text{ g NH}_3$$

Since 1 L has a mass of 900 g and 252 g is NH_3, the difference (900-252) is solvent. So 1

L has 648 g water.

$$\frac{14.8 \text{ mol NH}_3}{1 \text{ L}} \cdot \frac{1 \text{ L}}{648 \text{ g H}_2\text{O}} \cdot \frac{1000 \text{ g H}_2\text{O}}{1 \text{ kg H}_2\text{O}} = 22.8 \text{ m or } 23\text{m (to 2sf)}$$

(2) The mole fraction of ammonia is:

Calculate the # of moles of water present:

$$648 \text{ g H}_2\text{O} \cdot \frac{1 \text{ mol H}_2\text{O}}{18.02 \text{ g H}_2\text{O}} = 35.96 \text{ mol H}_2\text{O (retaining 1 extra sf)}$$

The mf NH_3 is : $\dfrac{14.8 \text{ mol NH}_3}{(14.8 \text{ mol} + 35.96 \text{ mol})} = 0.29$

(3) Weight percentage of NH_3:

14.8 mol NH_3 has a mass of 252.0 g, and 1 L of the solution has a mass of 900 g.

$$\% \text{ NH}_3 = \frac{252.0 \text{ g NH}_3}{900 \text{ g solution}} \text{NH}_3 \cdot 100 = 28 \% \text{ (to 2sf)}$$

80. To make a 0.100 m solution, we need a ratio of #moles of ions/kg solvent that is 0.100.

$$0.100 \text{ m} = \frac{\# \text{ mol ions}}{0.125 \text{ kg solvent}} \text{ and solving for \# mol ions: } 0.0125 \text{ mol ions}$$

The salt will dissociate into 3 ions per formula unit (2 Na^+ and 1 SO_4^{2-}).

The amount of Na_2SO_4 is :

$$0.0125 \text{ mol ions} \cdot \frac{1 \text{ mol Na}_2\text{SO}_4}{3 \text{ mol ions}} \cdot \frac{142.04 \text{ g Na}_2\text{SO}_4}{1 \text{ mol Na}_2\text{SO}_4} = 0.592 \text{ g Na}_2\text{SO}_4$$

82. Solution properties:

(a) The solution with the higher boiling point:

Recall that $\Delta t = m \cdot Kfp \cdot i$. The difference in these solutions will be in the product $(m \cdot i)$. The products for these solutions are:

sugar $= 0.30 \cdot 1 = 0.30$ (the sugar molecule remains as one unit)

KBr $= 0.20 \cdot 2 = 0.40$ (KBr dissociates into K^+ and Br^- ions)

KBr will provide the larger Δt.

(b) The solution with the lower freezing point:

Using the same logic as in part (a), NH_4NO_3 provides 2 ions/formula unit with Na_2CO_3 provides 3. The product , $(m \cdot i)$, is larger for Na_2CO_3, so **Na_2CO_3** gives the greater Δt and the lower freezing point.

84. The change in temperature of the freezing point is: $\Delta t = m \cdot K_{fp} \cdot i$

Calculate the molality:

$$35.0 \text{ g } CaCl_2 \cdot \frac{1 \text{ mol } CaCl_2}{111.0 \text{ g } CaCl_2} = 0.315 \text{ mol } CaCl_2 \text{ in } 0.150 \text{ kg water.}$$

$$m = \frac{0.315 \text{ mol } CaCl_2}{0.150 \text{ kg}} = 2.10 \text{ molal } CaCl_2$$

$\Delta t = m \cdot K_{fp} \cdot i = (2.10 \text{ molal} \cdot -1.86 \text{ °C/molal} \cdot 2.7) = -10.6 \text{ °C.}(-11 \text{ to 2sf})$

The freezing point of the solution is $0.0°C - 11 °C = -11 °C$

86. The molar mass of hexachlorophene if 0.640 g of the compound in 25.0 g of $CHCl3$ boils at 61.93 °C:

Recalling the Δt equation: $\Delta t = m \cdot K_{bp} = m \cdot \frac{+3.63 \text{ °C}}{molal} = (61.93 - 61.70) \text{ °C}$

Solving for m: $\frac{0.23 \text{ °C}}{3.63 \text{ °C/m}} = 0.0634 \text{ m}$

Substitute into the definition for molality: m = #mol/kg solvent

$$0.0634 \text{ molal} = \frac{\dfrac{0.640 \text{ g hexachloraphene}}{MM}}{0.025 \text{ kg}} \text{ and solving for MM; } 4.0 \times 10^2 \text{ g/mol} = MM$$

88. Solubility of $N_2 = k \cdot P_{N_2}$

$$= (8.42 \times 10^{-7} \frac{M}{mmHg}) \cdot 585 \text{ mm Hg} = 4.93 \times 10^{-4} \text{ M } N_2$$

$$\text{and } 4.93 \times 10^{-4} \frac{mol}{L} \cdot \frac{28.01 \text{ g } N_2}{1 mol \text{ } N_2} = 1.38 \times 10^{-2} \frac{g \text{ } N_2}{L}$$

90.   (a) *Average* MM of starch if 10.0 g starch/L has an osmotic pressure=3.8 mm Hg at 25 °C.

$$3.8 \text{ mm Hg} \bullet \frac{1 \text{ atm}}{760 \text{ mm Hg}} = (M)(0.08205 \frac{L \bullet atm}{K \bullet mol})(298 \text{ K})$$

$$2.045 \times 10^{-4} = \text{Molarity or } 2.0 \times 10^{-4} \text{ (to 2 sf)}$$

The definition of molarity is #mol/L. Substituting into the definition we obtain:

$$2.0 \times 10^{-4} \frac{\text{mol bovine insulin}}{L} = \frac{\frac{10.0 \text{ g starch}}{MM}}{1L}$$

Solving for MM $= 4.9 \times 10^4$ g/mol

(b) Freezing point of the solution:

$$\Delta t = m \bullet K_{fp} \bullet i \quad \text{(assume that i=1and that Molarity = molality)}$$

$$\Delta t = m \bullet (-1.86 \text{ °C/molal})$$

and the M $= \dfrac{\frac{10.0 \text{ g starch}}{4.9 \times 10^4 \text{ g/mol}}}{1 \text{ L}} = 2.0 \times 10^{-4}$ so the

$\Delta t = 2.0 \times 10^{-4} \bullet (-1.86 \text{ °C/molal}) = 3.8 \times 10^{-4}$ °C. In essence the starch will boil at the temperature of pure water. From this data we can assume that *it will NOT be easy* to measure the molecular weight of starch using this technique.

92. The apparent molecular weight of acetic acid in benzene, determined by the depression of benzene's freezing point.

$$\Delta t = m \bullet K_{fp} \bullet i \; ; \quad (3.37 \text{ °C} - 5.50 \text{ °C}) = m(-5.12 \text{ °C/molal}) i$$

$$\text{and} \frac{-2.13 \text{ °C}}{-5.12 \text{ °C/molal}} = m \bullet i \quad \text{so } 0.416 \text{ molal} = m \bullet i \text{ (assume i =1)}$$

and the apparent molecular weight is:

$$0.416 \text{ molal} = \frac{\frac{5.00 \text{ g acetic acid}}{MM}}{0.100 \text{ kg}} \quad \text{and solving for MM;} \quad 120 \text{ g/mol} = MM$$

The apparent molecular weight of acetic acid in water

$$\Delta t = m \bullet K_{fp} \bullet i \; ; \quad (-1.49 \text{ °C} - 0.00 \text{ °C}) = m(-1.86 \text{ °C/molal}) i$$

$$\text{and} \frac{-1.49 \text{ °C}}{-1.86 \text{ °C/molal}} = m \bullet i \text{ so } 0.801 \text{ molal} = m \bullet i$$

(once again, momentarily i $= 1$) and the apparent molecular weight is:

$$0.801 \text{ molal} = \frac{\frac{5.00 \text{ g acetic acid}}{MM}}{0.100 \text{ kg}} \quad \text{and solving for MM;} \quad 62.4 \text{ g/mol} = MM$$

The accepted value for acetic acid's molecular weight is approximately 60 g/mol. Hence the value for i isn't much larger than 1, indicating that the degree of dissociation of acetic acid molecules in water is not great—a finding consistent with the designation of acetic acid as a weak acid. The apparently doubled molecular weight of acetic acid in benzene indicates that the acid must exist primarily as a dimer.

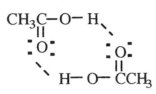

94. (a) The eggshell is predominantly calcium carbonate, and reacts with the acetic acid in the same way we expect any carbonate salt to react with an acid—liberating carbon dioxide and water *and* the water soluble calcium acetate. The membrane is however not reactive with the acetic acid and remains intact.

 (b) The egg—minus its shell—when placed in water has a higher solute concentration inside the egg membrane than in the water. The resulting osmotic pressure is an attempt to reduce the solute concentration—and the net transport of water from outside to inside the membrane results in a swelling of the egg.

 (c) When the egg is placed in a solution with a high solute concentration, the situation is the reverse of that in (b). The higher solute concentration is outside the membrane and the net transport of water from inside the membrane to outside the membrane occurs—with the concomitant shriveling of the egg.

96. Since hydrophilic colloids are those that "love water", we would expect starch to form a hydrophilic colloid since it contains the OH bonds that can hydrogen bond to water. Hydrocarbons on the other hand have non-polar bonds that should have little-to-no attraction to water molecules, and form a hydrophobic colloid.

98. The enthalpies of solution for Li_2SO_4 and K_2SO_4:
    The process is MX (s) → MX (aq)
    Using the data for **Li_2SO_4**:
    $\Delta H_{solution} = \Delta H_f (aq) - \Delta H_f (s) = (-1464.4 \text{ kJ/mol}) - (-1436.4 \text{ kJ/mol}) = -28.0 \text{ kJ/mol}$
    Using the data for **K_2SO_4**:
    $\Delta H_{solution} = \Delta H_f (aq) - \Delta H_f (s) = (-1414.0 \text{ kJ/mol}) - (-1437.7 \text{ kJ/mol}) = 23.7 \text{ kJ/mol}$
    Note that for Li_2SO_4 *the process is* **exothermic** *while for* K_2SO_4 *the process is*
        endothermic.

Similar data for LiCl and KCl:

For LiCl: ΔH_f (aq) - ΔH_f (s) = (-445.6 kJ/mol) - (-408.6 kJ/mol) = - 37.0 kJ/mol and

for KCl: ΔH_f (aq) - ΔH_f (s) = (-419.5 kJ/mol) - (-436.7 kJ/mol) =   17.2 kJ/mol

Note the similarities of the chloride salts, with the lithium salt being **exothermic** *while the*

potassium salt is **endothermic**.

100. Graham's law says that the pressure of a mixture of gases (benzene and toluene) is the sum of
the partial pressures. So, using Raoult's Law $P_{benzene}$ = mf benzene • $P°_{benzene}$ and
similarly for toluene. The total pressure is then:

P_{total} = $P_{benzene}$ + $P_{toluene}$

$$= (\frac{2 \text{ mol benzene}}{3 \text{ mol}} \cdot 75 \text{ mm Hg}) + (\frac{1 \text{ mol toluene}}{3 \text{ mol}} \cdot 22 \text{ mm Hg}) = 57 \text{mm Hg}$$

What is the mole fraction of each component in the liquid and in the vapor?

The **mf of the components in the liquid** are: benzene: 2/3 and toluene: 1/3

The **mf of the components in the vapor** are proportional to their pressures in the vapor

state.

The mf of benzene is: $\frac{50 \text{ mm Hg}}{57 \text{ mm Hg}}$ = 0.87; the mf of toluene would be (1-0.87) or 0.13.

102. A 2.0 % aqueous solution of novocainium chloride(NC) is also 98.0 % in water. Assume that
we begin with 100 g of solution. The molality of the solution is then:

$$\frac{2.0 \text{ g} \cdot \dfrac{1 \text{mol NC}}{272.8 \text{ g NC}}}{0.0980 \text{ kg water}} = 0.075 \text{ m}$$

Using the "delta T' equation:

Δt = m • K_{fp} • i , we can solve for i: $\frac{\Delta t}{m \cdot K_{fp}}$ = i

$$\frac{- 0.237 °C}{0.075 m \cdot -1.86 °C/m} = 1.7$$

So approximately **2 moles of ions are present per mole of compound.**

104. (a) We can calculate the freezing point of sea water if we calculate the molality of the solution.
Let's imagine that we have 1,000,000 (or 10^6) g of sea water. The amounts of the ions are
then equal to the concentration (in ppm). Calculating their concentrations we get:

Cl⁻ 1.95×10^4 g Cl⁻ • $\dfrac{1 \text{mol Cl}^-}{35.45 \text{ g Cl}^-}$ = 550. mol Cl⁻

Na⁺ 1.08×10^4 g Cl⁻ • $\dfrac{1 \text{mol Na}^+}{22.99 \text{ g Na}^+}$ = 470. mol Na⁺

Mg^{+2} 1.29×10^3 g $Mg^{+2} \cdot \dfrac{1 mol\ Mg^{+2}}{24.31\ g\ Mg^{+2}} = 53.1$ mol Mg^{+2}

SO_4^{-2} 9.05×10^2 g $SO_4^{-2} \cdot \dfrac{1 mol\ SO_4^{-2}}{96.06\ g\ SO_4^{-2}} = 9.42$ mol SO_4^{-2}

Ca^{+2} 4.12×10^2 g $Ca^{+2} \cdot \dfrac{1 mol\ Ca^{+2}}{40.08\ g\ Ca^{+2}} = 10.3$ mol Ca^{+2}

K^+ 3.80×10^2 g $K^+ \cdot \dfrac{1 mol\ K^+}{39.10\ g\ K^+} = 9.72$ mol K^+

Br^- 67 g $Br^- \cdot \dfrac{1 mol\ Br^-}{79.90\ g\ Br^-}$ $\underline{= 0.84$ mol Br^-}

For a total of: 1103 mol ions

The concentration per gram is: $\dfrac{1103\ mol\ ions}{10^6\ g\ H_2O}$

The *change* in the freezing point of the sea water is:

$\Delta t = m \cdot K_{fp} = \dfrac{1103\ mol\ ions}{10^6\ g\ H_2O} \cdot \dfrac{1000\ g\ H_2O}{1\ kg\ H_2O} \cdot -1.86\ °C/molal = -2.05\ °C$

So we expect this sea water to begin freezing at -2.05 °C.

(b) The osmotic pressure (in atmospheres) can be calculated if *we assume the density of sea water is 1.00 g/mL.*

$\Pi = MRT = \dfrac{1.103\ mol}{1\ L} \cdot 0.082057\ \dfrac{L \cdot atm}{K \cdot mol} \cdot 298\ K = 27.0$ atm

The pressure needed to purify sea water by reverse osmosis would then be a pressure greater than 27.0 atm.

106. A 2.00 % aqueous solution of sulfuric acid is also 98.00 % in water.

Assume that we begin with 100 g of solution.

(a) We can calculate the van't Hoff factor by first calculating the molality of the solution:

$$\dfrac{2.00\ g \cdot \dfrac{1 mol\ H_2SO_4}{98.06\ g\ H_2SO_4}}{0.09800\ kg\ water} = 0.208\ m$$

Using the "delta T" equation:

$\Delta t = m \cdot K_{fp} \cdot i$, we can solve for i: $\dfrac{\Delta t}{m \cdot K_{fp}} = i$

$\dfrac{-0.796\ °C}{0.208m \cdot -1.86\ °C/m} = 2.06 = i$

(b) Given the van't Hoff factor of 2 (above), the best representation of a dilute solution of sulfuric acid in water has to be: $H^+ + HSO_4^-$.

108. The vapor pressure data should permit us to calculate the molar mass of the boron compound.

$$P_{benzene} = X_{benzene} \cdot P^°_{benzene}$$

94.16 mm Hg = $X_{benzene} \cdot$ 95.26 mm Hg , and rearranging: $X_{benzene} = \dfrac{94.16 \text{ mm Hg}}{95.26 \text{ mm Hg}}$

$X_{benzene} = 0.9885$

Now we need to know the # of moles of the boron compound, so let's use the mf of benzene to find that:

10.0 g benzene $\cdot \dfrac{1 \text{ mol benzene}}{78.11 \text{ g benzene}} = 0.128$ mol benzene

$$0.9885 = \dfrac{0.128 \text{ mol benzene}}{0.128 \text{ mol benzene} + x \text{ mol } B_xF_y}$$

$0.9885(0.128 + x) = 0.128$ $x = 0.00147$ mol B_xF_y

Knowing that this # of moles of compound has a mass of 0.146 g, we can calculate the molar mass:

$$\dfrac{0.146 \text{ g}}{0.00147 \text{ mol } B_xF_y} = 99.3 \text{ g/mol}$$

We can calculate the empirical formula, since we know that the compound is 22.1% boron and 77.9 % fluorine.

In 100 g of the compound there are   22.1 g B $\cdot \dfrac{1 \text{ mol B}}{10.81 \text{ g B}} = 2.11$ mol B and

77.9 g F $\cdot \dfrac{1 \text{ mol F}}{19.00 \text{ g F}} = 4.10$ mol F

The empirical formula then is BF_2, which would have a formula weight of 48.8

Dividing the molar mass (found from the vapor pressure experiment) by the mass of the empirical formula, we get: $\dfrac{99.3}{48.8} = 2.0$

(a) The molecular formula is then B_2F_4.

(b) A Lewis structure for the molecule:

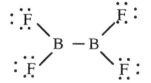

We know that the molecule is nonpolar (does not have a dipole moment), hence the proposed structure. The F-B-F bond angles are 120°, as are the F-B-B bond angles. Hence the molecule is planar (flat). The hybridization of the boron atoms is then sp^2.

Using Electronic Resources

110.   (a) Evidence for an unsaturated solution occurs when more solid dissolves in the solution.

(b) Testing for supersaturation can be accomplished by dropping a small crystal into the "supersaturated solution". If the solution is supersaturated, crystallization will typically begin immediately with the precipitate forming and settling to the bottom of the container.

112. *Dynamic Equilibrium* means that the condition is one of continuing change, much like the piece of playground equipment called a "see-saw"—with the continuing balance between two states. This process is a balancing act between two opposing processes—occurring at the same rate. The net result is *no obvious macroscopic changes*. Contrast dynamic changes to the "static" condition that would exist when you roll a ball down a hill. Once the ball reaches the lowest energy possible (the bottom of the hill), it becomes static—and no further action occurs (unless the ball is acted upon by some other force, e.g. someone picks it up).

114.   (a) Boiling point of a liquid is increased upon solution formation owing to the lowering of the vapor pressure, upon addition of a non-volatile solute. This decreased vapor pressure requires that a higher temperature be reached before the vapor pressure of the solution reaches normal atmospheric pressure (i.e. a higher boiling point temperature).

(b) The NH_4NO_3 will lower the freezing point to a greater degree than the ethylene glycol since the inorganic material will furnish two ions per formula unit while the ethylene glycol will furnish only one.

116.   (a) Surfactants help oil and water form colloids by having both polar and non-polar parts of a molecule. The polar "end" dissolves in water while the non-polar "end" dissolves in the oil forming a micelle. These micelles aggregate forming larger units, which don't form "true solutions"—a colloid.

(b) Fabric softeners are surfactants. When fabrics are coated with these fabric softener molecules (with their nonpolar "ends" sticking out into space), the nonpolar ends simply slip over one another—since their weak interparticle forces (dispersion forces) result in little attraction for other such molecules.

Chapter 15
Principles of Reactivity: Chemical Kinetics

Reviewing Important Concepts

2. From Figure 15.2, we see that after 2.0 h, the concentration of NO2 is
The initial concentration of N_2O_5 is 1.40 M. After 2.0 hours, the concentration has
dropped to 0.80M (a change of 0.60M). The stoichiometry of the equation shows that 4
molecules of NO_2 are created when 2 molecules of N_2O_5 decompose (a 2:1 ratio) The
decomposition of 0.60 M N_2O_5 results in the *formation of* 1.2 M NO_2.
The concentration of O_2 is similarly determined. The ratio of N_2O_5 to O_2 is 2:1, so the
decomposition of 0.60 M N_2O_5 results in the *formation* of 0.30 M O_2.

4. The manner in which the rate will change if [A] is tripled? If [A] is halved?
The experimental rate equation is *second order* in A. If [A] is tripled, the rate will change
by the factor $[3a]^2$. The rate will be changed by *a factor of 9 times*.
If [A] is halved, the new rate will be changed by the factor $[0.5A]^2$. The new rate is *1/4 the
original rate*.

8. Answering this question is a matter of viewing the integrated rate equations for zero, first, and
second-order reactions (page 642 has one listing of these equations). Equation 15.2 (the
equation for a *second-order reaction) indicates that a plot of 1/[A] vs t would give a
straight line*. Equation 15.1 (the equation for a *first-order reaction) indicates that a plot of
ln[A] vs t would give a straight line*.

Practicing Skills
Reaction Rates
16.   (a) $2\,O_3\,(g) \;\rightarrow\; 3\,O_2\,(g)$

$$\text{Reaction Rate} \;=\; -\frac{1}{2}\cdot\frac{\Delta[O_3]}{\Delta t} \;=\; +\frac{1}{3}\cdot\frac{\Delta[O_2]}{\Delta t}$$

   (b) $2\,HOF\,(g) \;\rightarrow\; 2\,HF\,(g) + O_2\,(g)$

$$\text{Reaction Rate} \;=\; -\frac{1}{2}\cdot\frac{\Delta[HOF]}{\Delta t} \;=\; +\frac{1}{2}\cdot\frac{\Delta[HF]}{\Delta t} \;=\; +\frac{\Delta[O_2]}{\Delta t}$$

18. For the reaction, $2\,O_3\,(g) \;\rightarrow\; 3\,O_2\,(g)$, the rate of formation of O_2 is 1.5×10^{-3} M/L•s.

SQ15.16(a) offers a clear assist, indicating that O_2 forms at a rate 1.5 times the rate that ozone decomposes. (2 ozones produce 3 oxygens!) Hence the rate of decomposition of O_3 is -1.0×10^{-3} M/L•s.

20. Plot the data for the hypothetical reaction $A \rightarrow 2\,B$

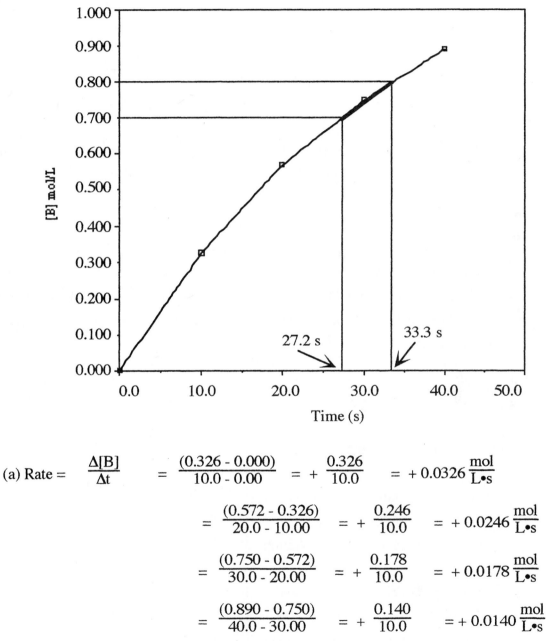

$$\text{(a) Rate} = \frac{\Delta[B]}{\Delta t} = \frac{(0.326 - 0.000)}{10.0 - 0.00} = +\frac{0.326}{10.0} = +0.0326\frac{\text{mol}}{\text{L•s}}$$

$$= \frac{(0.572 - 0.326)}{20.0 - 10.00} = +\frac{0.246}{10.0} = +0.0246\frac{\text{mol}}{\text{L•s}}$$

$$= \frac{(0.750 - 0.572)}{30.0 - 20.00} = +\frac{0.178}{10.0} = +0.0178\frac{\text{mol}}{\text{L•s}}$$

$$= \frac{(0.890 - 0.750)}{40.0 - 30.00} = +\frac{0.140}{10.0} = +0.0140\frac{\text{mol}}{\text{L•s}}$$

The rate of change decreases from one time interval to the next *due to a continuing decrease* in the amount of reacting material (A).

(b) Since each A molecule forms 2 molecules of B, the concentration of A will decrease at a rate that is **half** of the rate at which B appears. The negative signs here indicate a **decrease in [A]--not a negative concentration of A!**

T	[B]	$[A] = 1/2([B]_0 - [B])$
10.0 s	0.326	$-1/2(0.326) = -0.163$
20.0 s	0.572	$-1/2(0.572) = -0.286$

$$\text{Rate at which A changes} = \frac{\Delta[A]}{\Delta t} = \frac{(-0.286 - -0.163)}{20.0 - 10.00}$$

$$= \frac{-0.123}{10.0} \quad \text{or} - 0.0123 \frac{\text{mol}}{\text{L} \cdot \text{s}}$$

Note that the **negative sign** indicates a <u>reduction in the concentration of A</u> as the reaction proceeds. Compare this change with the change in [B] for the same interval above $(+ 0.0246 \frac{\text{mol}}{\text{L} \cdot \text{s}})$. The disappearance of A is half that of the appearance of B.

(c) The instantaneous rate when [B] = 0.750 mol/L:

The instantaneous rate can be calculated by noting the tangent to the line at the point , [B] = 0.750 mol/L. Taking points equidistant ([B] = 0.700 mol/L and [B] = 0.800 mol/L) and determining the times associated with those concentrations, we can calculate the instantaneous rate.

$$\frac{\Delta[B]}{\Delta t} = \frac{\left(0.800\frac{\text{mol}}{\text{L}} - 0.700\frac{\text{mol}}{\text{L}}\right)}{(33.3 \text{ s} - 27.2 \text{ s})} = \frac{0.100\frac{\text{mol}}{\text{L}}}{6.1 \text{ s}} = 0.0163 \frac{\text{mol}}{\text{L} \cdot \text{s}}$$

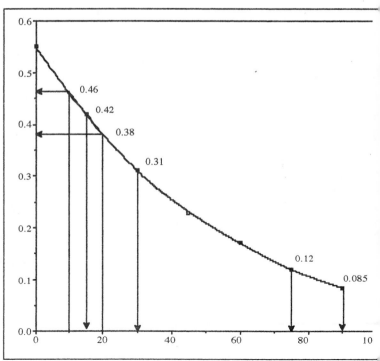

21. For the data concerning [Phenyl acetate] vs time:

   (a) The curved line indicates a changing rate—a fact that results from a continual decrease in the concentration of phenyl acetate.

   (b) As the graph indicates, the Δ[phenyl acetate] from 15 s to 30 s is - 0.11M while the concentration from 75 s to 90 s is –0.035M. The smaller value for the later data is an indication that the decreased concentration of reactants results in a decreased *rate of reaction*.

(c) From 60 s to 75 s, the [phenyl acetate] changes by -0.05 M (0.12M -0.17M) The stoichiometry of the reaction indicates that the [phenol] will change by $+0.05$M.

(d) The instantaneous rate at 15.0 s is determined by drawing a tangent to the curve at t=15.0 s. I picked a point at t =10.0 s and t =20.0 s. The concentrations are 0.46M and 0.38M. The rate is $\dfrac{(0.38 \text{ M} - 0.46\text{M})}{(20.0 \text{ s} - 10.0 \text{ s})} = 0.0080 \dfrac{\text{mol}}{\text{L} \cdot \text{s}}$:

Concentration and Rate Equations

22.   (a) The rate equation :   Rate $= k[NO_2][O_3]$

(b) Since k is constant, if $[O_3]$ is held constant, the rate would be tripled:
Let C represent the concentration of NO_2 . Substituting into the rate equation:

$Rate_1 = k[C][O_3]$

$Rate_2 = k[3C][O_3] = 3 \cdot k[C][O_3]$ or $3 \cdot Rate_1$

(c) Halving the concentration of O_3—assuming $[NO_2]$ is constant, would halve the rate.

$Rate_1 = k[NO_2][C]$

$Rate_2 = k[NO_2][1/2 \text{ C}] = 1/2[NO_2][C]$ or $1/2 \cdot Rate_1$

24.   (a) If we designate the three experiments (data sets in the table as i, ii, and iii respectively,

Experiment	[NO]	[O2]	$-\dfrac{\Delta[NO]}{\Delta t} (\dfrac{\text{mol}}{\text{L} \cdot \text{s}})$
i	0.010	0.010	2.5×10^{-5}
ii	0.020	0.010	1.0×10^{-4}
iii	0.010	0.020	5.0×10^{-5}

Note that experiment ii proceeds at a rate four times that of experiment i.

$$\frac{\text{experiment ii rate}}{\text{experiment i rate}} = \frac{1.0 \times 10^{-4} \frac{\text{mol}}{\text{L} \cdot \text{s}}}{2.5 \times 10^{-5} \frac{\text{mol}}{\text{L} \cdot \text{s}}} = 4$$

This rate change was the result of doubling the concentration of NO. The order of dependence of NO must be second order. Comparing experiments i and iii, we see that changing the concentration of O_2 by a factor of two, also affects the rate by a factor of two. The order of dependence of O_2 must be first order.

(b) Using the results above we can write the rate equation: Rate $= k[NO]^2[O_2]^1$

(c) To calculate the rate constant we have to have a rate. Note the data provided gives the rate of disappearance of NO. The relation of this concentration to the rate is:

$$Rate = -1/2 \cdot \frac{\Delta[NO]}{\Delta t}$$

Using experiment ii, the rate is $5.0 \times 10^{-5} \frac{mol}{L \cdot s}$

Substituting into the rate law

$$5.0 \times 10^{-5} \frac{mol}{L \cdot s} = k[0.020 \frac{mol}{L}]^2 [0.010 \frac{mol}{L}]$$

$$12 \frac{L^2}{mol^2 \cdot s} = k$$

(d) Rate when [NO] = 0.015 M and [O_2] = 0.0050 M

$$Rate = k[NO]^2[O_2]$$

$$= 12.5 \frac{L^2}{mol^2 \cdot s} (0.015 \frac{mol}{L})^2 (0.0050 \frac{mol}{L})$$

$$= -1.4 \times 10^{-5} \frac{mol}{L \cdot s}$$

(e) The relation between reaction rate and concentration changes:

$$Rate = -1/2 \cdot \frac{\Delta[NO]}{\Delta t} = -\frac{\Delta[O_2]}{\Delta t} = +1/2 \cdot \frac{\Delta[NO_2]}{\Delta t}$$

So when NO is reacting at $1.0 \times 10^{-4} \frac{mol}{L \cdot s}$ then O_2 will be reacting at

$5.0 \times 10^{-5} \frac{mol}{L \cdot s}$ and NO_2 will be forming at $1.0 \times 10^{-4} \frac{mol}{L \cdot s}$

26. For the reaction $2 NO(g) + O_2 (g) \rightarrow 2 NO_2 (g)$:

(a) The rate law can be determined by examining the effect on the rate by changing the concentration of *either* NO *or* O_2.

In Data sets 1 and 2, the [O_2] doubles, and the rate doubles—a first-order dependence.

In Data sets 2 and 3, the [NO] is halved, and the rate is quartered —a second-order dependence.
The rate law will be: Rate = $k[O_2][NO]^2$

(b) The rate constant is: $3.4 \times 10^{-8} = k[5.2 \times 10^{-3}][3.6 \times 10^{-4}]^2$

Solving for k: k = 50.45 (or 50. $L^2/mol^2 \cdot$h to 2sf).

Note that I selected the data from Experiment 1. Any of the data sets, (1, 2, or 3) would have provided the same value of k.

(c) The initial rate for Experiment 4 is determined by substitution into the rate law (with the value of k determined in (b):

Rate = $50[5.2 \times 10^{-3}][1.8 \times 10^{-4}]^2 = 8.4 \times 10^{-9}$ mol/L $\cdot$ h.

28. For the reaction: $2 CO(g) + O_2 (g) \rightarrow 2 CO_2 (g)$:

   (a) Determine m and n for the expression: Rate $= k[CO]^n[O_2]^m$

      Let's call the 3 experiments i, ii, and iii. Note that the Rate of experiment iii is 4 •Rate

      in experiment i. $\dfrac{1.47 \times 10^{-4}}{3.68 \times 10^{-5}} = 4$. This change in rate was caused by a change in [CO]

      by a factor of **2**. The order of dependence on CO must be second-order.

      Compare experiments i and iii: The ratio of the rates is $\dfrac{7.36 \times 10^{-5}}{3.68 \times 10^{-5}} = 2$. This doubling

      of the rate is occasioned by a change in [O_2] of 2. The order of dependence on O_2 must

      be first-order. So **n= 2** and **m = 1**.

   (b) The order with respect to CO is second order (m=2, yes?), and first order with respect to
      O_2 so the reaction is (2+1) third order overall.

   (c) The rate constant is determined by substituting into the rate expression:
      Rate $= k[CO]^2[O_2]^1$ so, using data from experiment i: $3.68 \times 10^{-5} = k(0.02)^2(0.02)^1$

$$\frac{3.68 \times 10^{-5}}{(0.02)^3} = 4.6\frac{L^2}{mol^2 \bullet min} \text{ or } 5\frac{L^2}{mol^2 \bullet min} \text{ (with 1sf)}$$

Concentration-Time Equations

30. Note that the reaction is first order. We can write the rate expression:
$$\ln\left(\frac{[C_{12}H_{22}O_{11}]}{[C_{12}H_{22}O_{11}]_0}\right) = -kt$$

   Substitute the concentrations of sucrose at $t = 0$ and $t = 2.57$ hr into the equation:

$$\ln\left(\frac{[0.0132 \text{ mol/L}]}{[0.0146 \text{ mol/L}]_0}\right) = -k(2.57 \text{ hr}) \text{ and solve for k to obtain k} = 0.0392 \text{ hr}^{-1}$$

32. Since the reaction is first order, we can write:
$$\ln\left(\frac{[SO_2Cl_2]}{[SO_2Cl_2]_0}\right) = -kt.$$ Given the rate constant, 2.8×10^{-3} min^{-1}, we can calculate the

   time required for the concentration to fall from 1.24×10^{-3} M to 0.31×10^{-3}M

$$\ln\left(\frac{[0.31 \times 10^{-3}\text{M}]}{[1.24 \times 10^{-3} \text{ M}]_0}\right) = -(2.8 \times 10^{-3} \text{ min}^{-1})t.$$

$$\frac{\ln(0.25)}{(-2.8 \times 10^{-3} \text{ min}^{-1})} = t = 495 \text{ min or } 5.0 \times 10^2 \text{ min (to 2 significant figures)}$$

34. The reaction is second order (the exponent for ammonium cyanate in the rate expression is 2). So we use the integrated form of the second-order rate law:

$$\frac{1}{[NH_4NCO]} - \frac{1}{[NH_4NCO]_0} = kt; \text{ and } \left[\frac{1}{[0.180M]}\right] - \left[\frac{1}{[0.229M]}\right] = (0.0113 \frac{L}{mol \bullet min})t.$$

$$\frac{1.189}{(0.0113 \frac{L}{mol \bullet min})} = t \text{ ; Solving for t gives 105 min.}$$

36. (a) Since the reaction is first order, we can write:

$$\ln\left(\frac{[H_2O_2]}{[H_2O_2]_0}\right) = -kt.$$ Given the rate constant, 1.06×10^{-3} min^{-1}, we can calculate the

time required for the concentration to fall from the original concentration to 85% of that value. Note that the concentrations per se are not that critical. Let's assume the initial concentration is 100.M and after the passage of t time the concentration is 85.0M (that's 15% decomposed, yes?)

$$\frac{\ln\left(\frac{85.0}{100}\right)}{1.06 \times 10^{-3} \, min^{-1}} = -t \text{ and solving for the fraction: } \frac{-1.897}{1.06 \times 10^{-3} \, min^{-1}} = -t$$

and t = 153 min (to 3sf).

(b) For 85% of the sample to decompose, we repeat the process, substituting 15.0 for the [H$_2$O$_2$] remaining:

$$\frac{\ln\left(\frac{15.0}{100}\right)}{1.06 \times 10^{-3} \, min^{-1}} = -t \text{ and solving for t = 1790 min (to 3sf).}$$

Half-Life

38. Given that the reaction is first order we can use the integrated form of the rate law:

$$\ln(\frac{[N_2O_5]}{[N_2O_5]_0}) = -kt.$$

(a) Since the **definition of half-life** is "the time required for half of a substance to react", the fraction on the left side = 1/2, and ln(0.50)= - 0.693

Given the rate constant 5.0×10^{-4} s^{-1} we can solve for t:

$- 0.693 = - (5.0 \times 10^{-4}$ s$^{-1})t$ and t = 1.4×10^3 seconds

(b) Time required for the concentration to drop to 1/10 of the original value:

Substitute the ratio 1/10 for the concentration of N$_2$O$_5$:

$\ln(0.10) = - (5.0 \times 10^{-4}$ s$^{-1})t$ and t = 4.6×10^3 seconds

40. Since the decomposition is first order: $\ln \dfrac{[\text{azomethane}]}{[\text{azomethane}]_0} = -kt$

Converting 2.00 g of azomethane to **moles** :

$$2.00 \text{ g azomethane} \cdot \dfrac{1 \text{ mol azomethane}}{58.08 \text{ g azomethane}} = 0.0344 \text{ mol}$$

$$\ln \dfrac{[\text{azomethane}]}{[0.0344 \text{ mol}]_0} = -(40.8 \text{ min}^{-1})(0.0500 \text{ min})$$

$$\ln \dfrac{[\text{azomethane}]}{[0.0344 \text{ mol}]_0} = -2.04$$

$$\dfrac{[\text{azomethane}]}{[0.0344 \text{ mol}]_0} = e^{-2.04} = 0.130$$

$$[\text{azomethane}] = 4.48 \times 10^{-3} \text{ mol}$$

Since 1 mol N_2 is produced when 1 mol of azomethane decomposes, the amount of N_2 formed is: (0.0344 mol - 0.00448 mol) = 0.0300 mol N_2 formed

42. Since this is a first-order process, $\ln \dfrac{[\text{Cu}^{2+}]}{[\text{Cu}^{2+}]_0} = -kt$ and $k = -\dfrac{0.693}{12.70 \text{ hr}}$

What fraction of the copper remains after time, t = 64 hr?

Radioactive decay is a first-order process so we use the equation:

$$\ln \dfrac{[\text{Cu}^{2+}]}{[\text{Cu}^{2+}]_0} = -\dfrac{0.693}{12.70 \text{ hr}} \cdot 64 \text{ hr}$$

$\ln \dfrac{[\text{Cu}^{2+}]}{[\text{Cu}^{2+}]_0} = -3.49$ and $\dfrac{[\text{Cu}^{2+}]}{[\text{Cu}^{2+}]_0} = e^{-3.49}$ or 0.030 or 3.0 % remains (to 2 sf)

44. For first-order kinetics, we know that $\ln\left(\dfrac{[\text{HCO}_2\text{H}]}{[\text{HCO}_2\text{H}]_0}\right) = -kt$.

Substituting into the equation, we solve for **k**.

$\ln\left(\dfrac{[25]}{[100]_0}\right) = -k(72 \text{ s})$ and rearranging to solve for k gives $\dfrac{-1.386}{-72 \text{ s}} = k = 0.01925 \text{ s}^{-1}$

and since we're pursing the $t_{1/2}$, $t_{1/2} = 0.693/k$ so $t_{1/2} = 0.693/0.01925 \text{ s}^{-1} = 36 \text{ s}$.

A **much simpler** route to this answer is to recognize that 1 half-life would consume 50% of the original sample, and the 2nd half-life would consume half of the remaining amount (25%). So two half-lives would result in the consumption of 75% of the original sample—or 1/2(72 s)!

Graphical Analysis of Rate Equations and k

46.   (a) Plot of ln[sucrose] and $\dfrac{1}{[\text{sucrose}]}$ versus time.

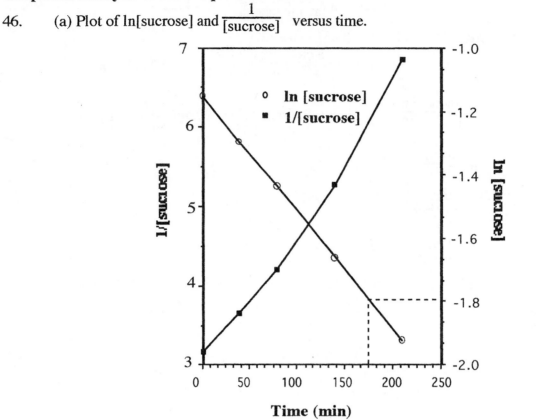

Time (min)

Since the plot of ln[sucrose] vs time gives a straight line, the reaction is first order.

(b) Since the reaction is first order with respect to sucrose(the plot of ln[sucrose] vs t is linear), the rate expression may be written: Rate = k [sucrose]. The rate constant can be calculated using two data points: Using the first two points yields:

$$\ln\left(\frac{[A]}{[A]_0}\right) = - kt \quad \text{and substituting}: \quad \ln\left(\frac{0.274}{0.316}\right) = - k(39 \text{ min})$$

$$\frac{\ln(0.867)}{39 \text{ min}} = - k \quad \text{and } 3.7 \times 10^{-3} \text{ min}^{-1} = k$$

(c) Using the graph of ln[sucrose] vs time, an estimate at 175 minutes yields:

ln[sucrose] = -1.8 corresponding to [sucrose] = 0.167 M

48. For the decomposition of N_2O:

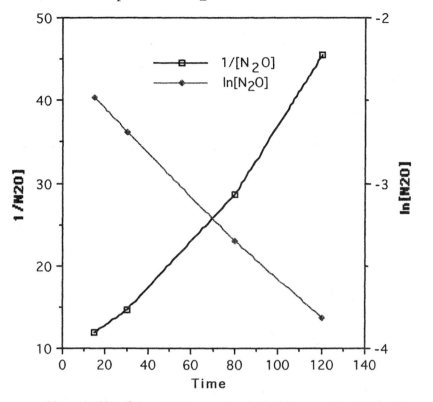

Since **ln[N$_2$O] vs t** gives a **straight line**, we know that the reaction is first order with respect to N_2O, and the line has a slope $= - k$. Taking the natural log (ln) of the concentrations at t = 120 min and t = 15.0 min gives ln(0.0220) = -3.8167; ln(0.0835) = -2.4829.

$$\text{slope} = -k = \frac{(-3.8167) - (-2.4829)}{(120.0 - 15.0)\text{min}} = \frac{1.3338}{105.0 \text{ min}} = 0.0127 \text{ min}^{-1}$$

The rate equation is: Rate $= k\ [N_2O]$ and

The rate of decomposition when $[N_2O] = 0.035$ mol/L :

$$\text{Rate} = (0.0127 \text{ min}^{-1})(0.035 \frac{\text{mol}}{\text{L}}) = 4.4 \times 10^{-4} \frac{\text{mol}}{\text{L} \cdot \text{min}}$$

50. Since the graph of reciprocal concentration gives a straight line, we know the reaction is **second-order** with respect to NO_2. Equation 15.2 (p 642) indicates that the **slope** of the line is k, so k = 1.1 L/mol •s.
    The rate law is Rate= k•$[NO_2]^2$

52. The straight line obtained when the reciprocal concentration of C_2F_4 is plotted vs t indicates that the reaction is second-order in C_2F_4.

The rate expression is: $\text{Rate} = \dfrac{-\Delta[C_2F_4]}{\Delta t} = 0.04 \dfrac{L}{mol \cdot s}[C_2F_4]^2$

Kinetics and Energy

54. The E* for the reaction $N_2O_5 (g) \rightarrow 2\,NO_2 (g) + 1/2\,O_2 (g)$

   Given k at 25 °C = 3.46×10^{-5} s^{-1} and k at 55 °C = 1.5×10^{-3} s^{-1}

   The rearrangement of the Arrhenius equation (in your text as Equation 15.7) is helpful here.

   $$\ln \frac{k_2}{k_1} = -\frac{E^*}{R}\left(\frac{1}{T_2} - \frac{1}{T_1}\right) \; ; \quad \ln \frac{1.5 \times 10^{-3}\,s^{-1}}{3.46 \times 10^{-5}\,s^{-1}} = -\frac{E^*}{8.31 \times 10^{-3}\,kJ/mol \cdot K}\left(\frac{1}{328} - \frac{1}{298}\right)$$

   and solving for E* yields a value of 102 kJ/mol for E*.

56. Using the Arrhenius equation: $\ln \dfrac{k_2}{k_1} = -\dfrac{E^*}{R}\left(\dfrac{1}{T_2} - \dfrac{1}{T_1}\right)$, $T_1 = 800K$, and $T_2 = 850$ K

   Given E* = 260 kJ/mol and $k_1 = 0.0315$ s^{-1}, we can calculate k_1.

   $$\ln \frac{k_2}{0.0315\,s^{-1}} = -\frac{260\,kJ/mol}{8.3145 \times 10^{-3}\,kJ/mol \cdot K}\left(\frac{1}{850\,K} - \frac{1}{800K}\right)$$

   $$\ln \frac{k_2}{0.0315\,s^{-1}} = 2.30 = \ln k_2 - \ln(0.0315\,s^{-1})$$

   $2.30 + \ln(0.0315\,s^{-1}) = \ln k_2 = 2.30 - 3.458 = -1.16$

   $k_2 = e^{-1.16} = 0.3$ s^{-1} (1 sf owing to a temperature (800K) with 1sf)

58. Energy progress diagram:

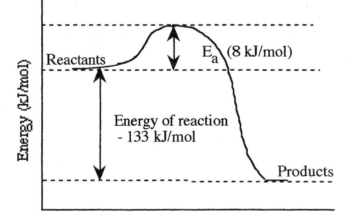

Reaction progress

Mechanisms

60. Elementary Step   Rate law

   (a) $NO (g) + NO_3 (g) \rightarrow 2\,NO_2 (g)$ Rate = k[NO][NO$_3$]

                                                      Reaction is bimolecular

(b) $Cl\ (g) + H_2\ (g) \rightarrow HCl\ (g) + H\ (g)$ Rate $= k[Cl][H_2]$

Reaction is bimolecular

(c) $(CH_3)_3CBr\ (aq) \rightarrow (CH_3)_3C^+\ (aq) + Br^-\ (aq)$ Rate $= k[(CH_3)_3CBr]$

Reaction is unimolecular

62. For the reaction reflecting the decomposition of ozone:

(a) The second step is the slow step, and therefore rate-determining.

(b) The rate equation involves *only* these substances that affect the rate, (since they participate in the rate determining step), Rate $= k[O_3][O]$.

64.   (a) Add the elementary steps:

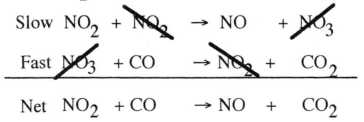

$$H_2O_2\ (aq)\ + I^- \longrightarrow H_2O\ (l)\ +\ \cancel{OI^-}(aq)$$

$$H^+\ (aq)\ +\ \cancel{OI^-}(aq) \longrightarrow \cancel{HOI}\ (aq)$$

$$\underline{\cancel{HOI}\ (aq) + H^+\ (aq) + I^-\ (aq) \rightarrow I_2\ (aq)\ +\ H_2O\ (l)}$$

$$H_2O_2\ (aq)\ +2\ I^-\ + 2H^+\ (aq) \rightarrow I_2\ (aq)\ +\ 2\ H_2O\ (l)$$

Note that (on the left side of the equations) the I⁻ and H⁺ ions "add". On the right side of the equations, H₂O "adds". HOI and OI⁻ "cancel", giving the overall stoichiometric equation.

(b) Molecularity: 1ˢᵗ eqn: bimolecular; 2ⁿᵈ eqn: bimolecular; 3ʳᵈ eqn: termolecular. In elementary steps, each species represents a molecule, so with two molecules as reactants—the equation is bimolecular, 3 molecules—termolecular, etc.

(c) To be consistent with kinetic data, the experimental rate equation should be: Rate $= k[H_2O_2][I^-]$—first order in both peroxide and iodide.(the **slow** step)

(d) Intermediates are species which are produced in one step and consumed in a subsequent step. In this mechanism, HOI and OI⁻ are intermediates.

66. For the reaction of NO₂ and CO:

$$\text{Slow}\quad NO_2\ +\ \cancel{NO_2} \rightarrow NO\ +\ \cancel{NO_3}$$

$$\underline{\text{Fast}\quad \cancel{NO_3}\ +\ CO\ \rightarrow \cancel{NO_2}\ +\ CO_2}$$

$$\text{Net}\quad NO_2\ +\ CO\ \rightarrow NO\ +\ CO_2$$

Note that when the two steps are added, the desired overall equation results.

(a) Classify the species:

NO$_2$(g)	Reactant (step 1); Product (step2)
CO (g)	Reactant (step 1)
NO$_3$ (g)	Intermediate (produced & consumed subsequently)
CO$_2$ (g)	Product
NO (g)	Product

(b) A reaction coordinate diagram

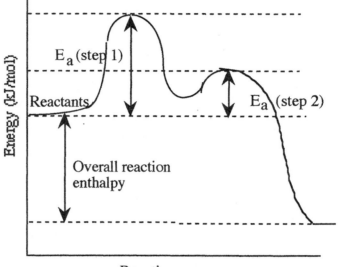

Reaction progress

General Questions

68. To determine second-order dependence, after acquisition of the pH vs time data, plot $1/[OH^-]$ versus time. The reaction is second-order in OH$^-$ if a straight line is obtained,

70. Regarding the following statements:
    (a) **Incorrect**; Reactions are faster at a higher T because the reactants possess more energy—leading to more frequent collisions.
    (b) **Correct**—See Figure 15.11
    (c) **Correct**—See Figure 15.12
    (d) **Incorrect**—Catalysts provide a different pathway with a lower Energy of activation

72. For the formation of ammonia from nitrogen and hydrogen:
    (a) Determine **n** and **m** for the equation: Rate = $k[N_2]^n[H_2]^m$

    Label the data in the table as experiments 1,2 and 3. Compare experiments 1 and 2. Note that while the concentration of N_2 doubles in experiment 2 compared to experiment 1. The Rate for experiment 2 is 4 times that of experiment 1. This is true if the order of dependence on nitrogen is second-order—**n=2**. Compare experiments 1 and 3. The concentration of N_2 is fixed, and the concentration of H_2 changes by a factor of 2. The rate for experiment 3 is 8 times the rate for experiment 1. This is true if the order of dependence on hydrogen is third-order—**m=3**.

    (b) With values of m and n known, substitute into *any* of the three experiments. I'm choosing experiment 1:
    Rate = $k[N_2]^n[H_2]^m$; $4.21 \times 10^{-5} = k(0.03)^2(0.01)^3$. Solve for k:
    $$\frac{4.21 \times 10^{-5}}{(0.03)^2(0.01)^3} = 5 \times 10^4 \frac{L^4}{mol^4 \cdot min} \text{ (to 1 sf)}$$

    (c) The order of dependence on H_2 is third order.

    (d) The overall order is the sum of the orders of dependence: 2+3= **5th** order overall.

74. For the formation of NOBr from NO and Br_2:
    (a) Determine the order of the reaction with respect to NO:
    Compare the rates and concentrations of NO in experiments 1 and 2. While the concentration of NO quadrupled (4 x) the rates (0.384/0.024) changed by a factor of 16. (16 x). The order of dependence on NO is second-order.

    (b) Determine the order of the reaction with respect to Br_2:
    To determine the order of dependence on bromine, we need two experiments in which the [NO] doesn't change, while the [Br_2] changes. Experiments 1 and 3 fit that requirement. The concentration of Br_2 changes by a factor of 2.5, the respective rates change by ($6.0 \times 10^{-2}/2.4 \times 10^{-2}$) or a factor of 2.5. The order of dependence on Br_2 is first-order.

    (c) The overall order is the sum of the orders of dependence: 2+1=3rd order overall.

76. For the reaction of cis-2-butene to trans-2-butene, the reaction coordinate diagram:

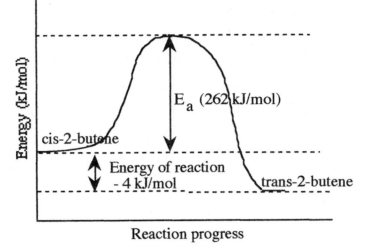

Reaction progress

78. For the dimerization to form octafluorocyclobutane, the following plot is obtained:

(a) The plot of reciprocal concentration vs time gives a straight line. Equation 15.2 tells us that such behavior is indicative of **a second-order process.**

The rate law is: Rate = $k[C_2F_4]^2$

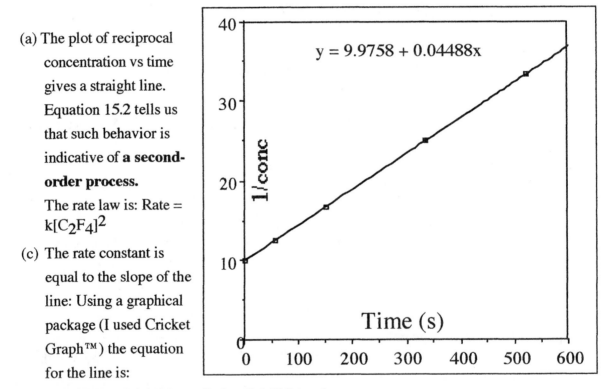

(c) The rate constant is equal to the slope of the line: Using a graphical package (I used Cricket Graph™) the equation for the line is:

y=9.9758 + 0.04488x.   So k = 0.045 L/mol•s

(d) The concentration after 600 s is found by using the integrated rate equation for second-order processes:

$$\frac{1}{[C_2F_4]_t} - \frac{1}{[(0.100)]_0} = 0.045\frac{L}{mol \cdot s}(600\ s)$$

$$\frac{1}{[C_2F_4]_t} = 0.045\frac{L}{mol \cdot s}(600\ s) + \frac{1}{0.100}$$

and the $[C_2F_4]_{600} = 0.027$ (or 0.03 M to 1sf)

(d) Time required for 90% completion: Using the same equation as in part (c), and substituting 10% of the initial concentration as our "concentration at time t" , we can solve for t.

$$\frac{1}{[(0.010)]_t} - \frac{1}{[(0.100)]_0} = 0.045 \frac{L}{mol \cdot s}(t)$$

$$(100 - 10) = 0.045 \frac{L}{mol \cdot s}(t) \text{ and } t = 2000 \text{ s}$$

80. For the formation of urea from ammonium cyanate:

(a) Plot the data to determine the order.

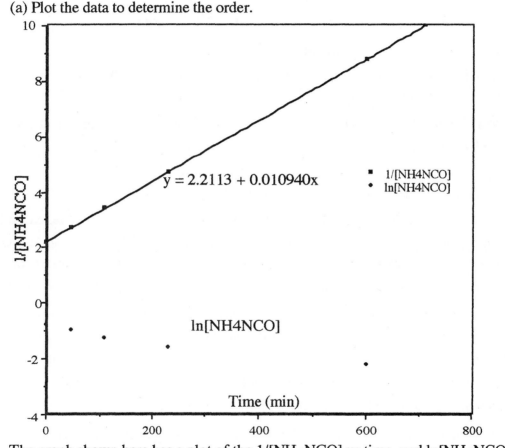

The graph shown here has a plot of the $1/[NH_4NCO]$ vs time, and $\ln[NH_4NCO]$ versus time. Note that the plot of reciprocal concentration gives a straight line, indicating the reaction is second-order in ammonium cyanate.

(b) k is the slope of the line, which is 0.0109 L/mol•min

(c) The half-life can be calculated using the integrated rate equation (as in SQ15.78)

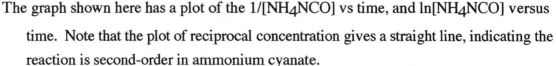

$$\frac{1}{[0.229]_t} - \frac{1}{[0.458]_0} = 0.0109 \frac{L}{mol \cdot min}(t)$$

The concentration at time t is 1/2 that of the original concentration of NH_4NCO (the definition of half-life).

$$4.367 - 2.183 = 0.0109 \frac{L}{mol \bullet min}(t) \text{ and solving for } t = 200. \text{ minutes.}$$

(d) The concentration of ammonium cyanate after 12.0 hours (720. min) is found by using the integrated rate equation. Since we know k and t, we can solve for "concentration at time t =(12.0 hours) "

$$\frac{1}{[NH_4NCO]_t} - \frac{1}{[0.458]_0} = 0.0109 \frac{L}{mol \bullet min}(t)$$

$$\frac{1}{[NH_4NCO]_t} = 0.0109 \frac{L}{mol \bullet min}(720. \text{ min}) + \frac{1}{[0.458]_0}$$

Solving for $[NH_4NCO]$ we obtain $[NH_4NCO] = 0.0997$ M

82. The reaction between carbon monoxide and nitrogen dioxide has a rate equation that is second-order in NO_2 This means that the *slowest step* in the mechanism involves **2** molecules of nitrogen dioxide. Mechanism 2 has a SLOW step that fulfills this requirement. Note that Mechanisms 1 and 3 are only 1st order in nitrogen dioxide.

84. For the formation of nitryl fluoride by nitrogen dioxide reacting with fluorine:

(a) The Rate equation is found by comparing experiments in which only one reactant concentration is varied between experiments and observing the differences in rates between the two experiments. Experiments 1 and 2 maintain a concentration of F_2 that is constant, while the concentration of NO_2 changes by a factor of 2. The rate for experiment 2 is two times that for experiment 1, indicating a first-order dependence on NO_2. Experiments 3 and 4 provide two data points which have a constant concentration for NO_2 while the $[F_2]$ changes by a factor of 2. The rate for experiment 4 is two times that for experiment 3, indicating a first-order dependence on F_2 as well.

The rate equation is : Rate = $k[NO_2] [F_2]$

(b) As found in part (a), the order of dependence is first order for both reactants.

(c) The rate constant can be found by substituting data from an experiment (your choice) into the Rate expression, and solving for k. I'll pick experiment 1—although any one of the experiments should yield the same value for k.

$2 \times 10^{-4} = k(0.001)^1 (0.005)^1$ for which k = 40 L/mol•s

86. The decomposition of dinitrogen pentaoxide has a first-order rate equation. Determine the rate constant and the half-life by substitution into the integrated rate equation for first-order reactions (Equation 15.1).

$$\ln\left(\frac{[N_2O_5]_t}{[N_2O_5]_0}\right) = -k \bullet t$$ The decomposition if 20% complete in 6.0 hours. The

amount of N2O5 remaining after 6.0 h is 80% of the original concentration. The left hand term is then 80/100.

$$\ln\left(\frac{80}{100}\right) = -k \bullet 6.0 \text{ h and } k = 0.037 \text{ h}^{-1}$$

Calculation of the half-life is accomplished by noting that the left-hand side of the equation has a value of 50% --and $\ln(0.50) = -0.693$. Substituting the value of k we calculated above:

$$\frac{-0.693}{-0.037 \text{ hr}^{-1}} = 19 \text{ h}$$

88. For the decomposition of dimethyl ether:

(a) The mass of dimethyl ether remaining after 125 min and after 145 min:

The half-life is 25.0 min. A period of 125 minutes is 5 half-lives. The fraction remaining after n half-lives is $\left(\frac{1}{2}\right)^n$ and with n = 5, the fraction remaining is 0.03125 (1/32 of the original amount). The mass remaining is $(0.03125)(8.00 \text{ g}) = 0.251$ g dimethyl ether. Note that 145 minutes is *almost* 6 half-lives. So you should be able to "guess" at a value for the amount remaining. The mass should be slightly greater than 1/64 of the original amount. The exact amount can be found by substitution into the first-order rate equation to solve for the rate constant:

$$\frac{-0.693}{25.0 \text{ min}} = -k \text{ and } k = 0.0277 \text{ min}^{-1}$$

Note that the "ln term" is simplified by remembering that a "half-life" is a time for which 50% decomposes. Then $\ln(0.50) = -0.693$

{A HANDY THING TO REMEMBER}

Substituting our value of k into the equation and solving for the fraction remaining:

$$\ln\left(\frac{[\text{dimethyl ether}]_t}{[\text{dimethyl ether}]_0}\right) = -0.0277 \text{ min}^{-1} \bullet 145 \text{ min} = -4.020$$

For simplicity, let's represent the fraction remaining as x. Solving for the ratio of concentrations gives $\ln(x) = -4.020$, for which x = 0.0179

The mass of dimethyl ether remaining is $(0.0179)(8.00 \text{ g}) = 0.144$ g

(b) The time required for 7.60 ng of ether to be reduced to 2.25 ng:

Substitute into the first order equation (as we've done above). Now we know the value for the "left side" and we know k; we can solve for t

$$\frac{\ln\left(\frac{[2.25\ ng]_t}{[7.60\ ng]_0}\right)}{-0.0277\ min^{-1}} = t = \frac{\ln(0.296)}{-0.0277\ min^{-1}} = \frac{-1.217}{-0.0277\ min^{-1}} = 43.9\ min$$

The fraction remaining after 150 minutes is easily calculated by noting that 150 minutes if *exactly* 6 half-lives. The fraction remaining is 1/64 or 0.016 (to 2sf)

90. For the reaction of NO with Br_2 to give BrNO:

(a) The balanced equation is: $2\ NO\ (g) + Br_2\ (g) \rightarrow 2\ BrNO\ (g)$

(b) The molecularity for each step in each mechanism:

Mechanism 1	Molecularity
	Termolecular (3 reactant species)
Mechanism 2	
Step 1	Bimolecular (2 reactant species)
Step 2	Bimolecular (2 reactant species)
Mechanism 3	
Step 1	Bimolecular (2 reactant species)
Step 2	Bimolecular (2 reactant species)

(c) Intermediates formed in Mechanisms 2 and 3:
In mechanism 2, Br_2NO is produced in step 1 and consumed in step 2—not appearing as reactant or product (the definition of intermediate). In mechanism 3, N_2O_2 fits that description.

(d) Comparison of rate laws from the three mechanisms:
For mechanisms 2 and 3, there are two steps and no indication of the "slow" step. For purposes of our discussion, assume that the first step in each mechanism is the "slow" step.

Mechanism 1	2nd order in NO and 1st order in Br_2
Mechanism 2	1st order in NO and 1st order in Br_2
Mechanism 3	2nd order in NO and 0 order in Br_2

92. Show consistency of mechanism with the rate law: Rate = $k[O_3]^2/[O_2]$:

The rate law for the **slow** step is Rate=$k[O_3][O]$, but the concentration of O is affected by the preceding equilibrium step. Solve for the [O] in that equilibrium step:

The equilibrium constant expression for the fast step is:

$$K = \frac{[O_2][O]}{O_3}; \text{solving for } [O] = \frac{K \cdot [O_3]}{[O_2]}$$

Substitute this concentration into the Rate law for the slow step above:

$$\text{Rate=}k[O_3][O], ; \text{Rate} = \frac{K \cdot k[O_3][O_3]}{[O_2]}$$

Combining $K \cdot k$ as k', $\text{Rate} = \frac{k'[O_3]^2}{[O_2]}$

94. Finely divided rhodium has a larger surface area than a block of the metal with the same mass. Since hydrogenation reactions depend on adsorption of H_2 on the catalyst surface, the greater the surface area, the greater the locations for such adsorption to occur.

96. Assume that you begin with labeled oxygen in $CH_3O^{18}H$. If we represent the labeled oxygen with a ●, you can see from the equation below that labeled oxygen (in the water) results **if the O originated from the methanol.**

$$CH_3\overset{\overset{O}{\|}}{C}O\underline{H} + \underline{H}\overset{●}{O}CH_3 \longrightarrow CH_3\overset{\overset{O}{\|}}{C}OCH_3 + \underline{H}\overset{●}{O}\,H$$

98. Equation 15.7 (page 642) allows us to calculate Ea if we know values of rate constants at two temperatures: $\ln\dfrac{k2}{k1} = -\dfrac{Ea}{R}\left(\dfrac{1}{T_2} - \dfrac{1}{T_1}\right)$

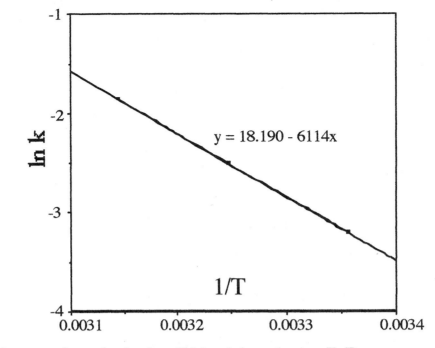

The slope was determined to be -6114 and since slope $= -Ea/R$,

$$-6114 = -\frac{Ea}{8.314 \times 10^{-3}\ \dfrac{kJ}{K \bullet mol}} = 51\ kJ/mol$$

100. $Ea = 103$ kJ, and $k = 0.0900\ min^{-1}$ at 328.0 K. k at 318.0 K is:

Using one form of the Arrhenius equation: $\ln\dfrac{k2}{k1} = -\dfrac{Ea}{R}\left(\dfrac{1}{T_2} - \dfrac{1}{T_1}\right)$

Substituting: $\ln\dfrac{k2}{k1} = -\dfrac{103\ kJ}{8.314 \times 10^{-3}\ \dfrac{kJ}{K \bullet mol}}\left(\dfrac{1}{318.0\ K} - \dfrac{1}{328.0\ K}\right)$

Solving for $\ln k_2/k_1 = -1.188$ and $k_2/k_1 = 0.3049$

With $k_1 = 0.0900$ then $k_2 = 0.0274\ min^{-1}$

102. The net reaction (obtained by adding the 3 steps and removing intermediates is:

$$HA + X \rightarrow A^- + products$$

Step 3 is the slow step and the Rate equation should be: Rate $= k[XH^+]$

Write equilibrium expressions for Step 1 and Step 2:

$$K_1 = \frac{[H^+][A^-]}{[HA]} \text{ and } K_2 = \frac{[XH^+]}{[H^+][X]}$$

Since $[XH^+]$ is controlled by K_2, solve for that concentration :

$$K_2[H^+][X] = [XH^+] \qquad \text{(expression 1)}$$

$[H^+]$ is controlled by the equilibrium controlled by K_1

Solve the K_1 expression for $[H^+]$: $\dfrac{K_1[HA]}{[A^-]} = [H^+]$;

Substitute for $[H^+]$ in expression 1: $K_2 \dfrac{K_1[HA]}{[A^-]}[X] = [XH^+]$

Substitute this expression into the rate equation to obtain: Rate $= kK_2 \dfrac{K_1[HA]}{[A^-]}[X]$

The order of the reaction with respect to HA is first-order.

Doubling [HA] would *double the rate*, since the reaction is first-order dependent on HA.

104. Since 245 min is the half-life of this decomposition, we can quickly determine the partial pressures of each of the gases without a calculator! The stoichiometry of the reaction indicates that 1 mole of SO_2 and 1 mole of Cl_2 is produced when 1 mole of SO_2Cl_2 decomposes.

Substance	Initial Pressure	Pressure after 245 minutes
SO_2Cl_2	25 mm Hg	12.5 mm Hg
SO_2	0 mm Hg	12.5 mm Hg
Cl_2	0 mm Hg	12.5 mm Hg

The pressure after 12.0 hrs (720 minutes) will require a calculation, since this time is not a multiple of the half-life.

The rate constant is determined by the equation: $\ln\left(\dfrac{[SO_2Cl_2]_t}{[SO_2Cl_2]_0}\right) = -k \bullet t$

We know that the half-life is 245 min so we can substitute 245 minutes for t, and the fraction 1/2 for the left-hand term, to obtain: $\ln(0.50) = -k \bullet 245$ min

Solving for $k = 2.83 \times 10^{-3}$ min^{-1}

Use k to solve for the fraction remaining (the left hand side) at a time = 720 minutes.

$\qquad \ln(x) = -(2.83 \times 10^{-3} \text{ min}^{-1})720 \text{ min} = -2.04$ and $x = 0.130$

So 13% of the SO_2Cl_2 remains, and the partial pressure is $0.13 \bullet 25$ mm Hg = 3.3 mm Hg

The partial pressure of the products is then 87% of 25mm Hg or 21.7 mm Hg.

The total pressure is $(21.7 + 21.7 + 3.3)$mm Hg = 46.7 mm Hg.

Using Electronic Resources

106. (a) The difference between an instantaneous rate and an average rate is similar to the situation you encounter when making a trip. The instantaneous rate is the reading of your speedometer as you glance down at the meter. The average rate is obtained by deciding what distance you traveled over several hours. With chemical reactions, the

situation is identical. The instantaneous rate begins (with the inception of the reaction) at its maximum value, and continues to decrease as reactants are consumed. The average rate is a measure of how long the reaction requires to consume a certain amount of the reactant(s).

(b) The steepness of the plot is an indication of the *instantaneous rate*, with that rate decreasing as the reaction proceeds.

(c) As noted above, as the concentration of dye decreases, the reaction rate slows. The rate of a reaction is proportional to the concentration of reactants!

108. (a) The necessity for collecting data at the *beginning* of the reaction is owing to the fact that the rate slows as the concentrations of reactants decrease. Additionally products may interfere with the reaction mechanism.

(b) The rate law is: Rate=$k[NH_4NCO]^2$ If the concentration of NH_4NCO is 0.18M and

$k = 0.011$ L/mol•min, the rate is $(0.011$ L/mol•min$)(0.18$mol/L$)^2$

$= 3.6 \times 10^{-4}$ mol/L•min .

110. (a) 3270 minutes corresponds to 5 half-lives. So the fraction remaining is $(1/2)^5$ or 1/32 of the original amount. The concentration at that time is then $(0.03125)• 0.020$ M

or 6.25×10^{-4} M (or 6.3×10^{-4} M to 2 sf) After 3924 minutes (6 half-lives) there will be 1/64 of the original amount or 3.1×10^{-4} M to 2 sf.

(b) The fractions remaining are 1/32 and 1/64 respectively.

112. The difference between these two reactions is the presence (for the two-step reaction) of two transition states (with their respective energies of activation) The one-step (or bimolecular substitution reaction) reaction has but a single transition state.

114. (a) The difference between an overall mechanism and an elementary step resides in the fact that the overall mechanism contains the reactants and the products (with all intermediates omitted). The elemental steps include such transitory species.

(b) The stoichiometric coefficients for an elementary step correspond one-to-one with the *order of dependence* for that specie. If, for example, the coefficient of specie X is 2, then the order of dependence of X is second-order.

(c) The rate law for Step 2 of Mechanism 2 is: Rate = $k[NO_3][CO]$

(d) No such $N^{16}O^{18}O$ would be formed. Only if isotopic scrambling of O atoms occurred would this isotope be formed. If CO reacted with the NO_2 molecules directly, then the NO produced would be either $N^{16}O$ or $N^{18}O$.

Chapter 16
Principles of Reactivity: Chemical Equilibria

Reviewing Important Concepts

2. Will $PbCl_2$ or PbF_2 have a greater concentration of Pb^{2+} ion in an aqueous solution?

   The equilibrium expressions for the solids are shown. Recall the concept of the equilibrium constants. You won't have to do the calculation to answer this question! Note that K for $PbCl_2$ has a larger value than for the fluoride. The result is that the concentrations of the "right hand species" will be greater for the chloride than for the fluoride. So the beaker with $PbCl_2$ will have a greater concentration of Pb^{2+}.

3. The decomposition of $CaCO_3$ is endothermic. Heat is *absorbed* as the carbonate decomposes.

   The addition of heat (increasing the temperature) will favor the absorption of that heat (favors the endothermic process) and shift the equilibrium *toward the right*.
   The addition of more solid will *not* affect the position of equilibrium (recall that solids do not appear in the equilibrium expression).
   If additional CO_2 is added to the flask, the equilibrium will accelerate the reaction which absorbs CO_2—shift the equilibrium *to the left*.

4. For the oxidation of NO to NO_2, heat is evolved upon formation of the dioxide.

   (a) LeChatelier's principle states that removal of heat will tend to move the equilibrium position *toward* the production of heat (formation of the dioxide). The equilibrium will shift *to the right* when the temperature is lowered.
   (b) LeChatelier's principle also states that a decrease in pressure favors the side of the equilibrium with the *greater total number* of moles of gaseous substances. An increase in the volume of the flask (and a concomitant reduction in pressure) will favor the *left* side of the equilibrium—the system will shift to the left.

6. Characterize the reactions as product-favored or reactant-favored:

   Product-favored reactions are those with $K > 1$, reactant-favored reactions have a $K < 1$.
   (a) $CO (g) + 1/2\ O_2 (g) \rightleftharpoons CO_2 (g)$ $K = 1.2 \times 10^{45}$ *Product-favored*
   (b) $H_2O (g) \rightleftharpoons H_2 (g) + 1/2\ O_2 (g)$ $K = 9.1 \times 10^{-41}$ *Reactant-favored*
   (c) $CO (g) + Cl_2 (g) \rightleftharpoons COCl_2 (g)$ $K = 6.5 \times 10^{11}$ *Product-favored*

Practicing Skills

Writing Equilibrium Constant Expression

8. Equilibrium constant expressions:

(a) $K = \dfrac{[H_2O]^2[O_2]}{[H_2O_2]^2}$ (b) $K = \dfrac{[CO_2]}{[CO][O_2]^{1/2}}$ (c) $K = \dfrac{[CO]^2}{[CO_2]}$ (d) $K = \dfrac{[CO_2]}{[CO]}$

Note that **solids** in the equations are *omitted* in the equilibrium constant expressions.

The Reaction Quotient

10. The equilibrium expression for the reaction: $I_2\,(g) \rightleftharpoons 2\,I\,(g)$ has a $K = 5.6 \times 10^{-12}$

Substituting the molar concentrations into the equilibrium expression:

$$Q = \frac{[I]^2}{[I_2]} = \frac{(2.00 \times 10^{-8})^2}{(2.00 \times 10^{-2})} = 2.0 \times 10^{-14}$$

Since $Q < K$, the system is not at equilibrium and will move to the right to make more product (and reach equilibrium).

12. Is the system at equilibrium?

[SO2]	5.0 x 10-3 M
[O2]	1.9 x 10-3 M
[CO3]	6.9 x 10-3 M

Substituting into the equilibrium expression: $\dfrac{[SO_3]^2}{[SO_2]^2[O_2]} = \dfrac{\left[6.9 \times 10^{-3}\right]^2}{\left[5.0 \times 10^{-3}\right]^2\left[1.9 \times 10^{-3}\right]} = 1000$

Since $Q > K$, the system is **not** at equilibrium, and will move "to the left" to make more reactants to reach equilibrium.

Calculating an Equilibrium Constant

14. For the equilibrium: $PCl_5\,(g) \rightleftharpoons PCl_3\,(g) + Cl_2\,(g)$, calculate K

The equilibrium concentrations are: $[PCl_5] = 4.2 \times 10^{-5}$ M

$[PCl_3] = 1.3 \times 10^{-2}$ M

$[Cl_2] = 3.9 \times 10^{-3}$ M

The equilibrium expression is : $\dfrac{[PCl_3][Cl_2]}{[PCl_5]} = \dfrac{[1.3 \times 10^{-2}][3.9 \times 10^{-3}]}{[4.2 \times 10^{-5}]} = 1.2$

16. The equilibrium expression is : $K = \dfrac{[CO]^2}{[CO_2]}$

The quantities given are **moles**, so first calculate the **concentrations at equilibrium**.

$$[CO] = \frac{0.10 \text{ moles}}{2.0 \text{ L}} = 0.050 \text{ M} \qquad [CO_2] = \frac{0.20 \text{ moles}}{2.0 \text{ L}} = 0.10 \text{ M}$$

(a) $K = \dfrac{[CO]^2}{[CO_2]} = \dfrac{[0.050]^2}{[0.10]} = 2.5 \times 10^{-2}$

(b & c) The only change here is in the amount of carbon. Since C does not appear in the equilibrium expression, K would be the same as in (a).

18. To keep track of concentrations, construct a table:

The reaction is $CO \ (g) + Cl_2 \ (g) \rightleftharpoons COCl_2(g)$

	CO	Cl_2	$COCl_2$
Initial	0.0102 M	0.00609 M	0 M
Change	- 0.00308 M	- 0.00308 M	+ 0.00308 M
Equilibrium	0.00712 M	0.00301M	+ 0.00308 M

(a) Equilibrium concentrations of CO and $COCl_2$ is found by noting that 0.00308 M Cl_2 is consumed, an equimolar amount of CO is consumed and an equal amount of $COCl_2$ is produced.

(b) $K = \dfrac{[+0.00308]}{[0.00712][0.00301]} = 144$

Using Equilibrium Constants

20. For the system butane $\rightleftharpoons$ isobutane K = 2.5

Equilibrium concentrations may be found using a table. First note that the amount of butane must be converted to molar concentrations. So 0.017mol butane/0.50 L = 0.034 M.

	butane	isobutane
Initial	0.034	0
Change	- x	+ x
Equilibrium	0.034 - x	x

Substituting these equilibrium concentrations into the equilibrium expression:

$$K = \frac{x}{0.034 - x} = 2.5 \quad \text{and } x = 2.4 \times 10^{-2}$$

and [isobutane] $= 2.4 \times 10^{-2}$ M, [butane] $= 0.034 - x = 1.0 \times 10^{-2}$ M

22. For the equilibrium, the equilibrium constant is $\dfrac{[I]^2}{[I_2]} = 3.76 \times 10^{-3}$

The initial concentration of I_2 is $\dfrac{0.105 \text{mol}}{12.3 \text{ L}}$ or 8.54×10^{-3} M

The equation indicates that 2 mol of I form for each mol of I_2 that reacts.

If some amount, say x M, of I_2 reacts, then the amount of I that forms is 2x, and the amount of I_2 remaining at equilibrium is then $(8.54 \times 10^{-3} - x)$

The equilibrium concentrations can then be substituted into the equilibrium expression.

$$\frac{(2x)^2}{8.54 \times 10^{-3} - x} = 3.76 \times 10^{-3}$$

or $4x^2 = 3.76 \times 10^{-3}(8.54 \times 10^{-3} - x)$ rearranging we get:

$$4x^2 + 3.76 \times 10^{-3} x - 3.21 \times 10^{-5} = 0.$$

Solve this using the quadratic equation. Using the positive solution to that equation we find that $x = 2.40 \times 10^{-3}$.

So at equilibrium:

$$[I_2] = 8.54 \times 10^{-3} - 2.40 \times 10^{-3} = 6.14 \times 10^{-3} \text{ M}$$

$$[I] = 2(2.40 \times 10^{-3}) = 4.79 \times 10^{-3} \text{ M}$$

24. Given at equilibrium : $K = \dfrac{[CO][Br_2]}{[COBr_2]} = 0.190$ at $73°C$

First, calculate concentrations: $[COBr_2] = \dfrac{0.500 \text{ mol}}{2.00 \text{ L}} = 0.250$ M

Substituting the $[COBr_2]$ into the equilibrium expression we get: $\dfrac{[CO][Br_2]}{(0.250)} = 0.190$

Note that the stoichiometry of the equation tells us that for **each CO, we obtain 1 Br$_2$**.

	COBr$_2$	CO	Br2
Initial	0.250 M	0 M	0 M
Change	-x	+ x	+ x
Equilibrium	0.250 -x	+ x	+ x

We can rewrite the expression to read: $\dfrac{[CO][Br_2]}{0.250 - x} = \dfrac{[x]^2}{0.250 - x} = 0.190$

Solve this using the quadratic equation. $x^2 + 0.190x - 0.0475 = 0$

Using the positive solution to that equation we find that $x = 0.143$ M $= [CO]$ and $[COBr_2] = (0.250-0.143) = 0.107$ M

The percentage of COBr$_2$ that has decomposed is: $\dfrac{0.143 \text{ M}}{0.250 \text{ M}} \cdot 100 = 57.1 \%$

Balanced Equations and Equilibrium Constants

26. To compare the two equilibrium constants, write the equilibrium expressions for the two.

$$\frac{[C]^2}{[A][B]} = K_1 \quad \text{and} \quad \frac{[C]^4}{[A]^2[B]^2} = K_2 \quad \text{Note that the first expression is equal to the second}$$

expression *squared*. So (b) $K_2 = K_1{}^2$

28. Comparing the two equilibria:

(1) SO_2 (g) + $\frac{1}{2}$ O_2 (g) $\rightleftharpoons$ SO_3 (g) K_1

(2) $2\,SO_3$ (g) $\rightleftharpoons$ $2\,SO_2$ (g) + O_2 (g) K_2

the expression that relates K_1 and K_2 is (c) : $K_2 = \dfrac{1}{K_1^2}$

Reversing equation 1 gives: SO_3 (g) $\rightleftharpoons$ SO_2 (g) + $\frac{1}{2}$ O_2 (g) with the eq. constant $\dfrac{1}{K_1}$

Multiplying the modified equation 1 x 2 gives: $2\,SO_3$ (g) $\rightleftharpoons$ $2\,SO_2$ (g) + O_2 (g) and the

modified eq. constant $(\dfrac{1}{K_1})^2$

30. Calculate K for the reaction:
SnO_2 (s) + 2 CO (g) $\rightleftharpoons$ Sn (s) + 2 CO_2 (g) given:

(1) SnO_2 (s) + 2 H_2 (g) $\rightleftharpoons$ Sn (s) + 2 H_2O (g)   K = 8.12
(2) H_2 (g) + CO_2 (g) $\rightleftharpoons$ H_2O (g) + CO (g)   K = 0.771

Take equation 2 and reverse it, then multiply by 2 to give:

$2H_2O$ (g) + 2 CO (g) $\rightleftharpoons$ 2 H_2 (g) + 2 CO_2 (g) $K = (\dfrac{1}{0.771})^2 = 1.68$

add equation 1 : SnO_2 (s) + 2 H_2 (g) $\rightleftharpoons$ Sn (s) + 2 H_2O (g)   K = 8.12

_____

SnO_2 (s) + 2 CO (g) $\rightleftharpoons$ Sn (s) + 2 CO_2 (g) K_{net} = 8.12 • 1.68 = 13.7

Disturbing a Chemical Equilibrium

32. The equilibrium may be represented :
N_2O_3 (g) + heat $\rightleftharpoons$ NO_2 (g)+ NO (g)

Since the process is **endothermic** (ΔH = +), heat is absorbed in the "left to right" reaction.
The effect of:

(a) Adding more N_2O_3 (g): An increase in the Pressure of N_2O_3 (adding more N_2O_3)
   will shift the equilibrium to the **right**, producing more NO_2 and NO.

(b) Adding more NO_2 (g): An increase in the Pressure of NO_2 (adding more NO_2) will
   shift the equilibrium to the **left**, producing more N_2O_3 .

(c) Increasing the volume of the reaction flask: If the volume of the flask is increased, the
   pressure will drop. (Remember that P for gases is inversely related to volume.) A drop
   in pressure will favor that side of the equilibrium with the "larger total number of moles
   of gas" — so this equilibrium will shift to the **right**, producing more NO_2 and NO.

(d) Lowering the temperature: You should note that a change in T (up or down) will result in a **change in the equilibrium constant**. (None of the three changes mentioned above change K!) However, the same principle applies. The removal of heat (a decrease in T) favors the exothermic process (shifts the equilibrium to the **left**) producing more N_2O_3

34. K for butane $\rightleftharpoons$ iso-butane is 2.5.

(a) Equilibrium concentration if 0.50 mol/L of iso-butane is added:

	butane	iso-butane
Original concentration	1.0	2.5
Change immediately after addition	1.0	2.5 + 0.50
Change (going to equilibrium)	+ x	- x
Equilibrium concentration	1.0 + x	3.0 - x

Substituting into the equilibrium expression: $K = \dfrac{[\text{iso-butane}]}{[\text{butane}]} = \dfrac{3.0 - x}{1.0 + x} = 2.5$

$$3.0 - x = 2.5 (1.0 + x) \quad \text{and } 0.14 = x$$

The equilibrium concentrations are:

$$[\text{butane}] = 1.0 + x = 1.1 \text{ M and } [\text{iso-butane}] = 3.0 - x = 2.9 \text{ M}$$

(b) Equilibrium concentrations if 0.50 mol /L of butane is added:

	butane	iso-butane
Original concentration	1.0	2.5
Change immediately after addition	1.0 + 0.50	2.5
Change (going to equilibrium)	- x	+ x
Equilibrium concentration	1.5 - x	2.5 + x

$$K = \dfrac{[\text{iso-butane}]}{[\text{butane}]} = \dfrac{2.5 + x}{1.5 - x} = 2.5$$

$$2.5 + x = 2.5 (1.5 - x) \quad \text{and } x = 0.36$$

The eq. concentrations are:$[\text{butane}] = 1.5 - x = 1.1 \text{ M and } [\text{iso-butane}] = 2.5 + x = 2.9 \text{ M}$

General Questions

36. For the equilibrium: $Br_2 (g) \rightleftharpoons 2Br (g)$, we must first calculate the concentration of Br_2

$$[Br_2] = \dfrac{0.086 \text{ mol}}{1.26 \text{ L}} = 0.068 \text{ M}$$

Then we can complete an equilibrium table:

	Br_2	Br
Initial	0.068 M	0
Change	- (0.037)(0.068) M	+2 (0.037)(0.068) M
Equilibrium	0.066 M	+0.0051 M

Substituting into the K expression: $K = \dfrac{[Br]^2}{[Br_2]} = \dfrac{(0.0051)^2}{(0.066)} = 3.9 \times 10^{-4}$

38. K_p is 6.5×10^{11} at 25°C for $CO(g) + Cl_2(g) \rightleftharpoons COCl_2(g)$

What is the value of K_p for $COCl_2(g) \rightleftharpoons CO(g) + Cl_2(g)$

Writing the equilibrium expressions for these two processes will show that one is the *reciprocal* of the other. Then K_p for the two will be mathematically related by the reciprocal relationship. So K_p for the second reaction is $1/K_p$ for the first reaction or $1/6.5 \times 10^{11}$ or 1.5×10^{-12}.

40. Calculate K for the system $CS_2(g) + 3\ Cl_2 \rightleftharpoons S_2Cl_2(g) + CCl_4(g)$

An equilibrium table will help keep track of amounts:

Note that this occurs in a 1.00 L flask. The **molar concentrations** and the **number of moles** will be numerically identical.

	CS_2	Cl_2	S_2Cl_2	CCl_4
Initial	1.2	3.6	0	0
Change	- 0.90	- 3 • 0.90	+0.90	+0.90
Equilibrium	0.3	0.9	0.90	0.90

Equilibrium concentrations are easily determined by noting the stoichiometry of the reaction. If 0.90 mol of CCl_4 are formed (to reach equilibrium), an equal amount of S_2Cl_2 is formed and an equal amount of CS_2 reacted. The amount of Cl_2 that reacts is found by noting the 1:3 ratio of CS_2 to Cl_2. If 0.90 mol of CS_2 react then 3 x 0.90 mol of Cl_2 react.

Substituting into the equilibrium expression gives: $\dfrac{[S_2Cl_2][CCl_4]}{[CS_2][Cl_2]^3} = \dfrac{[0.90][0.90]}{[0.3][0.9]^3} = 4$ (1 sf)

42. The equilibrium 2C (a colorless compound) $\rightleftharpoons$ B (a blue compound) is disturbed:

(a) Color change immediately upon reducing the flask size by half:

LeChatelier's Principle indicates that an increase in P (the V is halved) favors the side of the equilibrium with the *smaller total number* of moles of gas. The equilibrium should shift to the right, producing more B, and a greater(more intense) blue color.

(b) The equilibrium shift to the right should result in a darker blue color.

44. For the system $N_2 (g) + O_2 (g) \rightleftharpoons 2 NO (g)$

(a) The system is at equilibrium **if** the product concentrations when inserted into the K expression, give a value of 1.7×10^{-3}. Substitute into the K expression to determine Q:

$$\frac{[NO]^2}{[N_2][O_2]} = \frac{[0.0042]^2}{[0.25][0.25]} = 2.8 \times 10^{-4}$$

The system is **not at equilibrium** since Q is < K.

(b) To reach equilibrium Q needs to "grow". This is accomplished by producing more NO—the system will shift to **right**.

(c) The equilibrium concentrations will be:

	$[N_2]$	$[O_2]$	$[NO]$
Initial	0.25 M	0.25 M	0.0042 M
Change	- x	- x	+ 2x
Equilibrium	0.25 -x	0.25 - x	0.0042 + 2x

Substituting into the K expression:

$$\frac{[NO]^2}{[N_2][O_2]} = \frac{[0.0042 + 2x]^2}{[0.25 - x][0.25 - x]} = 1.7 \times 10^{-3}$$ Mathematically it's easier to solve by

noting that both sides are perfect squares. Take the square root of both sides to obtain:

$$\frac{[0.0042 + 2x]}{[0.25 - x]} = 4.1 \times 10^{-2}$$ solve for x,   x = 0.0030

Using our expressions from the equilibrium table, $[N_2] = [O_2] = 0.25$ M (to 2 sf)

Then [NO] = 0.0042 + 2(0.0030) = 0.0102 M

46. For the equilibrium: $2 SO_2 + O_2 \rightleftharpoons 2 SO_3$, $K = \dfrac{[SO_3]^2}{[SO_2]^2[O_2]} = 279$

This problem can be approached as in preceding examples.

Convert mass of reactants into moles and express them as molar concentrations (easily done since the reaction is in a 1.0- L flask.).

3.00 g SO_2 = 0.0468 mol SO_2 and 5.00g O_2 = 0.156 mol O_2

There is excess oxygen (2 moles of SO_2 are required for 1 mol O_2

K is very large, so the equilibrium lies very far to the right—although not totally (279 isn't that large!) The exact solution is as follows:

$$\frac{[0.0468 - x]^2}{[x]^2[0.5(0.156 - 2(0.0468 - x))]} = 279.$$ The equation is a cubic equation, and will have three

roots. A simpler treatment might suffice—since we've been given 4 answers from which to choose. *Assume* that *all* the SO_2 reacts to form SO_3 The amount of SO_3 formed would be 0.0468 mol SO_3, with a mass of (0.0468 mol • 80.06 g/mol) 3.75g. Given this *approximation*, we can assume that (c) **3.61 g is the reasonable answer**.

48. Given at equilibrium : $\dfrac{[CO][Br_2]}{[COBr_2]}$ = 0.190 at 73 °C

First, calculate concentrations: $[COBr_2] = \dfrac{0.500 \text{ mol}}{2.00 \text{ L}} = 0.250$ M

	$[COBr_2]$	$[CO]$	$[Br_2]$
Original concentration	0.250 M	0	0
Concentration at equilibrium: See SQ16.24 for the full solution	0.107 M	0.143 M	0.143 M
Concentration immediately after addition	0.107 M	0.143 M + 1.00 M	0.143 M
Change (going to equilibrium)	+ x	- x	- x
Equilibrium concentration	0.107 + x	1.143 -x	0.143 -x

(a) Note that the change predicts that more $COBr_2$ will form (in accordance with LeChatelier's Principle.

Substituting into the equilibrium expression we get: $\dfrac{[1.143 - x][0.143 - x]}{(0.107 + x)} = 0.190$

This will generate an equation needing the quadratic equation to solve:

$x^2 - 1.476x + 0.1431 = 0$ The two roots that results are x = 1.37M and x = 0.104 M. The first root makes no sense as it would require more CO to react than is present immediately after the addition of to 2 moles of CO.

(b) So x = 0.104M and the equilibrium concentrations are:

$$[COBr_2] = 0.107 + 0.104 = 0.211 \text{ M}$$
$$[CO] = 1.143 - 0.104 = 1.039 \text{ M}$$
$$[Br_2] = 0.143 - 0.104 = 0.039 \text{ M}$$

(c) The % of $COBr_2$ remaining is (0.211/0.250) x 100 or 84.4 %. The % decomposing is roughly 16%. Compare this to the percentage of 57% (SQ16.24)

50. For the equilibrium: PCl_5 (g) $\rightleftharpoons$ PCl_3 (g) + Cl_2 (g), calculate K

	[PCl_5]	[PCl_3]	[Cl_2]
Equilibrium (expressed as mol)	3.120g/208.24 = 0.01498 mol	3.845/137.33 = 0.02800 mol	1.787/70.91 = 0.02520 mol
Concentration immediately after addition	0.01498 mol	0.02800 mol	0.02520 mol + 0.02000 mol
Change (going to equilibrium).	+x	-x	-x
New equilibrium	0.01498 + x	0.02800 - x	0.04520 - x

(a) The addition of the chlorine will shift the equilibrium to the left.

(b) The new equilibrium concentrations will be found using the values from the table, but we must know the value of K, which we can get by using the equilibrium concentrations given in the problem data.

$$K = \frac{[PCl_3][Cl_2]}{[PCl_5]} = \frac{[0.02800][0.02520]}{[0.01498]} = 0.0470$$

Substituting the values from our table above: $\dfrac{[0.02800 - x][0.04520 - x]}{[0.01498 - x]} = 0.0470$

The resulting quadratic equation can be solved for x. The "sensible" root for x = 0.00485 M.

(c) The resulting equilibrium concentrations are:

$[PCl_5] = 0.01498 + x = 0.01498 + 0.00485 = 0.0199$ M

$[PCl_3] = 0.02800 - x = 0.02800 - 0.00485 = 0.0231$ M

$[Cl_2] = 0.04520 - x = 0.04520 - 0.00485 = 0.0403$ M

52. The K expression for the decomposition is: $K_P = P_{NH_3} \cdot P_{H_2S} = 0.11$.

[Recall that solids are omitted from K expressions.]

Noting that the two products are produced in equal amounts, so the $P(NH_3) = P(H_2S)$.

So $P(NH_3) = P(H_2S) = (0.11)^{1/2} = 0.33$ atm.

P_{total} will be the sum of pressure from NH_3 and H_2S = 0.33 + 0.33 = 0.66 atm

54. The reaction indicates 2 NH_3 and 1 CO_2 form from each ammonium carbamate molecule

We know the total pressure to be 0.116 atm, and that this results from $P(NH_3) + P(CO_2)$.

Since we get 2 ammonias, the pressure that ammonia contributes will be 2x the pressure from CO_2. This can be stated: $P_T = P(NH_3) + P(CO_2)$.

Since $P(NH_3) = 2 \cdot P(CO_2)$, we can write $P_T = 3 \cdot P(CO_2) = 0.116 \, atm$, so $P(CO_2) = 0.116/3$ or $0.0387 \, atm$ and $P(NH_3) = 2 \cdot (0.0387 \, atm) = 0.0773 \, atm$..

K_p can be calculated: $K_p = P^2_{NH_3} \cdot P_{CO_2} = (0.0773)^2(0.0387) = 2.31 \times 10^{-4}$

56. For the gas phase equilibrium of acetic acid : $K = \dfrac{[dimer]}{[monomer]^2} = 3.2 \times 10^4$

	monomer	dimer
Initial pressure	5.4×10^{-4}	0
Equilibrium	x	$\frac{1}{2}(5.4 \times 10^{-4} - x)$

Substituting into the K expression: $K = \dfrac{\frac{1}{2}(5.4 \times 10^{-4} - x)}{x^2} = 3.2 \times 10^4$

or $2.7 \times 10^{-4} - 0.5 \, x = 3.2 \times 10^4 \, x^2$

rearranging $3.2 \times 10^4 \, x^2 + 0.5 \, x - 2.7 \times 10^{-4} = 0$

and solving via the quadratic equation yields $x = 8.4 \times 10^{-5}$

(a) The % of acetic acid converted to the dimer is

$\dfrac{5.4 \times 10^{-4} - 8.4 \times 10^{-5}}{5.4 \times 10^{-4}} \times 100 = 84\%$

(b) As the temperature increases, the equilibrium would **shift to the left**, producing more monomeric acetic acid. This change reflects the move to **favor the endothermic process** as T increases.

58. $P_{total} = P_{NO_2} + P_{N_2O_4} = 1.5 \, atm$; $P_{N_2O_4} = 1.5 - P_{NO_2}$

$K_p = \dfrac{P_{N_2O_4}}{P^2_{NO_2}} = 6.75 = \dfrac{(1.5 - P_{NO_2})}{P^2_{NO_2}} = 6.75$ or

$1.5 - P_{NO_2} = 6.75 \, P^2_{NO_2}$ rearranging $6.75 \, P^2_{NO_2} + P_{NO_2} - 1.5 = 0$

and solving for P_{NO_2} (with the quadratic equation) yields:

$P_{NO_2} = 0.40 \, atm$ and $P_{N_2O_4} = 1.1 \, atm$

60. As the liquid water evaporates, the process $H_2O \, (l) \rightarrow H_2O \, (g)$ proceeds. As more gaseous water is formed, the process $H_2O \, (g) \rightarrow H_2O \, (l)$ (also known as condensation)—initially slow owing to a scarcity of gaseous water—begins to increase (since more gaseous water is

present as the evaporation of water continues). The $l \rightarrow g$ transition (evaporation) slows as the $g \rightarrow l$ transition (condensation) increases. When these two opposing processes are proceeding **at the same rate**, equilibrium is achieved. Once equilibrium has been achieved, the two *opposing processes continue* (it's dynamic) to occur—at the same rate. The net effect is that "nothing appears to be happening".

<u>62</u>. Consider the equilibrium : O_3 (g) + NO (g) $\rightleftharpoons$ O_2 (g) + NO_2 (g) for which K = 6.0 x 10^{34}.

(a) If a system is at equilibrium, the ratios of "reactants" and "products" , when substituted into the equilibrium expression will have a value equal to K.

Substituting we get: $\dfrac{[O_2][NO_2]}{[O_3][NO]}$ = $\dfrac{[8.2 \times 10^{-3}][2.5 \times 10^{-4}]}{[1.0 \times 10^{-6}][1.0 \times 10^{-5}]}$ = 2.1 x 10^5

Q is considerably smaller than K, so the reaction will proceed to the right (forming more O_2 and NO_2)

(b) Determine if the "left to right" process is endo- or exothermic:

$\Delta H^{\circ}_{reaction}$ = [ΔH°_f O_2 + ΔH°_f NO_2] - [ΔH°_f O_3 + ΔH°_f NO]

= [0 kJ/mol (1mol) + 33.1 kJ/mol(1mol)] -

0   [142.7 kJ/mol(1mol) +90.25 kJ/mol(1mol)]

= - 199.9 kJ

The "left to right" process is exothermic. On a warm day (when T is increased), the equilibrium would shift to the left , making more O_3 (g) + NO (g).

<u>64</u>. For the decomposition of lanthanum oxalate, we have:

$La_2(C_2O_4)_3$ (s) $\rightleftharpoons$ La_2O_3 (s) + 3 CO(g) + 3 CO_2 (g)

(a) The K_p expression would be $K_p = P_{CO}^3 \cdot P_{CO_2}^3$ With equal # of molecules of CO and CO_2 forming, the pressure at equilibrium will be an equal contribution from the two types of molecules. If P_T = 0.200 atm, then P_{CO} = 0.100 atm and P_{CO2} = 0.100 atm K_P = $(0.100)^6$. or 1.00 x 10^{-6}

(b) To calculate the amount of the solid unreacted we need the gas law: (Assume T=373 K)

$$ n = \dfrac{P \cdot V}{R \cdot T} = \dfrac{(0.100 \text{ atm}) \cdot (10.0 \text{ L})}{(0.08205 \dfrac{L \cdot torr}{K \cdot mol}) \cdot (373 \text{ K})} = 0.0327 \text{ mol of } CO_2 $$

The stoichiometry of the reaction tells us that 1/3 this # of moles of the lanthanum oxalate reacted: The #mol remaining unreacted is: 0.100 mol – (0.333(0.0327))= 0.089 mol.

66. For the decomposition of sulfuryl chloride: SO_2Cl_2 (g) $\rightleftharpoons$ SO_2 (g) + Cl_2(g)   K = 0.045

The concentration of sulfuryl chloride is $\dfrac{6.70 \text{ g}}{1.00 \text{ L}} \cdot \dfrac{1 \text{ mol } SO_2Cl_2}{134.97 \text{ } SO_2Cl_2}$ = 0.0496 M

	[SO_2Cl_2]	[SO_2]	[Cl_2]
Initial	0.0496 M	0	0
Change	-x	+ x	+ x
Equilibrium	0.0496 - x	+ x	+ x

(a) Substitute into the K expression: $K = \dfrac{[x][x]}{[0.0496 - x]} = 0.045$

Since the concentration of the chloride and K are of the same magnitude, it won't be judicious to make any simplifying assumptions!

(e.g. 0.0496-x $\approx$ 0.0496). Using the quadratic equation to solve: $x^2 = 0.045(0.0496 - x)$

$x^2 + 0.045x - 0.00223 = 0$ for which x = 2.98 x 10^{-2} or 3.0 x 10^{-2} (to 2sf)

The equilibrium concentrations are then:

$$[SO_2Cl_2] = 0.0496 - 0.0298 = 0.020 \text{ M}$$

$$[SO_2] = 0.030 \text{ M}$$

$$[Cl_2] = 0.030 \text{ M} \quad \text{(limited to 2 sf owing to the value of K)}$$

The fraction dissociated is (0.0296/0.0496) = 0.60

(0.0296 = initial value(0.0496) – equilibrium value (0.020))

(b) Equilibrium concentrations if the initial mixture also contains 1.00 atm Cl_2:

To solve this, we have to express the concentration of chlorine in units of M, just as sulfuryl chloride. The gas law should work:

$$\frac{n}{V} = \frac{P}{R \cdot T} = \frac{1.00 \text{ atm}}{\left(0.08205 \dfrac{L \cdot torr}{K \cdot mol}\right)(273 + 375)} = 0.0188 \text{ M}$$

Now we can use an equilibrium table (as before):

	[SO_2Cl_2]	[SO_2]	[Cl_2]
Initial	0.0496 M	0	0.0188 M
Change	-x	+ x	+ x
Equilibrium	0.0496 - x	+ x	0.0188 + x

Substituting into the K expression:

$K = \dfrac{[x][0.0188 + x]}{[0.0496 - x]} = 0.045$. The quadratic equation that results is:

$x^2 + 0.0638x - 0.002232 = 0$ and the solution of the equation gives x = 0.025 (to 2 sf)

Solving for equilibrium concentrations using the table:

$[SO_2Cl_2] =$ $0.0496 - x$ $= 0.025$ M

$[SO_2] =$ x $= 0.025$ M

$[Cl_2] =$ $0.0188 + x$ $= 0.044$ M

The fraction dissociated is (0.025/0.0496) $= 0.50$

(c) The reduction in dissociation of sulfuryl chloride from 0.60 to 0.50 is consistent with LeChatelier's Principle, which would predict that the addition of Cl_2 to the flask would shift the equilibrium to the left (preserving more SO_2Cl_2).

68. For the reaction of hemoglobin (Hb) with CO we can write: $K = \dfrac{[HbCO][O_2]}{[HbO_2][CO]}$ $= 2.0 \times 10^2$

If $\dfrac{[HbCO]}{[HbO_2]}$ $= 1$ then $\dfrac{[O_2]}{[CO]}$ $= 2.0 \times 10^2$

If $[O_2] = 0.20$ atm then $\dfrac{0.20 \text{ atm}}{[CO]}$ $= 2.0 \times 10^2$ and solving for $[CO] = 1.0 \times 10^{-3}$ atm.

So a partial pressure of $[CO] = 1.0 \times 10^{-3}$ atm would likely be fatal.

70. For the reaction:

	NOBr	⇌	NO	1/2 Br$_2$
Initial	x		0	0
Change	-0.34x		+ 0.34x	+0.17x
Equilibrium	0.66x		+ 0.34x	+ 0.17x

P(total) is given in units of mm Hg. Since Kp uses units of atmospheres, convert the P(total) to atmospheres: $190 \text{ mm Hg} \bullet \dfrac{1 \text{ atm}}{760 \text{ mm Hg}} = 0.250$ atm

The P(total) = P(NOBr) + P(NO) + P(Br$_2$) = 0.66x + 0.34x + 0.17x = 0.250 atm

Adding the terms: 1.17x = 0.250 atm and x = 0.214 atm

The equilibrium pressures are :

NOBr = 0.66x = 0.141 atm

NO = 0.34x = 0.0726 atm

Br$_2$ =0.17x = 0.0363 atm

$$K_p = \dfrac{P_{NO} \bullet P^{1/2}_{Br2}}{P_{NOBr}} = \dfrac{(0.0726) \bullet (0.0363)^{1/2}}{(0.141)} = 0.098 \text{ (to 2sf)}$$

Using Electronic Resources

72. Adding more SCN⁻ to the solution at equilibrium would shift the equilibrium to the right, increasing the red color of the $Fe(SCN)^{2+}$. As a result, the concentration of the complex would increase, the $[Fe^{3+}]$ would decrease (as more of the iron(III) iron reacted to form the complex. The $[SCN^-]$ would increase—in response to the addition of the ion.

74. The value for the equilibrium constants for the hypothetical reactions:

    For reactions (a),(b),and (c), I chose an initial concentration of the reactant =0.800.
    For reaction (d), I chose $[M]_i = 0.250$ and $[L]_i = 1.00$

    Experimentation with these systems will prove that the initial concentration is *not critical*. The ratio of reactants to products will always be the same for any system at any one temperature.

    (a) $A \rightleftharpoons B$ $\qquad K = \dfrac{[B]}{[A]} = \dfrac{0.655}{0.145} = 4.5$

    (b) $B \rightleftharpoons 2B$ $\qquad K = \dfrac{[B]^2}{[B]} = \dfrac{[0.898]^2}{[0.351]} = 2.3$

    (c) $HA \rightleftharpoons H^+ + A^-$ $\quad K = \dfrac{[H^+][A^-]}{[HA]} = \dfrac{[0.715][0.715]}{[8.516 \times 10^{-2}]} = 6.0$

    (d) $M + 4L \rightleftharpoons ML_4$ $\quad K = \dfrac{[ML_4]}{[M][L]^4} = \dfrac{[0.102]}{[0.148][0.592]^4} = 5.6$

Chapter 17
Principles of Reactivity: Chemistry of Acids and Bases

Reviewing Important Concepts

1. Water can be both a Bronsted base and a Lewis base. The Bronsted system requires that a base be able to accept a H^+ ion. Water does that in the formation of the H_3O^+ ion. The Lewis system defines a base as an electron pair donor. The two lone pairs of electrons on the O atom provide a source for those electrons. Examine the reaction of H_2O with H^+. The hydrogen ion accepts the electron pairs from the O atom of the water molecule—fulfilling water's role as a Lewis base and the H^+'s role as a Lewis acid.

   A Bronsted acid furnishes H^+ to another specie. The autoionization of water provides just one example of water acting as a Bronsted acid. Water however cannot function as a Lewis acid, as it has no capacity to accept an electron pair.

5. (a) As hydrogen atoms are successively replaced by the very electronegative chlorine atoms, an increase in acidity is seen. This is quite understandable if you remember that an increasing "pull" on the electrons in the "carboxylate" end of the molecule--brought about by an increasing number of chlorine atoms--will weaken the O-H bond and increase the acidity of the specie.

   (b) The acid with the largest K_a (Cl_3CCO_2H) would be the strongest acid and would have the lowest pH. The acid with the smallest K_a (CH_3CO_2H) would have the highest pH.

6. H_2SeO_4 should be the stronger acid. The higher oxidation state of Se in this acid (+6) compared to the selenous acid (+4) means that the H-O bonds in the selenic acid will be weaker, and the hydrogen ions more ready to leave—a stronger acid. Preparing equimolar solutions of both acids in water should result in a lower pH for the stronger acid.

9. This puzzle is a great way to test your chemical knowledge. Organize what we know.

Cations	Anions
Na^+	Cl^-
NH_4^+	OH^-
H^+	

Experimentally we observe: $B + Y \rightarrow$ acidic solution

                                             $B + Z \rightarrow$ basic solution

                                             $A + Z \rightarrow$ neutral solution

If $A + Z$ give a neutral solution, then $A + Z$ must be acid and base (in some order).

Since $B + Z$ gives a basic solution, $B + Z$ cannot be acid and base (as $A + Z$), otherwise B + Z would be neutral (as $A + Z$). Z must be basic (OH^-), meaning Y must be neutral (Cl^-). Since $B + Y$ gives an acid solution, then B must contain NH_4^+. If B contains ammonium, then A must contain H^+, and C must contain Na^+. In summary then:

$$A = H^+; B = NH_4^+ ; C = Na^+ ; Y = Cl^-; Z = OH^- .$$

Pairing these cations with chloride and potassium ions, we have:

$$A = HCl; B = NH_4Cl ; C = NaCl ; Y = KCl; Z = KOH .$$

Practicing Skills
The Bronsted Concept

10. | Conjugate Base of: | Formula | Name |
|---|---|---|
| (a) HCN | CN^- | cyanide ion |
| (b) HSO_4^- | SO_4^{2-} | sulfate ion |
| (c) HF | F^- | fluoride ion |

12. Products of acid-base reactions:

(a) $HNO_3 (aq)$ + $H_2O(\ell)$ $\rightarrow$ $H_3O^+(aq)$ + $NO_3^-(aq)$

         acid   base   conjugate acid   conjugate base

(b) $HSO_4^- (aq)$ + $H_2O(\ell)$ $\rightarrow$ $H_3O^+(aq)$ + $SO_4^{2-}(aq)$

         acid   base   conjugate acid   conjugate base

(c) $H_3O^+(aq)$ + $F^-(aq)$ $\rightarrow$ $HF(aq)$ + $H_2O(\ell)$

         acid   base   conjugate acid   conjugate base

14. Hydrogen oxalate acting as Bronsted acid and Bronsted base:

Bronsted acid: $HC_2O_4^- + OH^- \rightarrow C_2O_4^{2-} + H_2O$

Bronsted base: $HC_2O_4^- + HNO_3 \rightarrow H_2C_2O_4 + NO_3^-$

The action of the hydrogen oxalate ion is determined by the substance in solution. If that substance is a *stronger acid than hydrogen oxalate*(HNO_3), then hydrogen oxalate acts *as a base*. If the substance is a *weaker acid (or stronger base, e.g. hydroxide ion)*, then hydrogen oxalate acts *as an acid*.

16.   (a) HCO_2H (aq) + H_2O (ℓ) $\rightarrow$ HCO_2^- (aq) + H_3O^+ (aq)

        acid   base   conjugate   conjugate

                                   of HCO_2H of H_2O

(b) NH_3 (aq)   + H_2S (aq) $\rightleftharpoons$ NH_4^+ (aq)   + HS^- (aq)

      base   acid   conjugate   conjugate

                                   of NH_3 of H_2S

(c) HSO_4^- (aq) + OH^- (aq) $\rightleftharpoons$ SO_4^{2-} (aq) + H_2O (ℓ)

      acid   base   conjugate   conjugate

                                   of HSO_4^- of OH^-

pH Calculations

18. Since pH = 3.75, $[H_3O^+] = 10^{-pH}$ or $10^{3.75}$ or 1.8×10^{-4} M

Since the $[H_3O^+]$ is greater than 1×10^{-7} (pH< 7), the solution is acidic.

20. pH of a solution of 0.0075 M HCl:

Since HCl is considered a strong acid, a solution of 0.0075 M HCl has

$[H_3O^+] = 0.0075$ or 7.5×10^{-3}.

$$pH = -\log[H_3O^+] = -\log[7.5 \times 10^{-3}] = 2.12$$

The hydroxide ion concentration is readily determined since $[H_3O^+] \cdot [OH^-] = 1.0 \times 10^{-14}$

$$[OH^-] \;\; = \;\; \frac{1.0 \times 10^{-14}}{7.5 \times 10^{-3}} = 1.3 \times 10^{-12} \text{ M}$$

22. pH of a solution of 0.0015 M $Ba(OH)_2$:

Soluble metal hydroxides are strong bases. To the extent that it dissolves ($Ba(OH)_2$ is not

that soluble), $Ba(OH)_2$ gives two OH^- for each formula unit of $Ba(OH)_2$.

0.0015 M $Ba(OH)_2$ would provide 0.0030 M OH^-. Since $[H_3O^+] \cdot [OH^-] = 1.0 \times 10^{-14}$

$[H_3O^+]$ would then be:

$$[H_3O^+] = \frac{1.0 \times 10^{-14}}{3.0 \times 10^{-3}} = 3.3 \times 10^{-12} \text{ M and } pH = -\log[3.3 \times 10^{-12}] = 11.48.$$

Equilibrium Constants for Acids and Bases

24. Concerning the following acids:

Phenol		Oxalic acid		Hydrogen oxalate ion	
C_6H_5OH	1.3×10^{-10}	HCO_2H	1.8×10^{-4}	$HC_2O_4^-$	6.4×10^{-5}

(a) The strongest acid is oxalic. The weakest acid is phenol. Acid strength is proportional to
the magnitude of K_a.

(b) Recall the relationship between acids and their conjugate base: $K_a \cdot K_b = K_w$

Since K_w is a constant, The greater the magnitude of K_a, the smaller the value of the K_b for the conjugate base. The strongest acid (oxalic) has the weakest conjugate base.

(c) The weakest acid (phenol) has the strongest conjugate base.

26. The substance which has the smallest value for K_a will have the strongest conjugate base.
One can prove this quantitatively with the relationship: $K_a \cdot K_b = K_w$. An examination of Appendix H or Table 17.3 shows that--of these three substances--HClO has the smallest K_a, and ClO$^-$ will therefore be the strongest conjugate base.

28. The equation for potassium carbonate dissolving in water:
$$K_2CO_3 \text{ (aq)} \rightarrow 2\,K^+ \text{ (aq)} + CO_3^{2-} \text{ (aq)}$$
Soluble salts--like K_2CO_3--dissociate in water. The carbonate ion formed in this process is a base, and reacts with the acid, water.
$$CO_3^{2-} \text{ (aq)} + H_2O \text{ (}\ell\text{)} \rightleftharpoons HCO_3^- \text{ (aq)} + OH^- \text{ (aq)}$$
The production of the hydroxide ion, a strong base, in this second step is responsible for the basic nature of solutions of this carbonate salt.

30. Most of the salts shown are sodium salts. Since Na^+ does not hydrolyze, we can estimate the acidity (or basicity) of such solutions by looking at the extent of reaction of the anions with water (hydrolysis).

The Al^{3+} ion is acidic, as is the $H_2PO_4^-$ ion. From Table 17.3 we see that the K_a for the hydrated aluminum ion is greater than that for the $H_2PO_4^-$ ion, making the Al^{3+} solution more acidic —lower pH than that of $H_2PO_4^-$. All the other salts will produce basic solutions and since the S^{2-} ion has the largest K_b, we will anticipate that the Na_2S solution will be most basic—i.e. have the highest pH.

pKa , A Logarithmic Scale of Acid Strength
32. The pK_a for an acid with a K_a of 6.5×10^{-5}. $pK_a = -\log(6.5 \times 10^{-5}) = 4.19$

34. K_a for epinephrine, whose $pK_a = 9.53$. $K_a = 10^{-pK_a}$ so $K_a = 10^{-9.53}$ or 3.0×10^{-10}

36. *2-chlorobenzoic* acid has a smaller pK_a than benzoic acid, so it *is the stronger acid.*

Ionization Constants for Weak Acids and Their Conjugate Bases

38. The K_b for the chloroacetate ion:

Recall the relationship between acids and their conjugate base: $K_a \cdot K_b = K_w$

K_b for the chloroacetate ion will be $\dfrac{1.0 \times 10^{-14}}{1.36 \times 10^{-3}} = 7.4 \times 10^{-12}$

40. The K_b for $(CH_3)_3NH^+$ is 10^{-pK_b} or $10^{-9.80} = 1.6 \times 10^{-10}$

Then $K_b = \dfrac{1.0 \times 10^{-14}}{1.6 \times 10^{-10}} = 6.3 \times 10^{-5}$

Predicting the Direction of Acid-Base Reactions

42. $CH_3CO_2H \ (aq) + HCO_3^- \ (aq) \rightleftharpoons CH_3CO_2^- \ (aq) + H_2CO_3 \ (aq)$

Since acetic acid is a stronger acid that carbonic acid, the equilibrium lies predominantly to the right.

44. Predict whether the equilibrium lies predominantly to the left or to the right:

(a) $NH_4^+ \ (aq) + Br^- \ (aq) \rightleftharpoons NH_3 \ (aq) + HBr \ (aq)$

HBr is a stronger acid than NH_4^+; equilibrium lies to left.

(b) $HPO_4^{2-} \ (aq) + CH_3CO_2^- \ (aq) \rightleftharpoons PO_4^{3-} \ (aq) + CH_3CO_2H \ (aq)$

CH_3CO_2H is a stronger acid than HPO_4^{2-}; equilibrium lies to left,

(c) $Fe(H_2O)_6^{3+} \ (aq) + HCO_3^- \ (aq) \rightleftharpoons Fe(H_2O)_5(OH)^{2+} \ (aq) + H_2CO_3 \ (aq)$

$Fe(H_2O)_6^{3+}$ is a stronger acid than H_2CO_3; equilibrium lies to right.

Types of Acid-Base Reactions

46. (a) The net ionic equation for the reaction of NaOH with Na_2HPO_4:

$OH^-(aq) + HPO_4^{2-}(aq) \rightleftharpoons H_2O(\ell) + PO_4^{3-}(aq)$

(b) The equilibrium lies to the right (since HPO_4^{2-} is a stronger acid than H_2O and OH^- is a stronger base than PO_4^{3-}), but phosphate and hydroxide ions aren't too different in basic strength, so the position of equilibrium does not lie very far to the right. The result is that remaining OH^- will result in a basic solution.

48. (a) The net ionic equation for the reaction of CH_3CO_2H with Na_2HPO_4:

$CH_3CO_2H \ (aq) + HPO_4^{2-}(aq) \rightleftharpoons CH_3CO_2^-(aq) + H_2PO_4^-(aq)$

(b) The equilibrium lies to the right (since CH_3CO_2H is a stronger acid than $H_2PO_4^-$ and HPO_4^{2-} is a stronger base than $CH_3CO_2^-$. The result is that the weak acetic acid and the $H_2PO_4^-$ ion will result in a weakly acidic solution.

Using pH to Calculate Ionization Constants

50. For a 0.015 M HOCN, pH = 2.67:

(a) Since pH = 2.67, $[H_3O^+] = 10^{-pH}$ or $10^{-2.67}$ or 2.1×10^{-3} M

(b) The Ka for the acid is in general: $K_a = \dfrac{[H^+][A^-]}{[HA]}$

The acid is monoprotic, meaning that for each H^+ ion, one also gets a OCN^-
The two numerator terms are then equal. The concentration of the molecular acid is the
(original concentration – concentration that dissociates).

$$K_a = \frac{[2.1 \times 10^{-3}][2.1 \times 10^{-3}]}{[0.015 - 2.1 \times 10^{-3}]} = 3.5 \times 10^{-4}$$

52. With a pH = 9.11, the solution has a pOH of 4.89 and $[OH^-] = 10^{-4.89} = 1.3 \times 10^{-5}$ M.

The equation for the base in water can be written:

$$H_2NOH\ (aq) + H_2O\ (\ell) \rightleftharpoons H_3NOH^+\ (aq) + OH^-\ (aq)$$

At equilibrium, $[H_2NOH] = [H_2NOH] - [OH^-] = (0.025 - 1.3 \times 10^{-5})$ or
approximately 0.025 M.

$$K_b = \frac{[H_3NOH^+][OH^-]}{[H_2NOH]} = \frac{(1.3 \times 10^{-5})^2}{0.025} = 6.6 \times 10^{-9}$$

54. (a) With a pH = 3.80 the solution has a $[H_3O^+] = 10^{-3.80}$ or 1.6×10^{-4} M

(b) Writing the equation for the unknown acid, HA, in water we obtain:

$$HA\ (aq) + H_2O\ (\ell) \rightleftharpoons H_3O^+\ (aq) + A^-\ (aq)$$

$[H_3O^+] = 1.6 \times 10^{-4}$ implying that $[A^-]$ is also 1.6×10^{-4}. Therefore the
equilibrium concentration of acid, HA, is $(2.5 \times 10^{-3} - 1.6 \times 10^{-4})$ or $\approx 2.3 \times 10^{-3}$.

$$K_a = \frac{[H_3O^+][A]}{[HA]} = \frac{(1.6 \times 10^{-4})^2}{2.3 \times 10^{-3}} = 1.1 \times 10^{-5}$$

We would classify this acid as a **moderately weak acid.**

Using Ionization Constants

56. For the equilibrium system: $CH_3CO_2H\ (aq) + H_2O(\ell) \rightleftharpoons CH_3CO_2^- + H_3O^+\ (aq)$

Initial	0.20 M			
Change	- x		+ x	+ x
Equilibrium	0.20 - x		+ x	+ x

Substituting into the K_a expression:

$$K_a = \frac{[H_3O^+][A]}{[HA]} = \frac{(x)^2}{0.20 - x} = 1.8 \times 10^{-5} \text{ and solving for } x = 1.9 \times 10^{-3}$$

So $[CH_3CO_2H] \cong 0.20$ M; $[CH_3CO_2^-] = [H_3O^+] = 1.9 \times 10^{-3}$ M

58. Using the same logic as in question 56 above, we can write:

$$HCN \text{ (aq)} + H_2O \text{ (}\ell\text{)} \rightleftharpoons H_3O^+ \text{ (aq)} + CN^- \text{ (aq)}$$

Initial	0.025M			
Change	- x		+ x	+ x
Equilibrium	0.025 - x		+ x	+ x

Substituting these values into the K_a expression for HCN:

$$K_a = \frac{[CN^-][H_3O^+]}{[HCN]} = \frac{(x)^2}{(0.025 - x)} = 4.0 \times 10^{-10}$$

Assuming that the denominator may be approximated as 0.025 M, we obtain:

$$\frac{x^2}{0.025} = 4.0 \times 10^{-10}$$

and $x = 3.2 \times 10^{-6}$. The equilibrium concentrations of $[H_3O^+] = [CN^-] = 3.2 \times 10^{-6}$ M.

The equilibrium concentration of $[HCN] = (0.025 - 3.2 \times 10^{-6})$ or 0.025 M.

Since x represents $[H_3O^+]$ the pH $= -\log(3.2 \times 10^{-6})$ or 5.50.

60. The equilibrium of ammonia in water can be written:

$$NH_3 + H_2O \rightleftharpoons NH_4^+ + OH^- \qquad K_b = 1.8 \times 10^{-5}$$

Initial	0.15 M		0	0
Change	- x		+ x	+ x
Equilibrium	0.15 - x		+ x	+ x

Substituting into the K_b expression:

$$K_b = \frac{[NH_4^+][OH^-]}{[NH_3]} = \frac{(x)^2}{(0.15 - x)} = 1.8 \times 10^{-5}$$

With K_b small, the extent to which ammonia reacts with water is slight. We can

approximate the denominator (0.15 - x) as 0.15 M, and solve the equation:

$$\frac{(x)^2}{0.15} = 1.8 \times 10^{-5} \text{ and } x = 1.6 \times 10^{-3}$$

So $[NH_4^+] = [OH^-] = 1.6 \times 10^{-3}$ M and $[NH_3] = (0.15 - 1.6 \times 10^{-3}) \approx 0.15$ M

The pH of the solution is then:

$$pOH = -\log(1.6 \times 10^{-3}) = 2.80 \text{ and the pH} = (14.00 - 2.80) = 11.21$$

62. For CH_3NH_2 (aq) + $H_2O(\ell)$ $\rightleftharpoons$ $CH_3NH_3^+$(aq) + OH^-(aq) K_b = 4.2 x 10^{-4}

Using the approach of problem 60, the equilibrium expression can be written:

$$K_b = \frac{x^2}{0.25 - x} = 4.2 \times 10^{-4} \approx \frac{x^2}{0.25}$$

Solving the expression **with approximations** gives x = 1.0 x 10^{-2}

[OH$^-$] = 1.0 x 10^{-2} M

pOH would then be 1.99 and pH 12.01 (to 2 sf)

64. pH of 1.0 x 10^{-3} M HF; K_a = 7.2 x 10^{-4}

Using the same method as in question 56, we can write the expression:

$$K_a = \frac{x^2}{(1.0 \times 10^{-3} - x)} = 7.2 \times 10^{-4}$$

The concentration of the HF and K_a preclude the use of our usual approximations

$(1.0 \times 10^{-3} - x \approx 1.0 \times 10^{-3})$

So we multiply both sides of the equation by the denominator to get:

$x^2 = 7.2 \times 10^{-4}(1.0 \times 10^{-3} - x)$

$x^2 = 7.2 \times 10^{-7} - 7.2 \times 10^{-4} x$

Using the quadratic equation we solve for x:

and x = 5.6 x 10^{-4} M = [F$^-$] = [H$_3$O$^+$] and pH = 3.25

Acid-Base Properties of Salts

66. Hydrolysis of the NH_4^+ produces H_3O^+ according to the equilibrium:

NH_4^+ (aq) + H_2O (l) $\rightleftharpoons$ H_3O^+ (aq) + NH_3 (aq)

With the ammonium ion acting as an acid, to donate a proton, we can write the K_a expression:

$$K_a = \frac{[NH_3][H_3O^+]}{[NH_4^+]} = \frac{K_w}{K_b} = \frac{1.0 \times 10^{-14}}{1.8 \times 10^{-5}} = 5.6 \times 10^{-10}$$

The concentrations of both terms in the numerator are equal, and the concentration of ammonium ion is 0.20 M. (Note the approximation for the **equilibrium** concentration of NH_4^+ to be equal to the **initial** concentration.)

Substituting and rearranging we get

$$[H_3O^+] = \sqrt{0.20 \cdot 5.6 \times 10^{-10}} = 1.1 \times 10^{-5} \text{ and the pH} = 4.98.$$

243

68. The hydrolysis of CN^- produces OH^- according to the equilibrium:

$$CN^- (aq) + H_2O (l) \rightleftharpoons HCN (aq) + OH^- (aq)$$

Calculating the concentrations of Na^+ and CN^- :

$$[Na^+]_i = [CN^-]_i = \frac{10.8 \text{ g NaCN}}{0.500 \text{ L}} \cdot \frac{1 \text{ mol NaCN}}{49.01 \text{ g NaCN}} = 4.41 \times 10^{-1} \text{ M}$$

Then $K_b = \dfrac{K_w}{K_a} = \dfrac{1.0 \times 10^{-14}}{4.0 \times 10^{-10}} = 2.5 \times 10^{-5} = \dfrac{[HCN][OH^-]}{[CN^-]}$

Substituting the $[CN^-]$ concentration into the K_b expression and noting that:

$$[OH^-]_e = [HCN]_e \quad \text{we may write}$$

$$[OH^-]_e = [(2.5 \times 10^{-5})(4.41 \times 10^{-1})]^{\frac{1}{2}} = 3.3 \times 10^{-3} \text{ M}$$

$$[H_3O^+] = \frac{1.0 \times 10^{-14}}{3.3 \times 10^{-3}} = 3.0 \times 10^{-12} \text{ M}$$

pH After an Acid-Base Reaction

70. The net reaction is:

$$CH_3CO_2H (aq) + NaOH (aq) \rightleftharpoons CH_3CO_2^- (aq) + Na^+ (aq) + H_2O(l)$$

The addition of 22.0 mL of 0.15 M NaOH (3.3 mmol NaOH) to 22.0 mL of 0.15 M CH_3CO_2H (3.3 mmol CH_3CO_2H) produces water and the soluble salt, sodium acetate (3.3 mmol $CH_3CO_2^-$ Na^+). The acetate ion is the anion of a weak acid and reacts with water according to the equation:

$$CH_3CO_2^- (aq) + H_2O (l) \rightleftharpoons CH_3CO_2H (aq) + OH^- (aq)$$

The equilibrium constant expression is: $K_b = \dfrac{[CH_3CO_2H][OH^-]}{[CH_3COO^-]} = 5.6 \times 10^{-10}$

The concentration of acetate ion is: $\dfrac{3.3 \text{ mmol}}{(22.0 + 22.0) \text{ ml}} = 0.075 \text{ M}$

	$CH_3CO_2^-$	CH_3CO_2H	OH^-
Initial concentration	0.075		
Change	-x	+x	+x
Equilibrium	0.075 - x	+x	+x

$$K_b = \frac{[CH_3CO_2H][OH^-]}{[CH_3CO_2^-]} = \frac{x^2}{0.075 - x} = 5.6 \times 10^{-10}$$

Simplifying $(100 \cdot K_b \ll 0.075)$ we get $\frac{x^2}{0.075} = 5.6 \times 10^{-10}$; $x = 6.5 \times 10^{-6} = [OH^-]$

The hydrogen ion concentration is related to the hydroxyl ion concentration by the equation:

$$K_w = [H_3O^+][OH^-] = 1.0 \times 10^{-14}$$

$$[H_3O^+] = \frac{1.0 \times 10^{-14}}{[OH^-]} = \frac{1.0 \times 10^{-14}}{6.5 \times 10^{-6}} = 1.5 \times 10^{-9}$$

$$pH = 8.81$$

The pH is greater than 7, as we expect for a salt of a strong base and weak acid.

72. Equal numbers of moles of acid and base are added in each case, leaving only the salt of the acid and base. The reaction (if any) of that salt with water (hydrolysis) will affect the pH.

pH of solution	Reacting Species	Reaction controlling pH
(a) >7	CH3CO2H/KOH	Hydrolysis of CH3CO2$^-$
(b) <7	HCl/NH3	Hydrolysis of NH4$^+$
(c) =7	HNO3/NaOH	No hydrolysis

Polyprotic Acids and Bases

74.   (a) pH of 0.45 M H2SO3:

The equilibria for the diprotic acid are:

$$K_{a1} = \frac{[HSO_3^-][H_3O^+]}{[H_2SO_3]} = 1.2 \times 10^{-2} \quad \text{and} \quad K_{a2} = \frac{[SO_3^{2-}][H_3O^+]}{[HSO_3^-]} = 6.2 \times 10^{-8}$$

For the first step of dissociation:

	H2SO3	HSO3$^-$	H3O$^+$
Initial concentration	0.45 M		
Change	-x	+x	+x
Equilibrium	0.45 - x	+x	+x

Substituting in to the K_{a1} expression: $K_{a1} = \frac{(x)(x)}{(0.45-x)} = 1.2 \times 10^{-2}$

We must solve this expression with the quadratic equation since $(0.45 < 100 \cdot K_{a1})$.

The equilibrium concentrations for HSO3$^-$ and H3O$^+$ ions are found to be 0.0677 M.

The further dissociation is indicated by K_{a2}.

Using the equilibrium concentrations obtained in the first step, we substitute into the K_{a2}

expression.

	HSO_3^-	SO_3^{2-}	H_3O^+
Initial concentration	0.0677	0	0.0677
Change	-x	+x	+x
Equilibrium	0.0677 - x	+x	0.0677 + x

$$K_{a2} = \frac{[SO_3^{2-}][H_3O^+]}{[HSO_3^-]} = \frac{(+x)(0.0677 + x)}{(0.0677 - x)} = 6.2 \times 10^{-8}$$

We note that x will be small in comparison to 0.0677, and we simplify the expression:

$$K_{a2} = \frac{(+x)(0.0677)}{(0.0677)} = 6.2 \times 10^{-8}$$

In summary, the concentrations of HSO_3^- and H_3O^+ ions have been virtually unaffected

by the second dissociation.

Then $[H_3O^+] = 0.0677$ M and pH = 1.17

(b) The equilibrium concentration of SO_3^{2-} :

From the K_{a2} expression above: $[SO_3^{2-}] = 6.2 \times 10^{-8}$ M

76.   (a) Concentrations of OH^-, $N_2H_5^+$, and $N_2H_6^{2+}$ in 0.010 M N_2H_4:

The K_{b1} equilibrium allows us to calculate $N_2H_5^+$ and OH^- formed by the reaction of

N_2H_4 with H_2O. $K_{b1} = \frac{[N_2H_5^+][OH^-]}{[N_2H_4]} = 8.5 \times 10^{-7}$

	N_2H_4	$N_2H_5^+$	OH^-
Initial concentration	0.010	0	
Change	-x	+x	+x
Equilibrium	0.010 - x	+x	0.0677 + x

Substituting into the K_{b1} expression : $\frac{(x)(x)}{0.010 - x} = 8.5 \times 10^{-7}$

We can simplify the denominator $(0.010 > 100 \cdot K_{b1})$.

$$\frac{(x)(x)}{0.010} = 8.5 \times 10^{-7} \text{ and } x = 9.2 \times 10^{-5} \text{ M} = [N_2H_5^+] = [OH^-]$$

The second equilibrium (K_{b2}) indicates further reaction of the $N_2H_5^+$ ion with water. The

step should consume some $N_2H_5^+$ and produce more OH^-. The magnitude of K_{b2}

indicates that the equilibrium "lies to the left" and we anticipate that not much $N_2H_6^{2+}$ (or additional OH^-) will be formed by this interaction.

	$N_2H_5^+$	$N_2H_6^{2+}$	OH^-
Initial concentration	9.2×10^{-5}	0	9.2×10^{-5}
Change	-x	+x	+x
Equilibrium	$9.2 \times 10^{-5} - x$	+x	$9.2 \times 10^{-5} + x$

$$K_{b2} = \frac{[N_2H_6^{2+}][OH^-]}{[N_2H_5^+]} = 8.9 \times 10^{-16} = \frac{x \cdot (9.2 \times 10^{-5} + x)}{(9.2 \times 10^{-5} - x)}$$

Simplifying yields $\dfrac{x \cdot (9.2 \times 10^{-5})}{(9.2 \times 10^{-5})} = 8.9 \times 10^{-16}$; $x = 8.9 \times 10^{-16} M = [N_2H_6^{2+}]$

In summary we see that the second stage produces a negligible amount of OH^- and consumes very little $N_2H_5^+$ ion. The equilibrium concentrations are:

$[N_2H_5^+]$: 9.2×10^{-5} M; $[N_2H_6^{2+}]$: 8.9×10^{-16} M; $[OH^-]$: 9.2×10^{-5} M

(b) The pH of the 0.010 M solution: $[OH^-] = 9.2 \times 10^{-5}$ M so pOH = 4.04
        and pH = 14.0 - 4.04 = 9.96

Lewis Acid and Bases

78.   (a) H_2NOH electron rich (accepts H^+)   Lewis base
        (b) Fe^{2+} electron poor   Lewis acid
        (c) CH_3NH_2 electron rich (accepts H^+)   Lewis base

80. CO is a Lewis base (donates electron pairs) in complexes with nickel and iron.

Molecular Structure, Bonding, and Acid-Base Behavior

82. HOCN will be the stronger acid. In HOCN the proton is attached to the very electronegative O atom. This great electronegativity will provide a very polar bond, weakening the O-H bond, and making the H more acidic than in HCN.

84. Benzenesulfonic acid is a Bronsted acid owing to the inductive effect of three oxygen atoms attached to the S (and through the S to the benzene ring).These very electronegative O atoms will—through the inductive effect—remove electron density between the O and the H—weakening the OH bond, and making the H acidic.

General Questions on Acids and Bases

86. For the equilibrium: $HC_9H_7O_4$ (aq) + H_2O (ℓ) $\rightleftharpoons$ $C_9H_7O_4^-$ (aq) + H_3O^+ (aq)

we can write the K_a expression:

$$K_a = \frac{[C_9H_7O_4^-][H_3O^+]}{[HC_9H_7O_4]} = 3.27 \times 10^{-4}$$

The initial concentration of aspirin is:

$$2 \text{ tablets} \cdot \frac{0.325 \text{ g}}{1 \text{ tablet}} \cdot \frac{1 \text{ mol } HC_9H_7O_4}{180.2 \text{ g } HC_9H_7O_4} \cdot \frac{1}{0.225 \text{ L}} = 1.60 \times 10^{-2} \text{ M } HC_9H_7O_4$$

Substituting into the K_a expression:

$$K_a = \frac{[H_3O^+]^2}{1.60 \times 10^{-2} - x} = 3.27 \times 10^{-4} = \frac{x^2}{1.60 \times 10^{-2} - x}$$

Since $100 \cdot K_a \approx$ [aspirin], the quadratic equation will provide a "good" value.
Using the quadratic equation, $[H_3O^+] = 2.13 \times 10^{-3}$ M and pH $= 2.671$.

88. 4-chlorobenzoic acid has a larger K_a than benzoic acid, so it is the stronger acid. A 0.010 M

solution of *benzoic acid* would produce a lower concentration of hydronium ions than would a

similar solution of 4-chlorobenzoic acid, and would be therefore less acidic—*have a higher pH.*

90. The reaction between H_2S and $NaCH_3CO_2$:

$$H_2S \text{ (aq)} + CH_3CO_2^- \text{ (aq)} \rightleftharpoons HS^- \text{ (aq)} + CH_3CO_2H \text{ (aq)}$$

An examination of Table 17.3 (page 702) reveals that CH_3CO_2H is a stronger acid than

H_2S, so the equilibrium will lie *to the left.*

92. Monoprotic acid has $K_a = 1.3 \times 10^{-3}$. Equilibrium concentrations of HX, H_3O^+ and pH for

0.010 M solution of HX:

	HX	H^+	X^-
Initial concentration	0.010	0	0
Change	-x	+x	+x
Equilibrium	0.010 - x	+x	+ x

$$K_a = \frac{[H_3O^+][A]}{[HA]} = \frac{(x)^2}{0.010 - x} = 1.3 \times 10^{-3}$$

Since Ka and the concentration are of the same order of magnitude, the quadratic equation

will be of help: $x^2 + 1.3 \times 10^{-3}x - 1.3 \times 10^{-5} = 0$

For which the "reasonable solution" is $x = 3.0 \times 10^{-3}$

Concentrations are then: $[HX] = 0.010 - x = 0.010 - 0.0030 = 7.0 \times 10^{-3}$ M

$[H^+] = +x = 3.0 \times 10^{-3}$ M $[X^-] = +x = 3.0 \times 10^{-3}$ M

pH $= -\log(3.0 \times 10^{-3}) = 2.52$

94. The pKa of m-Nitrophenol:

The Ka expression for this weak acid is: $K_a = \dfrac{[H^+][A^-]}{[HA]}$

The pH of a 0.010 M solution of nitrophenol is 3.44, so we know: $[H^+] = 3.63 \times 10^{-4}$

Since we get one " A-" for each " H+" the concentrations of the two are equal.
The $[HA] = (0.010 - 0.000363)$. Substituting those values into the K_a expression yields:

$K_a = \dfrac{[H^+][A^-]}{[HA]} = \dfrac{[3.63 \times 10^{-4}][3.63 \times 10^{-4}]}{[0.010 - 3.63 \times 10^{-4}]} = 1.4 \times 10^{-5}$ (to 2sf)

$pK_a = -\log(1.4 \times 10^{-5}) = 4.86$

96. For Novocain, the $pK_a = 8.85$. The pH of a 0.0015M solution is:

First calculate K_a. $K_a = 10^{-8.85}$ so $K_a = 1.41 \times 10^{-9}$

Now we can treat Novocain as any weak acid, with the appropriate K_a expression:

$K_a = \dfrac{[H^+][A^-]}{[HA]} = \dfrac{[x][x]}{[0.0015 - x]} = 1.4 \times 10^{-9}$

Since Ka is so small, we can assume that the denominator is approximated by 0.0015 M.

$\dfrac{[x]^2}{0.0015} = 1.4 \times 10^{-9}$ and $x^2 = 1.45 \times 10^{-6}$

Since $[H^+] = 1.45 \times 10^{-6}$; pH $= -\log(1.45 \times 10^{-6}) = 5.84$

98. Regarding ethylamine and ethanolamine:

(a) Since ethylamine has a large K_b, ethylamine is the stronger base.

(b) The pH of 0.10 M solution of ethylamine:

$C_2H_5NH_2 + H_2O \rightleftharpoons C_2H_5NH_3^+ + OH^-$ $K_b = 4.3 \times 10^{-4}$

Initial	0.10 M	0	0
Change	- x	+ x	+ x
Equilibrium	0.10 - x	+ x	+ x

Substituting into the K_b expression:

$K_b = \dfrac{[NH_4^+][OH^-]}{[NH_3]} = \dfrac{x^2}{(0.10 - x)} = 4.3 \times 10^{-4}$

So $x = 6.56 \times 10^{-3}$ so pOH $= -\log(6.56 \times 10^{-3}) = 2.18$ and pH $= 14.00 - 2.18 = 11.82$

<u>100</u>. K_b for pyridine = 1.5 x 10^{-9} (from Appendix I) ; K_a = 1.0 x 10^{-14}/1.5 x 10^{-9} = 6.7 x 10^{-6}

Treating the hydrochloride as a weak acid (K_a = 6.7 x 10^{-6}), we can write:

$$C_5H_5NH^+ + H_2O \rightleftharpoons C_5H_5N + H_3O^+$$

And the equilibrium concentrations are:

	$C_5H_5NH^+$	H^+	C_5H_5N
Initial concentration	0.025	0	0
Change	-x	+x	+x
Equilibrium	0.025 - x	+x	+ x

So $K_a = \dfrac{[C_5H_5N][H^+]}{[C_5H_5NH^+]} = \dfrac{[x][x]}{[0.025 - x]} = 6.7 \times 10^{-6}$

Solving for x = 4.05 x 10^{-4} and pH = -log(4.05 x 10^{-4}) = 3.39

102. pH of aqueous solutions of

	reaction	pH
(a) $NaHSO_4$	hydrolysis of HSO_4^- produces H_3O^+	< 7
(b) NH_4Br	hydrolysis of HSO_4^- produces H_3O^+	< 7
(c) $KClO_4$	no hydrolysis occurs	= 7
(d) Na_2CO_3	hydrolysis of CO_3^{2-} produces OH^-	> 7
(e) $(NH_4)_2S$	hydrolysis of S^{2-} produces OH^-	> 7
(f) $NaNO_3$	no hydrolysis occurs	= 7
(g) Na_2HPO_4	hydrolysis of HPO_4^{2-} produces OH^-	> 7
(h) LiBr	no hydrolysis occurs	= 7
(i) $FeCl_3$	hydrolysis of Fe^{3+} produces H_3O^+	< 7

104. Arrange the following 0.1M solution in order of increasing pH:

Examine the nature of each solute.

(a) NaCl—no hydrolysis of cation or anion

(b) NH_4Cl—hydrolysis of NH_4^+-- to give H_3O^+

(c) HCl—a strong acid

(d) $NaCH_3CO_2$—hydrolysis of the acetate anion—to produce OH^-

(e) KOH—a strong base

From low pH (acidic) to high pH (basic): HCl < NH_4Cl < NaCl < $NaCH_3CO_2$ < KOH

106. For oxalic acid $K_{a1} = 5.9 \times 10^{-2}$ and $K_{a2} = 6.4 \times 10^{-5}$

Representing oxalic as H_2A. The first step can be written as:

$$H_2A + H_2O \rightleftharpoons HA^- + H_3O^+ \qquad K_{a1} = 5.9 \times 10^{-2}$$

We can write the second step:

$$HA^- + H_2O \rightleftharpoons A^{2-} + H_3O^+ \qquad K_{a2} = 6.4 \times 10^{-5}$$

If we add the two equations we get:

$$H_2A + 2 H_2O \rightleftharpoons A^{2-} + 2 H_3O^+ \qquad \overline{}$$
$$K_{net} = (5.9 \times 10^{-2})(6.4 \times 10^{-5})$$
$$K_{net} = 3.8 \times 10^{-6}$$

108. (a) As the K_a values increase, the concentration of Hydronium ion that the acid

produces increases—reducing the pH of the solution. So the nitropyridinium hydrochloride ($x = NO_2$) solution would have the lowest pH and the methylpyridinium hydrochloride ($x = CH_3$) solution would provide the highest pH.

(b) Strongest and weakest Brønsted base:
Remembering the relationship that $K_a \cdot K_b = K_w$, it follows that the conjugate acid with the largest K_a (nitropyridinium hydrochloride) will have the conjugate base (nitropyridine) with the smallest K_b (weakest base) and similarly methyl pyridine would be the strongest base.

110. Confirm that $K = 1.8 \times 10^9$ is valid for the reaction of hydrochloric acid with ammonia:

Begin by writing the equation:

$$NH_3 (aq) + H_3O^+(aq) \rightleftharpoons NH_4^+(aq) + H_2O(\ell)$$

Careful examination of this equation reveals that it is the *reverse* of the reaction for ammonium ions in water. So K for this process is the *reciprocal* of the K_a for ammonium ion. The K_a for that ion is 5.6×10^{-10}

Calculated by: $K' = \dfrac{K_w}{K_b \, conj} = \dfrac{1.0 \times 10^{-14}}{1.8 \times 10^{-5}} = 5.6 \times 10^{-10}$

The reciprocal of 5.6×10^{-10} is 1.8×10^9.

112. From the table of pK_a value for the compounds:

(a) The strongest acid has the smallest pK_a, so Chloroacetic acid is the strongest acid.
Recall that the smallest pK_a corresponds to the largest K_a, which would represent the acid with the greatest production of H^+.

(b) The acid with the strongest conjugate base would be the weakest acid, Benzylammonium ion.

(c) The acids in order of increasing strength:

benzylammonium	<	conjugate acid	<	benzoic	<	thioacetic	<	chloroacetic
ion		of cocaine		acid		acid		acid

<u>114.</u>   (a) BF_3 is electron-deficient and is the **Lewis acid**. Dimethyl ether, $(CH_3)_2O$, has

electron pairs (on the oxygen atom) and is the **Lewis base.**

(b) Calculate the # of moles of complex:

$$0.100 \text{ g complex} \cdot \frac{1 \text{ mol complex}}{113.9 \text{ g complex}} = 0.000878 \text{ mol complex}$$

Since the complex is a gas, we can calculate the initial pressure of the complex using the

Ideal Gas Law:

$$P = \frac{n \cdot R \cdot T}{V} = \frac{0.000878 \text{ mol} \cdot 0.082057 \frac{L \cdot atm}{K \cdot mol} \cdot 298 \text{ K}}{0.565 \text{ L}} = 0.0380 \text{ atm}$$

For the equilibrium we can write:

$(CH_3)_2O\text{-}BF_3 \text{ (g)} \rightleftharpoons BF_3 \text{ (g)} + (CH_3)_2O \text{ (g)} \quad K_p = 0.17$

	$(CH_3)_2O\text{-}BF_3$ (g)	BF_3 (g)	$(CH_3)_2O$ (g)
Initial	0.0380	0	0
Change	- x	+ x	+ x
Equilibrium	0.0380 - x	x	x

$$K_p = \frac{P_{BF_3} \cdot P_{(CH_3)_2O}}{P_{(CH_3)_2O - BF_3}} = \frac{x^2}{0.0380 - x} = 0.17$$

Given the magnitude of K_p and the pressure of the complex (0.0380 atm), our usual

simplifying assumptions won't suffice here. Solve the equation, with the quadratic equation.

$x^2 = 0.17(0.0380 - x)$ and $x^2 + 0.17x - 6.46 \times 10^{-3} = 0$

$x = 0.032$ atm (to 2 significant figures)

So the pressure of: $BF_3 = 0.032$ atm ; $(CH_3)_2O = 0.032$ atm;

$(CH_3)_2O\text{-}BF_3 = (0.0380 - 0.032) = 0.006$ atm

and the total pressure is: $P_{BF_3} + P_{(CH_3)2O} + P_{(CH_3)2O\text{-}BF_3}$

0.032 + 0.032 + 0.006 = 0.0700 atm

116. The pH of a solution of 1.25 g sodium sulfanilate in 125 mL: Sulfanilic acid is a weak acid,
     so it's conjugate base should provide a slightly basic solution. The $pK_a = 3.23$
     so $K_a = 10^{-3.23} = 5.9 \times 10^{-4}$.

     The formula for sulfanilic acid is complex ($H_2NC_6H_4SO_3H$), so let's use HSA to

     represent the acid and SA to represent the anion.

     The concentration of the sodium salt (represented as NaSA) will be necessary:

     $$\frac{1.25 \text{ g}}{0.125 \text{ L}} \cdot \frac{1 \text{ mol NaSA}}{195.15 \text{ g NaSA}} = 0.0512 \text{ M}$$

     In water, the NaSA salt will exist predominantly as sodium cations and SA anions, so we
     can ignore the sodium cations.

     The reaction that occurs with the conjugate base and water is: $SA + H_2O \rightleftharpoons HSA + OH^-$

     (Note the production of hydroxyl ion –the reason we expect the solution to be basic.)

     As the anion is acting as a base, we'll need to calculate an appropriate K for the anion.

     K_b (conjugate) $= K_w/K_a$ so $\dfrac{1.0 \times 10^{-14}}{5.9 \times 10^{-4}} = 1.7 \times 10^{-11}$

     The K_b expression is: $\dfrac{[HSA][OH^-]}{[SA]} = \dfrac{[x][x]}{0.0512 - x} = 1.7 \times 10^{-11}$

     The size of the K tells us that we can safely approximate the denominator of the fraction as
     0.0512. The resulting equation is: $x^2 = (0.0512)(1.7 \times 10^{-11})$ and $x = 9.33 \times 10^{-7}$
     Since x represents the concentration of OH^- ion, the pOH = $-\log(9.33 \times 10^{-7}) = 6.03$
     The pH = 14.00-6.03 or 7.97.

Chapter 18
Principles of Reactivity: Other Aspects of Aqueous Equilibria

Reviewing Important Concepts

2. pH after equal molar amounts of the acids and bases indicated are mixed:
    (a) weak base + strong acid. (e.g. NH_3 + HCl) will produce an acidic solution (pH < 7) as the NH_4^+ ion produced during the initial reaction reacts with water to produce additional hydronium ions.
    (b) strong base + strong acid (e.g. NaOH +HNO_3) will produce a neutral solution (pH = 7) as the ions remaining at the equivalence point *do not* interact with water to any appreciable extent, leaving the solution at a neutral pH.
    (c) strong base + weak acid (e.g. KOH + CH_3CO_2H) will produce a basic solution (pH >7) as the conjugate base of the weak acid ($CH_3CO_2^-$ in our example) reacts with water to produce additional hydroxide ions.

4. Indicators are weak acids or bases, and have the characteristic that one form of the indicator(the acidic form) has one color while another form of the indicator(the conjugate of the first form) has a different color. The color change occurs at a pH that is a function of the acidic or basic components of the indicator molecule. When following an acid/base reaction, one chooses an indicator that changes color at a pH that is near the equivalence point of the acid and base reacting. It has no exact correspondence to the equivalence point.

7. The Henderson-Hasselbalch equation: $pH = pKa + log\dfrac{[conjugate\ base]}{[acid]}$
    (a) pH change when K_a increases: Arbitrarily suppose we have an acid with a $K_a = 1 \times 10^{-5}$, the pK_a would be 5. Again arbitrarily pick another acid with a $K_a = 1 \times 10^{-3}$, with a pK_a of 3. Assuming that the "log term" remains constant, the pH would **decrease**.
    (b) pH change when the acid concentration is decreased relative to its conjugate base: *Arbitrarily* pick a concentration for the two species. Suppose we begin with 0.25M base and 0.50 M acid. Assume that 0.10M acid reacts to produce 0.35 M base with 0.40 M acid (decreasing the amount of acid relative to the base).
    $$log\frac{0.25}{0.50} = -0.30 \quad and\ log\frac{0.35}{0.40} = -0.06$$

The log term *becomes less negative* as the base/acid ratio increases. The pH will then **increase** accordingly. [NOTE: There is nothing special about the values chosen. Pick two of your own and see if the trend is consistent]

13. Silver phosphate can be more soluble in water than calculated from Ksp data owing to competing reactions. The phosphate anion is also the conjugate base of a weak acid, and will undergo a reaction with water (hydrolysis), which lowers the $[PO_4^{3-}]$, and causes the solubility equilibrium to shift to the right, increasing the solubility of the phosphate salt. This is another example of LeChatelier's Principle in operation.

17. Solution producing precipitate:

For the six solutes listed, note that in water each would exist as the individual cations and anions. NaBr (aq) more accurately may be represented: Na^+ (aq) and Br^-(aq). The net equations that result are shown:

(a) Ag^+ (aq) $+$ Br^- (aq) $\rightarrow$ AgBr (s)

(b) Pb^{2+} (aq) $+$ 2 Cl^- (aq) $\rightarrow$ $PbCl_2$ (s)

(c) No precipitate is expected.

One can make these decisions in one of two ways: (1) Recalling the solubility tables—probably learned earlier or (2) Reviewing Appendix J (which contains compounds that we normally classify as insoluble—and precipitate from a solution containing the appropriate pairs of cations and anions (Ag^+ and Br^- for example).

Practicing Skills

Common Ion Effect

18. To determine how pH is expected to change, examine the equilibria in each case:

(a) NH_3 (aq) $+ H_2O$ (ℓ) $\rightleftharpoons$ NH_4^+ (aq) $+ OH^-$ (aq)

As the added NH_4Cl dissolves, ammonium ions are liberated—increasing the ammonium ion concentration and shifting the position of equilibrium to the left—reducing OH^-, and decreasing the pH.

(b) CH_3CO_2H (aq) $+$ $H_2O(\ell)$ $\rightleftharpoons$ $CH_3CO_2^-$ (aq) $+ H_3O^+$ (aq)

As sodium acetate dissolves, the additional acetate ion will shift the position of equilibrium to the left—reducing H_3O^+, and increasing the pH.

(c) NaOH (aq) $\rightarrow$ Na^+ (aq) $+$ OH^- (aq)

NaOH is a strong base, and as such is totally dissociated. Since the added NaCl does not hydrolyze to any appreciable extent—no change in pH occurs.

20. The pH of the buffer solution is:

$$K_b = \frac{[NH_{4^+}][OH^-]}{[NH_3]} = \frac{(0.20)[OH^-]}{(0.20)} = 1.8 \times 10^{-5}$$

solving for hydroxyl ion yields: $[OH^-] = 1.8 \times 10^{-5}$ M ; pOH = 4.74   pH = 9.26

22. pH of solution when 30.0 mL of 0.015 M KOH is added to 50.0 mL of 0.015 M benzoic acid:

The equilibrium affected is that of benzoic acid in water:

$$C_6H_5CO_2H + H_2O \rightleftharpoons C_6H_5CO_2^- + H_3O^+ \quad K_a = 6.3 \times 10^{-5}$$

$$K_a = \frac{[C_6H_5CO_2^-][H_3O^+]}{[C_6H_5CO_2H]} = 6.3 \times 10^{-5}$$

The KOH will *consume benzoic acid* and *produce* the conjugate *benzoate anion*.,

The base and benzoic acid react: $KOH + C_6H_5CO_2H \rightarrow C_6H_5CO_2^-K^+ + H_2O$

The pH of the solution will depend on the ratio of the conjugate pairs (acid and anion).

Determine the concentrations of the two:

Amount of benzoic acid present: 50 mL • 0.015 mol/L = 0.75 mmol of benzoic acid.

Amount of KOH added: 30 mL • 0.015 mol/L = 0.45 mmol of KOH

Amount of benzoic acid remaining: (0.75 mmol -0.45 mmol of KOH)= 0.30 mmol benzoic acid

Amount of benzoate anion produced: *0.45 mmol of benzoate anion.*

We can rearrange the K_a expression to solve for H_3O^+:

$$[H_3O^+] = 6.3 \times 10^{-5} \cdot \frac{[C_6H_5CO_2H]}{[C_6H_5CO_2^-]} = 6.3 \times 10^{-5} \cdot \frac{0.30 \text{ mmol}}{0.45 \text{ mmol}} = 4.2 \times 10^{-5}$$

pH = $-\log(4.2 \times 10^{-5})$ = 4.38

Buffer Solutions

24. The original pH of the 0.12M NH_3 solution will be:

$$\frac{[NH_4^+][OH^-]}{[NH_3]} = 1.8 \times 10^{-5} \quad \text{and} \quad \frac{x^2}{0.12 - x} = 1.8 \times 10^{-5}$$

Assuming $(0.12 - x \approx 0.12)$, $x = 1.47 \times 10^{-3}$

$[OH^-] = 1.47 \times 10^{-3}$ and pOH = 2.83 with pH = 11.17

Adding 2.2 g of NH_4Cl (0.041 mol) to 250 mL will produce an immediate increase of 0.16 M NH_4^+ (0.041 mol/0.250 L).

Substituting into the equilibrium expression as we did earlier, we get:

$$\frac{(x + 0.16)(x)}{0.12 - x} = 1.8 \times 10^{-5}$$

Assuming ($x + 0.16 \approx 0.16$) and ($0.12 - x \approx 0.12$)

$$\frac{0.16(x)}{0.12} = 1.8 \times 10^{-5} \quad x = 1.35 \times 10^{-5} = [OH^-]$$

Note the hundred-fold decrease in $[OH^-]$ over the initial ammonia solution, as predicted by LeChatelier's principle.

So pOH $= 4.88$ and pH $= 9.12$ (lower than original)

26. Mass of sodium acetate needed to change 1.00 L solution of 0.10 M CH_3CO_2H to pH = 4.5:

The equilibrium affected is that of acetic acid in water:

$$CH_3CO_2H + H_2O \rightleftharpoons CH_3CO_2^- + H_3O^+ \quad K_a = 1.8 \times 10^{-5}$$

The equilibrium expression is:

$$K_a = \frac{[CH_3CO_2^-][H_3O^+]}{[CH_3CO_2H]} = 1.8 \times 10^{-5}$$

We know the concentration of acetic acid (0.10M), and we know the desired $[H_3O^+]$:

pH $= 4.5$ so $[H_3O^+] = 10^{-4.5} = 3.2 \times 10^{-5}$

Substituting these values into the equilibrium expression gives:

$$\frac{[CH_3CO_2^-][H_3O^+]}{[CH_3CO_2H]} = 1.8 \times 10^{-5} = \frac{[CH_3CO_2^-][3.2 \times 10^{-5}]}{[0.10]}$$

We can solve for $[CH_3CO_2^-] = 0.057$ M

What mass of $NaCH_3CO_2$ would give this concentration of $CH_3CO_2^-$?

$$\frac{0.057 \text{ mol}}{1 \text{ L}} \cdot \frac{1 \text{L}}{1} \cdot \frac{82.0 \text{ g } NaCH_3CO_2}{1 \text{ mol } NaCH_3CO_2} = 4.7 \text{ g } NaCH_3CO_2$$

Using the Henderson-Hasselbalch Equation

28. The pH of a solution with 0.050 M acetic acid and 0.075 M sodium acetate:

$$pH = pK_a + \log \frac{[\text{conjugate base}]}{[\text{acid}]}$$

The pK_a for acetic acid is $= -\log(K_a) = -\log(1.8 \times 10^{-5})$ or 4.74

$$pH = 4.74 + \log \frac{[0.075]}{[0.050]} = 4.74 + 0.176 = 4.92$$

30.   (a) The pK_a for formic acid is $= -\log(K_a) = -\log(1.8 \times 10^{-4})$ or 3.74

$$pH = pK_a + \log \frac{[\text{conjugate base}]}{[\text{acid}]}$$

$$= 3.74 + \log \frac{0.035}{0.050} = 3.74 - 0.15 \text{ or } 3.59$$

(b) Ratio of conjugate pairs to increase pH by 0.5 (to pH= 4.09)

Substituting into the Henderson-Hasselbalch equation

$$4.09 = 3.74 + \log \frac{[\text{conjugate base}]}{[\text{acid}]}$$

$$+0.35 = \log \frac{[\text{conjugate base}]}{[\text{acid}]} \quad \text{so} \quad -0.35 = \log \frac{[\text{acid}]}{[\text{conjugate base}]}$$

$$0.4 = \frac{[\text{acid}]}{[\text{conjugate base}]}$$

Preparing a Buffer Solution

32. The best combination to provide a buffer solution of pH 9 is (b) the NH_3/NH_4^+ system. Note that K_a (for NH_4^+) is approximately 10^{-10}. Buffer systems are good when the desired pH is ± 1 unit from pK_a (10 in this case).

The HCl and NaCl don't form a buffer. The acetic acid/sodium acetate system would form an acidic buffer ($pK_a \approx 5$) in the pH range 4 - 6.

Adding an Acid or Base to a Buffer Solution

34.   (a) Initial pH

Need to know the concentrations of the conjugate pairs:

The equilibrium expression shows the *ratio of the conjugate pairs*, we can calculate **moles** of the conjugate pairs, and know that the ratio of the # of moles of the species will have the same value as the ratio of their concentrations!

$$CH_3CO_2H = 0.250 \text{ L} \cdot 0.150 \text{ M} = 0.0375 \text{ mol}$$

$$NaCH_3CO_2 = 4.95 \text{ g} \cdot \frac{1 \text{ mol}}{82.07 \text{ g}} = 0.0603 \text{ mol}$$

Substituting into the K_a expression:

$$\frac{[CH_3CO_2^-][H_3O^+]}{[CH_3CO_2H]} = 1.8 \times 10^{-5} = \frac{[0.0603][H_3O^+]}{[0.0375]}$$

and solving for $[H_3O^+] = 1.1 \times 10^{-5} \text{ M}$; pH $= 4.95$

(b) pH after 82. mg NaOH is added to 100. mL of the buffer. The amount of the conjugate pairs in 100/250 of the buffer is $(100/250)(0.0375 \text{ mol}) = 0.0150 \text{ mol } CH_3CO_2H$

and $(100/250)(0.0603 \text{ mol}) = 0.0241 \text{ mol } CH_3CO_2^-$

$$82 \text{ mg NaOH} \cdot \frac{1 \text{ mmol NaOH}}{40.0 \text{ mg NaOH}} = 2.05 \text{ mmol NaOH or } 0.00205 \text{ mol NaOH}$$

or 2.1 mmol NaOH (to 2 significant figures)

This base would consume an equivalent amount of CH_3CO_2H and produce an equivalent amount of $CH_3CO_2^-$.

After that process: (0.0150- 0.0021) or 0.0129 mol CH_3CO_2H and (0.0241 + 0.0021) or 0.0262 mol $CH_3CO_2^-$ are present. Substituting into the K_a expression as in part (a)

$$\frac{(0.0262)[H_3O^+]}{(0.0129)} = 1.8 \times 10^{-5} \text{ so } [H_3O^+] = 8.9 \times 10^{-6} \text{ and } pH = 5.05$$

36. (a) The pH of the buffer solution is:

$$K_b = \frac{[NH_4^+][OH^-]}{[NH_3]} = \frac{(0.250)[OH^-]}{(0.500)} = 1.8 \times 10^{-5}$$

Note : Here the data presented are given as moles (in the case of ammonium chloride) and molar concentration (in the case of ammonia). In SQ17.34 we substituted the # moles of the conjugate pairs into the K expression. Here we must first *decide* whether to substitute # moles or molar *concentrations* into the K_b expression. Either would work! What is critical to remember is that we have to have both species expression in one **or** the other form—not a mix of the two. Here I chose to convert moles of NH_4Cl into molar concentrations, and substitute.

Solving for hydroxyl ion in the K_b expression above yields: $[OH^-] = 3.6 \times 10^{-5}$ M and pOH = 4.45 so pH = 9.55

(b) pH after addition of 0.0100 mol HCl:

The basic component of the buffer (NH_3) will react with the HCl, producing more ammonium ion. The composition of the solution is then:

	NH_3	NH_4Cl
Moles present (before HCl added)	0.250	0.125
Change (reaction)	- 0.0100	+ 0.0100
Following reaction	0.240	+ 0.135

The amounts of NH_3 and NH_4Cl following the reaction with HCl are only slightly different from those amounts prior to reaction. Converting these numbers into molar concentrations (Volume is 500. mL) and substituting the concentrations into the K_b expression yields:

$$K_b = \frac{[NH_4^+][OH^-]}{[NH_3]} = \frac{(0.270)[OH^-]}{(0.480)} = 1.8 \times 10^{-5}$$

$[OH^-] = 3.2 \times 10^{-5}$; pOH = 4.50, and the new pH = 9.50.

More About Acid-Base Reactions: Titrations

38. We can calculate the amount of phenol present by converting mass to moles:

$$0.515 \text{ g C}_6\text{H}_5\text{OH} \cdot \frac{1 \text{ mol C}_6\text{H}_5\text{OH}}{94.11 \text{ g C}_6\text{H}_5\text{OH}} = 5.47 \times 10^{-3} \text{ mol phenol}$$

(a) The pH of the solution containing 5.47×10^{-3} mol phenol in 125 mL water:

With a $K_a = 1.3 \times 10^{-10}$, the K_a expression is:

$$K_a = \frac{[\text{C}_6\text{H}_5\text{O}^-][\text{H}_3\text{O}^+]}{[\text{C}_6\text{H}_5\text{OH}]} = 1.3 \times 10^{-10}$$

The stoichiometry of the compound indicates one phenoxide ion:one hydronium ion.

The initial concentration is 5.47×10^{-3} mol/ 0.125 L $= 0.0438$ M

The K_a expression becomes: $\dfrac{[x][x]}{[0.0438 - x]} = 1.3 \times 10^{-10}$

Given the magnitude of the K_a, we can safely approximate the denominator as 0.0438M.

$x^2 = 1.3 \times 10^{-10} \cdot 0.0438$ and x $= 2.39 \times 10^{-6}$ M and pH $= -\log(2.39 \times 10^{-6}) = 5.62$.

(b) At the equivalence point 5.47×10^{-3} mol of NaOH will have been added. Phenol is a monoprotic acid, so one mol of phenol reacts with one mol of sodium hydroxide. The volume of 0.123 NaOH needed to provide this amount of base is:

$$\text{moles} = \text{M} \times \text{V}$$

$$5.47 \times 10^{-3} \text{ mol NaOH} = \frac{0.123 \text{ mol NaOH}}{\text{L}} \times \text{V}$$

or 44.5 mL of the NaOH solution.

The total volume would be (125 + 44.5) or 170. mL solution.

Sodium phenoxide is a soluble salt hence the initial concentration of both sodium and phenoxide ions will be equal to:

$$\frac{5.47 \times 10^{-3} \text{ mol}}{0.170 \text{ L}} = 3.22 \times 10^{-2} \text{ M}$$

The phenoxide ion however is the conjugate ion of a weak acid and undergoes hydrolysis.

$$\text{C}_6\text{H}_5\text{O}^- \text{ (aq)} + \text{H}_2\text{O (}\ell\text{)} \rightleftharpoons \text{C}_6\text{H}_5\text{OH (aq)} + \text{OH}^- \text{ (aq)}$$

	$\text{C}_6\text{H}_5\text{O}^-$	$\text{C}_6\text{H}_5\text{OH}$	OH^-
Initial	3.22×10^{-2} M		
Change	-x	+x	+x
Equilibrium	3.22×10^{-2} M - x	x	x

$$K_b = \frac{[C_6H_5OH][OH^-]}{[C_6H_5O^-]} = \frac{1.0 \times 10^{-14}}{1.3 \times 10^{-10}} = \frac{(x)(x)}{3.22 \times 10^{-2} - x} = 7.7 \times 10^{-5}$$

Since $7.7 \times 10^{-5} \cdot 100 < 3.22 \times 10^{-2}$ we simplify: $\dfrac{(x)(x)}{3.22 \times 10^{-2}} = 7.7 \times 10^{-5}$

$$x = 1.5 \times 10^{-3}$$

At the equivalence point: $[OH^-] = 1.5 \times 10^{-3}$ M

and $[H_3O^+] = \dfrac{1.0 \times 10^{-14}}{1.5 \times 10^{-3}} = 6.5 \times 10^{-12}$ and pH $= 11.19$

While the phenoxide ion reacts with water (hydrolyzes) to some extent, the **Na$^+$** is a "spectator ion" and its concentration remains unchanged at 3.2×10^{-2} M. The concentration of phenoxide **is** reduced (albeit slightly) and at equilibrium is $(3.22 \times 10^{-2}$ M $- x)$ or 3.1×10^{-2} M.

40. (a) At the equivalence point the moles of acid $=$ moles of base.

$$(0.03678 \text{ L})(0.0105 \text{ M HCl}) = 3.86 \times 10^{-4} \text{ mol HCl}$$

If this amount of base were contained in 25.0 mL of solution, the concentration of NH$_3$ in the original solution was 0.0154 M.

(b) At the equivalence point NH$_4$Cl will hydrolyze according to the equation:

$$NH_4^+ \text{ (aq)} + H_2O \text{ (}\ell\text{)} \rightleftharpoons NH_3 \text{ (aq)} + H_3O^+ \text{ (aq)}$$

$$K_a = \frac{[NH_3][H_3O^+]}{[NH_4^+]} = \frac{1.0 \times 10^{-14}}{1.8 \times 10^{-5}} = 5.6 \times 10^{-10}$$

The salt (3.86×10^{-4} mol) is contained in $(25.0 + 36.78)$ 61.78 mL. Its concentration will

be $\dfrac{3.69 \times 10^{-4} \text{ mol}}{0.06178 \text{ L}}$ or 6.25×10^{-3} M.

Substituting into the K_a expression :

$$\frac{[H_3O^+]^2}{6.25 \times 10^{-3}} = 5.6 \times 10^{-10} \text{ and} : [H_3O^+] = 1.9 \times 10^{-6} \text{ and pH } = 5.73.$$

Since $[H_3O^+][OH^-] = 1.0 \times 10^{-14}$ then $[OH^-] = \dfrac{1.0 \times 10^{-14}}{1.9 \times 10^{-6}} = 5.3 \times 10^{-9}$ M

and the $[NH_4^+] = 6.25 \times 10^{-3}$ M

Titration Curves and Indicators

42. The titration of 0.10 M NaOH with 0.10 M HCl (a strong base vs a strong acid)

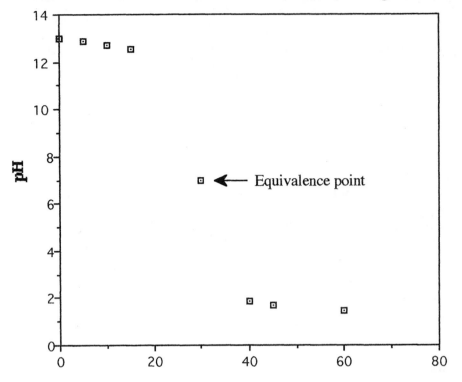

The initial pH of a 0.10 M NaOH would be pOH = - log[0.10] so pOH = 1.00
and pH = 13.00

When 15.0 mL of 0.10 M HCl have been added, one-half of the NaOH initially present will
be consumed, leaving 0.5 (0.030 L • 0.10 mol/L) or 1.50×10^{-3} mol NaOH in
45.0 mL—therefore a concentration of 0.0333 M NaOH. pOH = 1.48 and pH = 12.52.

At the equivalence point (30.0 mL of the 0.10 M acid are added) there is only NaCl present.
Since this salt does not hydrolyze, the pH at that point is exactly 7.0. The total volume
present at this point is 60.0 mL.

Once a total of 60.0 mL of acid are added, there is an excess of 3.0×10^{-3} mol of HCl.
Contained in a total volume of 90.0 mL of solution, the [HCl] = 0.0333 and the pH =1.5.

44. (a) pH of 25.0 mL of 0.11 M NH_3:

For the weak base, NH_3, the equilibrium in water is represented as:
$$NH_3 \text{ (aq)} + H_2O \text{ (}\ell\text{)} \rightleftharpoons NH_4^+ \text{ (aq)} + OH^- \text{ (aq)}$$

The slight dissociation of NH_3 would form equimolar amounts of NH_4^+ and OH^- ions.

$$K_b = \frac{[NH_4^+][OH^-]}{[NH_3]} = \frac{(x)(x)}{0.11 - x} = 1.8 \times 10^{-5}$$

Simplifying, we get : $\frac{x^2}{0.10} = 1.8 \times 10^{-5}; x = 1.4 \times 10^{-3} \text{ M} = [OH^-]$

$$pOH = 2.87 \text{ and } pH = 11.12$$

(b) Addition of HCl will consume NH_3 and produce NH_4^+ (the conjugate) according to the net equation: $NH_3 \text{ (aq)} + H^+ \text{ (ℓ)} \rightleftharpoons NH_4^+ \text{ (aq)}$

The strong acid will drive this equilibrium to the right so we will assume this reaction to be complete. Let us first calculate the moles of NH_3 initially present:

$$(0.0250 \text{ L}) (0.10 \frac{\text{mol } NH_3}{\text{L}}) = 0.00250 \text{ mol } NH_3$$

Reaction with the HCl will produce the conjugate acid, NH_4^+. The task is two-fold.

First calculate the amounts of the conjugate pair present. Second substitute the concentrations into the K_b expression. [One time-saving hint: The ratio of concentrations and the ratio of the amounts (moles) will have the same numerical value. One can substitute the amounts of the conjugate pair into the K_b expression.]

$$K_b = \frac{[NH_4^+][OH^-]}{[NH_3]} = 1.8 \times 10^{-5}$$

When 25.0 mL of the 0.10 M HCl has been added (total solution volume = 50.0 mL), the reaction is at the equivalence point. All the NH_3 will be consumed, leaving the salt, NH_4Cl. The NH_4Cl (2.50 millimol) has a concentration of 5.0×10^{-2} M. This salt, being formed from a weak base and strong acid, undergoes hydrolysis.

$$NH_4^+\text{(aq)} + H_2O \text{ (ℓ)} \rightleftharpoons NH_3 \text{ (aq)} + H_3O^+ \text{ (aq)}$$

	NH_4^+	NH_3	H_3O^+
Initial concentration	5.0×10^{-2}M		
Change	$-x$	$+x$	$+x$
Equilibrium	$5.0 \times 10^{-2} - x$	$+x$	$+x$

$$K_a = \frac{[NH_3][H_3O^+]}{[NH_4^+]} = 5.6 \times 10^{-10}$$

$$= \frac{x^2}{5.0 \times 10^{-2} - x} \approx \frac{x^2}{5.0 \times 10^{-2}} = 5.6 \times 10^{-10}$$

and $x = 5.3 \times 10^{-6} = [H_3O^+]$ and pH = 5.28 (equivalence point)

(c) The midpoint of the titration occurs when 12.50 mL of the acid have been added. At that point the amount of base and salt present are equal. An examination of the K_b expression will show that under these conditions the $[OH^-] = K_b$. So pOH of 4.75 and pH = 9.25.

(d) From the table of indicators in your text we see that one indicator to use is Methyl Red. This indicator would be yellow prior to the equivalence point and red past that point. Bromcresol green would also be suitable, being blue prior to the equivalence point and yellow-green after the equivalence point.

(e)

mL of 0.10 M HCl added	millimol HCl added	millimol NH$_3$ after reaction	millimol NH$_4^+$ after reaction	[OH$^-$] after reaction	pH
5.00	0.50	2.0	0.50	7.2×10^{-5}	9.85
15.0	1.5	1.0	1.5	1.2×10^{-5}	9.08
20.0	2.0	0.50	2.0	4.5×10^{-6}	8.65
22.0	2.2	0.30	2.2	2.5×10^{-6}	8.39

For the pH after 30.0 mL have been added: Addition of acid in excess of 25.00 mL will result in a solution which is essentially a strong acid. After the addition of 30.0 mL, substances present are: millimol HCl added: 3.00

millimol NH$_3$ present: 2.50

excess HCl present :   0.50 millimol

This HCl is present in a total volume of 55.0 mL of solution, hence the calculation for a strong acid proceeds as follows: $[H_3O^+] = 0.50$ mmol HCl/55.0 mL $= 9.1 \times 10^{-3}$ M and pH = 2.04.

A Summary

mL acid	pH
0.00	11.15
5.00	9.85
15.0	9.08
20.0	8.65
22.0	8.39
30.0	2.04

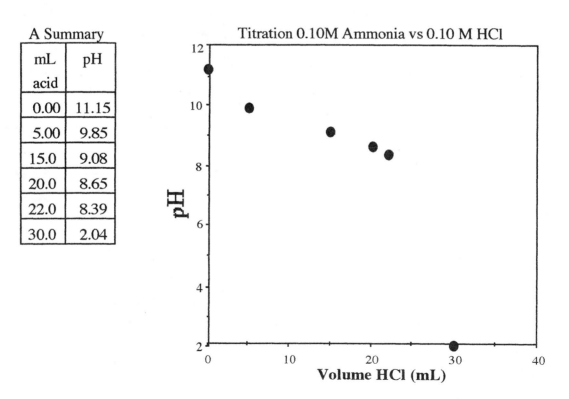

Titration 0.10M Ammonia vs 0.10 M HCl

46. Suitable indicators for titrations:

(a) HCl with pyridine: A solution of pyridinium chloride would have a pH of approximately 3. A suitable indicator would be thymol blue.

(b) NaOH with formic acid: The salt formed at the equivalence point is sodium formate. Hydrolysis of the formate ion would give rise to a basic solution (pH ≈ 8.5). Phenolphthalein would be a suitable indicator.

(c) Hydrazine and HCl: The salt, hydrazinium hydrochloride, hydrolyzes. The hydrazinium ion would produce an acidic solution (pH ≈ 4.5). A suitable indicator would be bromcresol green or methyl orange.

Solubility Guidelines

48. Two insoluble salts of

(a) Cl^-	$AgCl$ and $PbCl_2$
(b) Zn^{2+}	ZnS and $ZnCO_3$
(c) Fe^{2+}	FeS and $Fe(OH)_2$

50. Using the table of solubility guidelines, predict water solubility for the following:

(a) $(NH_4)_2CO_3$	Ammonium salts are **soluble.**
(b) $ZnSO_4$	Sulfates are generally **soluble.**
(c) NiS	Sulfides are generally **insoluble.**
(d) $BaSO_4$	Sr^{2+}, Ba^{2+}, and Pb^{2+} form **insoluble** sulfates.

Writing Solubility Product Constant Expressions

52.

Salt dissolving	K_{sp} expression	K_{sp} values
(a) $AgCN(s) \rightleftharpoons Ag^+(aq) + CN^-(aq)$	$K_{sp} = [Ag^+][CN^-]$	6.0×10^{-17}
(b) $NiCO_3(s) \rightleftharpoons Ni^{2+}(aq) + CO_3^{2-}(aq)$	$K_{sp} = [Ni^{2+}][CO_3^{2-}]$	1.4×10^{-7}
(c) $AuBr_3(s) \rightleftharpoons Au^{3+}(aq) + 3 Br^-(aq)$	$K_{sp} = [Au^{3+}][Br^-]^3$	4.0×10^{-36}

Calculating Ksp

54. Here we need only to substitute the equilibrium concentrations into the K_{sp} expression:

$$K_{sp} = [Tl^+][Br^-] = (1.9 \times 10^{-3})(1.9 \times 10^{-3}) = 3.6 \times 10^{-6}$$

56. What is K_{sp} for SrF_2?

The temptation is to calculate the number of moles of the solid that are added to water, but one must recall (a) not all that solid dissolves and (b) the concentration of the solid does not appear in the K_{sp} expression.

Note that the equilibrium concentration of $[Sr^{2+}] = 1.03 \times 10^{-3}$ M. The stoichiometry of the solid dissolving indicates two fluoride ions accompany the formation of one strontium ion. This tells us that $[F^-] = 2.06 \times 10^{-3}$ M. Substitution into the K_{sp} expression:

$K_{sp} = [Sr^{+2}][F^-]^2 = (1.03 \times 10^{-3})(2.06 \times 10^{-3})^2 = 4.37 \times 10^{-9}$.

58. For lead(II) hydroxide, the K_{sp} expression is $K_{sp} = [Pb^{+2}][OH^-]^2$.

Since we know the pH, we can calculate the $[OH^-]$.

pH = 9.15 and pOH = 14.00 - 9.15 = 4.85 so $[OH^-] = 1.4 \times 10^{-5}$.

For each mole of $Pb(OH)_2$ that dissolves, we get one mol of Pb^{+2} and two mol of OH^-.

Since $[OH^-]$ at equilibrium $= 1.4 \times 10^{-5}$ M, then $[Pb^{+2}] = 1/2 \cdot 1.4 \times 10^{-5}$ M

$K_{sp} = [Pb^{+2}][OH^-]^2 = K_{sp} = [7.0 \times 10^{-6}][1.4 \times 10^{-5}]^2 = 1.4 \times 10^{-15}$

Estimating Salt Solubility from K_{sp}

60. The solubility of AgI in water at 25°C in (a) mol/L and (b) g/L
    (a) The K_{sp} for AgI (from Appendix J) is 8.5×10^{-17}. The K_{sp} expression is:

$$K_{sp} = [Ag^+][I^-] = 8.5 \times 10^{-17}$$

Since the solid dissolves to give one silver ion/one iodide ion, the concentrations of Ag+ and I⁻ will be equal. We can write the K_{sp} as: $[Ag^+]^2 = [I^-]^2 = 8.5 \times 10^{-17}$.

Taking the square root of both sides we obtain: $[Ag^+] = [I^-] = 9.2 \times 10^{-9}$ and recognizing that each mole of AgI that dissolves per liter gives one mol of silver ion, the solubility of AgI will be 9.2×10^{-9} mol/L.

(b) The solubility in g/L: $9.2 \times 10^{-9} \dfrac{\text{mol AgI}}{\text{L}} \cdot \dfrac{234.77 \text{ g AgI}}{1 \text{ mol AgI}} = 2.2 \times 10^{-6}$ g AgI/L

62. The K_{sp} for $MgF_2 = 5.2 \times 10^{-11}$

$K_{sp} = [Mg^{2+}][F^-]^2 = 5.2 \times 10^{-11}$

if a mol/L of MgF_2 dissolve, $[Mg^{2+}] = a$ and $[F^-] = 2a$

$K_{sp} = (a)(2a)^2 = 4a^3 = 5.2 \times 10^{-11}$ and $a = 2.35 \times 10^{-4}$

(a) The molar solubility is then 2.4×10^{-4} M (to 2 sf)

(b) Solubility in g/L

$$2.4 \times 10^{-4} \dfrac{\text{mol MgF}_2}{\text{L}} \cdot \dfrac{62.3 \text{ g MgF}_2}{1 \text{ mol MgF}_2} = 1.5 \times 10^{-2} \text{ g MgF}_2 / \text{L}$$

64. $K_{sp} = [Ra^{2+}][SO_4^{2-}] = 3.7 \times 10^{-11}$ so $[Ra^{2+}] = [SO_4^{2-}] = 6.1 \times 10^{-6}$ M
    $RaSO_4$ will dissolve to the extent of 6.5×10^{-6} mol/L

Express this as grams in 100. mL (or 0.1 L)

$$\frac{6.1 \times 10^{-6} \text{ mol } RaSO_4}{1 \text{ L}} \cdot \frac{0.1 \text{ L}}{1} \cdot \frac{322 \text{ g } RaSO4}{1 \text{ mol } RaSO4} = 1.96 \times 10^{-4} \text{ g}$$

Expressed as milligrams: 0.196 mg or 0.20 mg (to 2 sf) of $RaSO_4$ will dissolve!

66. Which solute in each of the following pairs is more soluble.

	Compound	Ksp
(a)	**$PbCl_2$**	1.7×10^{-5}
	$PbBr_2$	6.6×10^{-6}
(b)	HgS (red)	4×10^{-54}
	FeS	6×10^{-19}
(c)	**$Fe(OH)_2$**	4.9×10^{-17}
	$Zn(OH)_2$	3×10^{-17}

To compare relative solubilities of two compounds of the same general formula, one can examine the Ksp value. The larger the K_{sp}, the more soluble the compound.

Common Ion Effect and Salt Solubility

68. The equilibrium for AgSCN dissolving is: $AgSCN \text{ (s)} \rightleftharpoons Ag^+ \text{ (aq)} + SCN^- \text{ (aq)}$.

As x mol/L of AgSCN dissolve in pure water, x mol/L of Ag^+ and x mol/L of SCN^- are produced.

$$K_{sp} = [Ag^+][SCN^-] = x^2 = 1.0 \times 10^{-12} \qquad \text{and } x = 1.0 \times 10^{-6} \text{ M}$$

So 1.0×10^{-6} mol AgSCN/L dissolve in pure water.

The equilibrium for AgSCN dissolving in NaSCN (0.010 M) is like that above. Equimolar amounts of Ag^+ and SCN^- ions are produced as the solid dissolves. However the $[SCN^-]$ is augmented by the soluble NaSCN.

$$K_{sp} = [Ag^+][SCN^-] = (x)(x + 0.010) = 1.0 \times 10^{-12}$$

We can simplify the expression by *assuming* that x + 0.010 ≈ 0.010.

$$(x)(0.010) = 1.0 \times 10^{-12} \quad \text{and } x = 1.0 \times 10^{-10} \text{ M}$$

The solubility of AgSCN in 0.010 M NaSCN is 1.0×10^{-10} M -- reduced by four orders of magnitude from its solubility in pure water. LeChatelier strikes again!

70. Solubility in mg/mL of AgI in (a) pure water and (b) water that is 0.020 M in Ag^+ (aq).

$$K_{sp} = [Ag^+][I^-] = 8.5 \times 10^{-17}$$

(a) In SQ 18.60 we found the solubility of AgI to be 9.2×10^{-9} mol/L or 2.2×10^{-6} g/L

Expressing this in mg/mL = 2.2×10^{-6} mg/mL

(b) The difference in solubility arises because of the presence of the 0.020 M in Ag^+. Recall that in SQ18.60 we knew that the silver and iodide ion concentrations would be equal. With the addition of the 0.020 M silver solution, this is no longer a valid assumption. If we let x mol/L of the AgI dissolve, the concentrations of Ag^+ and I^- (from the salt dissolving) will be x mol/L. The $[Ag^+]$ will be amended by 0.020 M, so we write: $K_{sp} = [x + 0.020][x] = 8.5 \times 10^{-17}$. This could be solved with the quadratic equation, but a bit of thought will simplify the process. In the (a) part we discovered that the $[Ag^+]$ was approximately 10^{-9} M. This value is small compared to 0.020. Let's use that approximation to convert the K_{sp} expression to: $[0.020][x] = 8.5 \times 10^{-17}$ and $x = 4.25 \times 10^{-15}$ M. This molar solubility translates into: $4.25 \times 10^{-15} \dfrac{mol\ AgI}{L} \cdot \dfrac{234.77\ g\ AgI}{1\ mol\ AgI} = 9.98 \times 10^{-13} g\ AgI/L$

or 1.0×10^{-12} mg/mL (to 2 sf).

Effect of Basic Anions on Salt Solubility

72. The soluble that should be more soluble in nitric acid than in pure water from the pairs:
    (a) $PbCl_2$ or PbS: PbS will be more soluble, since the S^{2-} ion will react with the nitric acid, reducing the $[S^{2-}]$, and increasing the amount of PbS that dissolves.

    (b) Ag_2CO_3 or AgI: Ag_2CO_3 will be more soluble. The CO_3^{2-} ion will react with the nitric acid, produce HCO_3^- and H_2CO_3 which will decompose to CO_2 and H_2O. The removal of carbonate will shift the equilibrium to the right, increasing the amount that dissolves.

    (c) $Al(OH)_3$ or AgCl: $Al(OH)_3$ will be more soluble. As in (b) above, the OH^- will react with the H^+ to form water. The reduction in $[OH^-]$ will increase the amount of the salt that dissolves.

Precipitation Reactions

74. Given the equation for $PbCl_2$ dissolving in water: $PbCl_2\ (s) \rightleftharpoons Pb^{+2}\ (aq) + 2\ Cl^-$ we can write the K_{sp} expression : $K_{sp} = [Pb^{+2}][Cl^-]^2 = 1.7 \times 10^{-5}$

Substituting the ion concentrations into the Ksp expression we get:
$Q = [Pb^{+2}][Cl^-]^2 = (0.0012)(0.010)^2 = 1.2 \times 10^{-7}$
Since Q is less than K_{sp} , no $PbCl_2$ precipitates.

76. If $Zn(OH)_2$ is to precipitate, the reaction quotient (Q) must exceed the K_{sp} for the salt.
    4.0 mg of NaOH in 10. mL corresponds to a concentration of:

$$[OH^-] = \frac{4.0 \times 10^{-3} \text{ g NaOH}}{0.0100 \text{ L}} \cdot \frac{1 \text{ mol NaOH}}{40.0 \text{ g NaOH}} = 0.01 \text{ M}$$

The value of Q is : $[Zn^{2+}][OH^-]^2 = (1.6 \times 10^{-4})(1.0 \times 10^{-2})^2 = 1.6 \times 10^{-8}$

The value of Q is greater than the K_{sp} for the salt (4.5×10^{-17}) , so $Zn(OH)_2$ precipitates.

78. The molar concentration of Mg^{2+} is:

$$\frac{1350 \text{ mg Mg}^{2+}}{1 \text{ L}} \cdot \frac{1 \text{ g Mg}^{2+}}{1000 \text{ mg Mg}^{2+}} \cdot \frac{1 \text{ mol Mg}^{2+}}{24.3050 \text{ g Mg}^{2+}} = 5.55 \times 10^{-2} \text{M}$$

For $Mg(OH)_2$ to precipitate, Q must be greater than the K_{sp} for the salt (5.6×10^{-12}).

$Q = [Mg^{2+}][OH^-]^2 = (5.55 \times 10^{-2})(OH^-)^2 = 5.6 \times 10^{-12}$ so

$(OH^-)^2 = 5.6 \times 10^{-12}/5.55 \times 10^{-2}$ or 1.01×10^{-10} , and $[OH^-] = 1.0 \times 10^{-5}$ M.

Solubility and Complex ions

80. Show that the equation: $AuCl(s) + 2 CN^- (aq) \rightleftharpoons Au(CN)_2^- (aq) + Cl^- (aq)$ is the sum of two

equations.

AuCl dissolving:	$AuCl(s) \rightleftharpoons Au^+ (aq) + Cl^- (aq)$	$K_{sp} = 2.0 \times 10^{-13}$
$Au(CN)_2^-$ formation:	$Au^+ (aq) + 2 CN^- (aq) \rightleftharpoons Au(CN)_2^- (aq)$	$K_f = 2.0 \times 10^{38}$
The net equation	$AuCl(s) + 2 CN^- (aq) \rightleftharpoons Au(CN)_2^- (aq) + Cl^-$	$K_{net} = K_{sp} \cdot K_f$

$K_{net} = 2.0 \times 10^{-13} \cdot 2.0 \times 10^{38} = 4.0 \times 10^{25}$

Separations

82. Separate the following pairs of ions:

(a) Ba^{2+} and Na^+ : Since most sodium salts are soluble, it is simple to find a barium salt

which is not soluble--e.g. the sulfate. Addition of dilute sulfuric acid should provide a

source of SO_4^{2-} ions in sufficient quantity to precipitate the barium ions, but not the

sodium ions.

[$K_{sp \ BaSO4} = 1.1 \times 10^{-10}$ $K_{sp \ Na2SO4}$ = not listed, owing to the large
solubility of sodium sulfate]

(b) Ni^{2+} and Pb^{2+}: The carbonate ion will serve as an effective reagent for selective

precipitation of the two ions. A solution of Na_2CO_3 can be added, and the less soluble

$PbCO_3$ will begin precipitating first. The 7 orders of magnitude difference in their

K_{sp}'s should provide satisfactory separation.

[$K_{sp \ PbCO3} = 7.4 \times 10^{-14}$ $K_{sp \ NiCO3} = 1.4 \times 10^{-7}$]

General Questions

84. Will $BaSO_4$ precipitate?

Calculate the concentrations of barium and sulfate ions (after the solutions are mixed).

$\dfrac{48 \text{ mL}}{72 \text{ mL}} \cdot 0.0012 \text{ M Ba}^{2+} = 0.0008 \text{ M Ba}^{2+}$ and

$\dfrac{24 \text{ mL}}{72 \text{ mL}} \cdot 1.0 \times 10^{-6} \text{ M SO}_4^{2-} = 3.3 \times 10^{-7} \text{ M SO}_4^{2-}$

Substituting in the K_{sp} expression: $Q = [Ba^{+2}][SO_4^{2-}] = (8.0 \times 10^{-4})(3.3 \times 10^{-7})$

$Q = 2.7 \times 10^{-10}$ Q is larger than the Ksp for the solid(1.1×10^{-10}) so $BaSO_4$

precipitates. Note that we are assuming that the sulfuric acid totally dissociates into protons and sulfate ions!

86. The pH and $[H_3O^+]$ when 50.0 mL of 0.40 M NH_3 is mixed with 50.0 mL of 0.40 M HCl:

The number of moles of each reactant: (0.40 mol HCl/L)(0.050 L) = 0.020 mol HCl and (0.40 mol NH_3/L)(0.050 L) = 0.020 mol HCl . This is the equivalence point. After the acid and base react, only NH_4Cl is present, in the form of NH_4^+ and Cl^- ions.

The concentration of the ammonium ions is 0.020 mol/0.100 L or 0.20 M

Since ammonium ions react with water to an appreciable extent (They are the conjugate acids of a weak base, after all!), we write. $NH_4^+(aq) + H_2O (\ell) \rightleftharpoons NH_3(aq) + H_3O^+(aq)$

Substituting into the Ka expression for the ammonium ions:

$$\frac{[NH_3][H_3O^+]}{[NH_4^+]} = \frac{[x][x]}{[0.20 - x]} = 5.6 \times 10^{-10}$$

The relative magnitudes of the ammonium ion concentration and the Ka for the ammonium ion will permit us to approximate the denominator of the fraction as 0.20. We solve:

$x^2 = (0.20)(5.6 \times 10^{-10})$ and $x = 1.1 \times 10^{-5}$ M (to 2 sf).

So and $[H_3O^+] = 1.1 \times 10^{-5}$ M and pH = 4.98

88.   Compounds in order of increasing solubility in H_2O:

Compound	K_{sp}
$BaCO_3$	2.6×10^{-9}
Ag_2CO_3	8.5×10^{-12}
Na_2CO_3	not listed: Na_2CO_3 is very soluble in water and is certainly the most soluble salt of the 3.

To determine the relative solubilities find the molar solubilities.

For $BaCO_3$: $K_{sp} = [Ba^{2+}][CO_3^{2-}] = (x)(x) = 2.6 \times 10^{-9}$ and $x = 5.1 \times 10^{-5}$

The molar solubility of $BaCO_3$ is 5.1×10^{-5} M.

and for Ag_2CO_3: $K_{sp} = [Ag^+]^2[CO_3^{2-}]$ $= (2x)^2(x) = 8.5 \times 10^{-12}$

$$4x^3 = 8.5 \times 10^{-12}$$

$$\text{and } x = 1.3 \times 10^{-4}$$

The molar solubility of Ag_2CO_3 is 1.3×10^{-4} M.

In order of increasing solubility: $BaCO_3 < Ag_2CO_3 < Na_2CO_3$

90. pH of a solution that contains 5.15 g NH_4NO_3 and 0.10 L of 0.15 M NH_3:

The concentration of the ammonium ion (from the nitrate salt) is:

$$5.15 \text{ g NH}_4NO_3 \cdot \frac{1 \text{ mol NH}_4NO_3}{80.04 \text{ g NH}_4NO_3} \cdot \frac{1}{0.10 \text{ L}} = 0.64 \text{ M}$$

We can calculate the pH using the K_b expression for ammonia:

$$K_b = \frac{[NH_4^+][OH^-]}{[NH_3]} = \frac{(0.64)(OH^-)}{(0.15)} = 1.8 \times 10^{-5} \text{ and solving for the hydroxide ion}$$

concentration $x = \dfrac{(1.8 \times 10^{-5})(0.15)}{(0.64)} = 4.2 \times 10^{-6}$ M. and pOH = 5.37 and pH = 8.63

Diluting the solution does not change the pH of the solution, since the dilution affects the concentration of both members of the conjugate pair. Since the pH of the buffer is a function of the *ratio* of the conjugate pair, the pH does not change.

92. For the titration of aniline hydrochloride with NaOH:

(a) initial pH: Recall that the hydrochloride (represented in this example as HA) is a weak acid: ($K_a = 2.4 \times 10^{-5}$) Use a K_a expression to solve for H_3O^+:

$$K_a = \frac{[A^-][H_3O^+]}{[HA]} = 2.4 \times 10^{-5} \cdot 0.10 = x^2 = 2.4 \times 10^{-6}$$

$x = 1.5 \times 10^{-3}$ M $= [H_3O^+]$ and pH = 2.81

(b) pH at the equivalence point:

At the equivalence point we know that the number of moles of acid - # moles base (by definition) Since we have 0.00500 mol of HA (M •V), we must ask the question, "How much 0.185 M NaOH contains 0.00500 mol NaOH?"

That volume is 0.00500 mol OH^- • 1L/ 0.185 mol NaOH = 27.0 mL of the NaOH solution.

The total volume is then 50 + 27 mL = 0.0770 L.

The [HA] = 0.00500 mol/0.0770 L = 0.0649 M

We can calculate the pH of the solution using the K_b expression for A:

$$K_b = \frac{[HA][OH^-]}{[A]} = \frac{1.0 \times 10^{-14}}{2.4 \times 10^{-5}} \text{ and knowing that the } [HA] = [OH^-]$$

$x^2 = 4.2 \times 10^{-12} \cdot 0.0649$ and $x = 5.2 \times 10^{-6}$ M; pOH = 5.28 and pH = 8.72

(c) pH at the midpont:

This portion is easy to solve if you examine the Ka expression or the Henderson-Hasselbalch expression for this system:

From the K_b expression above, note that the conjugate pairs are found in the numerator and denominator of the right-side of the K_b term (the same applies to Ka expressions).

At the mid-point you have reacted half the acid, (using 13.5 mL—see part (b))forming its conjugate base. The result is that the concentrations of the conjugate pairs are equal and the pOH = pK_b and pH = pK_a So pH = $-\log(2.4 \times 10^{-5}) = 4.62$.

(d) With a pH= 8.72 at the equivalence point, o-cresolphthalein or phenolphthalein would serve as an adequate indicator.

(e) The pH after the addition of 10.0, 20.0 and 30.0 mL base:

The volumes of base correspond to

10 mL (0.185 mol/L • 0.0100 L)	0.00185 mol NaOH
20.0 mL (0.185 mol/L • 0.0200 L)	0.00370 mol NaOH
30.0 mL(0.185 mol/L • 0.0300L)	0.00555 mol NaOH

Each mol of NaOH consumes a mol of aniline hydrochloride(HA) and produces an equal number of mol of aniline(A).

Recall that we began with 0.00500 mol of the acid (which we're representing as HA) Substitution into the Ka expression yields:

$$\frac{[A^-][H_3O^+]}{[HA]} = 2.4 \times 10^{-5} \text{ and rearranging them } [H_3O^+] = 2.4 \times 10^{-5} \cdot \frac{[HA]}{[A]}$$

After 10 mL of base are added, 0.00185 mol HA are consumed, and 0.00185 mol A produced. The acid remaining is (000500 - 0.00185) mol.

Substituting into the rearranged equation:

$$[H_3O^+] = 2.4 \times 10^{-5} \cdot \frac{[0.00315 \text{ mol}]}{[0.00185 \text{ mol}]} = 4.1 \times 10^{-5} \text{ and pH} = -\log (4.1 \times 10^{-5}) = 4.39$$

After 20 mL of base are added, 0.00370 mol HA are consumed, and 0.00370 mol A produced. The acid remaining is (000500 - 0.00370)mol.

Substituting into the rearranged equation:

$$[H_3O^+] = 2.4 \times 10^{-5} \cdot \frac{[0.00130 \text{ mol}]}{[0.00370 \text{ mol}]} = 8.4 \times 10^{-6} \text{ and pH} = -\log (8.4 \times 10^{-6}) = 5.07$$

After 30 mL of base are added, all the acid is consumed, and excess strong base is present (0.00555 mol − 0.00500)mol. We can treat this solution as one of a strong base. The volume of the solution is (50.0 + 30.0 or 80.0 mL.). The concentration of the NaOH is:

0.00055 mol/ 0.080 L = 0.006875 M so pOH = -log(0.006875) and pOH=2.16, the pH=11.84.

(f) The approximate titration curve:

Vol base	pH
0.0	2.81
10.0	4.39
13.5	4.62
20.0	5.07
27.0	8.72
30.0	11.84

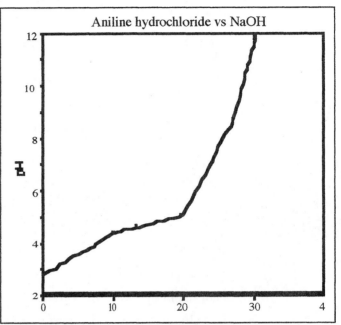

Aniline hydrochloride vs NaOH

94. Make a buffer of pH= 2.50 from 100. mL of 0.230 M H_3PO_4.

For the first ionization of the acid, K_a= 7.5 x 10^{-3} (Appendix H)

The pK_a of the acid = 2.12.The Henderson-Hasselbalch equation will help to determine the ratio of the conjugate pair to make the pH 2.50.

We'll assume that (since the second K_a is approximately 10^{-8}, that we're dealing only with the first ionization. This *is an approximation*.

$$pH = pK_a + \log \frac{[\text{conjugate base}]}{[\text{acid}]} \quad \text{so } 2.50 = 2.12 + \log \frac{[\text{conjugate base}]}{[\text{acid}]}$$

$$0.38 = \log \frac{[A]}{[HA]} \text{ and the ratio of } \frac{[A]}{[HA]} = 2.37$$

The amount of HA = (0.100 L • 0.230M HA) = 0.0230 mol HA.

Rearranging the ratio: [A] = 2.37[HA]. Knowing that the acid present is *either* the molecular acid, HA, or the conjugate base of the acid, A, we can write:

A + HA = 0.0230 mol and with A = 2.37 HA

2.37 HA + HA = 0.0230mol ; 3.37 HA = 0.0230 mol; and HA = 0.0068 mol.

We need to consume (0.0230 mol HA- 0.0068 mol HA) or 0.0162 mol HA.

This requires 0.0162 mol NaOH, and from 0.150M NaOH we need:

0.0162 mol NaOH = 0.150 mol/L • V and V = 0.108 L or 110 mL of NaOH (to 2 sf).

96. For the titration of 0.150 M ethylamine ($K_b = 4.27 \times 10^{-4}$) with 0.100 M HCl:

(a) pH of 50.0 mL of 0.0150 M $CH_3CH_2NH_2$:

For the weak base, $CH_3CH_2NH_2$, the equilibrium in water is represented as:

$$CH_3CH_2NH_2 \text{ (aq)} + H_2O \text{ (}\ell\text{)} \rightleftharpoons CH_3CH_2NH_3^+ \text{ (aq)} + OH^- \text{ (aq)}$$

Note : The K_b for ethylamine is given as 4.27×10^{-4}.

The slight dissociation of $CH_3CH_2NH_2$ would form equimolar amounts of $CH_3CH_2NH_3^+$ and OH^- ions.

$$K_b = \frac{[CH_3CH_2NH_3^+][OH^-]}{[CH_3CH_2NH_2]} = \frac{(x)(x)}{0.150 - x} = 4.27 \times 10^{-4}$$

The quadratic equation will be needed to find an exact solution.

$x^2 = 6.405 - 4.27 \times 10^{-4}x$

Rearranging: $x^2 + 4.27 \times 10^{-4}x - 6.405 \times 10^{-5} = 0$

and solving for $x = 7.79 \times 10^{-3}$; $pOH = -\log(7.79 \times 10^{-3}) = 2.11$

and $pH = 11.89$

(b) pH at the midpoint of the titration:

The volume of acid added isn't that important since the amounts of base and conjugate acid will be equal. In part (d), we find that 75.0 mL of acid are required for the equivalence point, so the volume of acid at this point is $(0.5 \cdot 75.0 \text{ mL})$ Substituting that fact into the equilibrium expression we obtain:

$$\frac{[CH_3CH_2NH_3^+][OH^-]}{[CH_3CH_2NH_2]} = \frac{(x)[OH^-]}{(x)} = 4.27 \times 10^{-4} \text{ so we can see that the}$$

$[OH^-] = 4.27 \times 10^{-4}$ and $pOH = 3.37$ and $pH = (14.00 - 3.37) = 10.63$

(c) pH when 75% of the required acid has been added:

Amount of base initially present is $(0.050 \text{ L} \cdot 0.150 \text{ M})$ or 0.0075 moles base.

So 75% of this amount is 0.00563 mol. requiring 0.00563 mol of HCl.

The volume of 0.100 M HCl containing that # mol is

0.00563 mol HCl/0.100 M = 0.0563 L or 56.3 mL

So 0.001875 mol base remain. Substituting into the K_b expression:

$$\frac{[CH_3CH_2NH_3^+][OH^-]}{[CH_3CH_2NH_2]} = \frac{0.00563 \text{ mol} \cdot [OH^-]}{0.001875 \text{ mol}} = 4.27 \times 10^{-4}$$

solving for hydroxyl ion concentration yields: 1.42×10^{-4} M and pOH=3.85 and pH = 10.15.

(d) pH at the equivalence point:

At the equivalence point, there are equal # of moles of acid and base. The number of moles of ethylamine = (0.050 L • 0.150 M) or 0.00750 moles ethylamine .

That amount of HCl would be:

$$7.50 \times 10^{-3} \text{ mol HCl} \cdot \frac{1 \text{ L}}{0.100 \text{ mol HCl}} = 0.0750 \text{ L (or 75.0 mL)}$$

This total amount of solution would be: 75.0 + 50.0 = 125.0 mL

Since we have added equal amounts of acid and base, the reaction between the two will result in the existence of only the salt.

and the concentration of salt would be: $\dfrac{7.50 \times 10^{-3} \text{ mol}}{0.1250 \text{ L}} = 6.00 \times 10^{-2} \text{ M}$

Recall that the salt will act as a weak acid and we can calculate the K_a:

$$CH_3CH_2NH_3^+ \text{ (aq)} + H_2O \text{ (}\ell\text{)} \rightleftharpoons CH_3CH_2NH_2 \text{ (aq)} + H_3O^+ \text{ (aq)}$$

The equilibrium constant (K_a) would be $\dfrac{K_w}{K_b} = \dfrac{1.0 \times 10^{-14}}{4.27 \times 10^{-4}} = 2.3 \times 10^{-11}$

The equilibrium expression would be:

$$\frac{[CH_3CH_2NH_2][H_3O^+]}{[CH_3CH_2NH_3^+]} = 2.3 \times 10^{-11}$$

Given that the salt would hydrolyze to form equal amount of ethylamine and hydronium ion (as shown by the equation above). If we represent the concentrations of those species as x, then we can write (using our usual approximation):

$$\frac{x^2}{6.00 \times 10^{-2}} = 2.3 \times 10^{-11} \text{ and solving for x} = 1.19 \times 10^{-6}$$

Since x represents [H_3O^+], then pH = $- \log(1.19 \times 10^{-6}) = 5.93$

(e) pH after addition of 10.0 mL of HCl more than required:

Past the equivalence point, the excess strong acid controls the pH. To determine the pH, we need only calculate the concentration of HCl. In part (d) we found that the total volume at the equivalence point was 125.0 mL. The addition of 10.0 mL will bring that total volume to 135.0 mL of solution. The # of moles of excess HCl is:

(0.100 mol HCl/L • 0.0100 L = 0.00100 mol HCl contained within 135.0 mL, for a concentration of 7.41×10^{-3} M HCl. Since HCl is a strong acid, the pH = $-\log(7.41 \times 10^{-3})$ or 2.13.

(f) The titration curve.

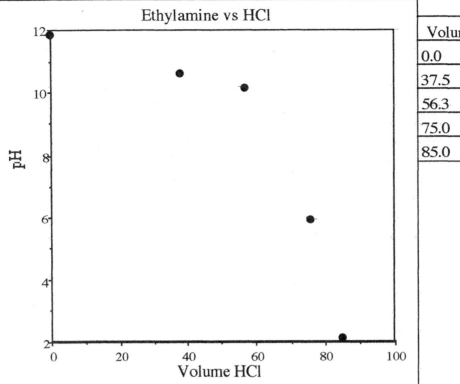

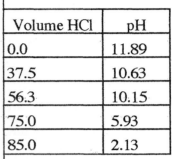

Volume HCl	pH
0.0	11.89
37.5	10.63
56.3	10.15
75.0	5.93
85.0	2.13

(g) Suitable indicator for endpoint: Alizarin or Bromcresol purple would be suitable indicators. (See page 700, Figure 17.2)

98. The effect on pH of:
   (a) Adding $CH_3CO_2^-Na^+$ to 0.100 M CH_3CO_2H:

   The equilibrium affected is: $CH_3CO_2H + H_2O \rightleftharpoons CH_3CO_2^- + H_3O^+$

   Addition of sodium acetate increases the concentration of acetate ion. The equilibrium will shift to the left—reducing hydronium ion, and increase the pH.
   (b) Adding $NaNO_3$ to 0.100 M HNO_3: No effect on pH. Since nitric acid is a strong acid, the equilibrium lies very far to the right. Addition of NO_3^- will not shift the equilibrium.

   (c) The effects differ owing to the nature of the two acids. Nitric acid (a strong acid) exists as hydronium and nitrate ions. Acetic acid exists as the molecular acid and acetate ion. The addition of conjugate base of both acids ($CH_3CO_2^-$ and NO_3^-) will affect the acetic acid equilibrium but not the nitric acid system.

100. Buffer (200.0 mL) of Na_3PO_4 and Na_2HPO_4 has pH=12.00

   (a) Component present in the larger amount:

   The Henderson-Hasselbalch equation will be of use here:

$$pH = pK_a + \log \frac{[\text{conjugate base}]}{[\text{acid}]}$$

The K_{a3} for the acid is 3.6×10^{-13}.

The pKa is then $-\log (3.6 \times 10^{-13}) = 12.44$. and solve for the log term.

$$12.00 = 12.44 + \log \frac{[PO_4^{3-}]}{[HPO_4^{2-}]} \quad \text{so} \quad -0.44 = \log \frac{[PO_4^{3-}]}{[HPO_4^{2-}]}$$

Solving for the ratio of the conjugates $10^{-0.44}$ gives a value $= 0.36$, so the acidic component is present in the greater amount (approximately 3 x the concentration of phosphate).

(b) With a concentration of Na_3PO_4 of 0.400 M, the mass of Na_2HPO_4

With the ratio of $0.36 = \dfrac{0.400 \text{ M}}{x}$ we can solve for the concentration of the Na_2HPO_4

$x = 0.400/0.36 = 1.10$ M Na_2HPO_4. The mass is then

(1.10 mol/L • 0.200 L • 141.96 g/mol) or 31.3 g of Na_2HPO_4

(c) Substance needed to change pH to 12.25:

Using Henderson-Hasselbalch as above we get:

$$12.25 = 12.44 + \log \frac{[PO_4^{3-}]}{[HPO_4^{2-}]} \quad \text{so} \quad -0.19 = \log \frac{[PO_4^{3-}]}{[HPO_4^{2-}]}$$

Solving for the ratio again, we get a ratio of phosphate to hydrogen phosphate of 0.646. Note that this is larger than the previous ratio, so we need to add phosphate ion to the solution to change the pH as desired.

Note that $[Na_2HPO_4] = 1.10$ M (from (b) above) We can solve for the phosphate ion concentration needed to provide a ratio of 0.646.

$$0.646 = \frac{x}{1.10 \text{ M}} \quad \text{and} \quad (0.646)(1.10) = 0.710 \text{ M } PO_4^{3-}$$

The # of moles we need is $(0.710 \text{ M } PO_4^{3-})(0.2000 \text{ L}) = 0.142$ mol Na_3PO_4.

We originally had 0.400 M phosphate in 200.0 mL, giving 0.0800 mol Na_3PO_4.

The amount we need to add is: (0.142 mol Na_3PO_4 _ 0.0800 mol Na_3PO_4)= 0.062 mol

The mass of sodium phosphate corresponding to this number of moles is:

0.062 mol • 163.94 g Na_3PO_4/mol = 10.1 g Na_3PO_4 (to 10. g Na_3PO_4 to 2sf).

102. These metal sulfates are 1:1 salts. The K_{sp} expression has the general form:

$$K_{sp} = [M^{2+}][SO_4^{2-}]$$

To determine the $[SO_4^{2-}]$ necessary to begin precipitation, we can divide the equation by the metal ion concentration to obtain:

$$\frac{K_{sp}}{[M^{2+}]} = [SO_4^{2-}]$$

The concentration of the three metal ions under consideration are each 0.10 M. Substitution of the appropriate K_{sp} for the sulfates and 0.10 M for the metal ion concentration yields the sulfate ion concentrations in the table below. As the soluble sulfate is added to the metal ion solution, the sulfate ion concentration increases from zero molarity. The lowest sulfate ion concentration is reached first, with higher concentrations reached later. The order of precipitation is listed in the last column of the table below.

Compound	K_{sp}	Maximum $[SO_4^{2-}]$	Order of Precipitation
$BaSO_4$	1.1×10^{-10}	1.1×10^{-9}	1
$SrSO_4$	3.4×10^{-7}	3.4×10^{-6}	2

104. The equation $AgCl(s) + I^-(aq) \rightleftharpoons AgI(aq) + Cl^-$

can be obtained by adding two equations:

1. $AgCl(s) \rightleftharpoons Ag^+(aq) + Cl^-(aq)$ $K_{sp1} = 1.8 \times 10^{-10}$

2. $Ag^+(aq) + I^-(aq) \rightleftharpoons AgI(s)$ $\dfrac{1}{K_{sp2}} = 1.2 \times 10^{16}$

The net equation: $AgCl(s) + I^-(aq) \rightleftharpoons AgI(aq) + Cl^-$

$K_{net} = K_{sp1} \cdot \dfrac{1}{K_{sp2}} = (1.8 \times 10^{-10})(1.2 \times 10^{16}) = 2.2 \times 10^6$

The equilibrium lies to the right. This indicates that AgI will form if I^- is added to a saturated solution of $AgCl$.

106. Regarding the solubility of barium and calcium fluorides:

(a) $[F^-]$ that will precipitate maximum amount of calcium ion:

BaF_2 begins to precipitate when the ion product just exceeds the K_{sp}.

$K_{sp} = [Ba^{2+}][F^-]^2 = 1.8 \times 10^{-7}$, and the concentration of barium ion = 0.10 M

$[F^-] = (1.8 \times 10^{-7}/0.10)^{0.5} = 1.3 \times 10^{-3}$ M

This would be the maximum fluoride concentration permissible.

(b) $[Ca^{2+}]$ remaining when BaF_2 just begins to precipitate:

The $[Ca^{2+}]$ ion remaining will be calculable from the K_{sp} expression for CaF_2.

$K_{sp} = [Ca^{2+}][F^-]^2 = 5.3 \times 10^{-11}$ and we know that the $[F^-]$ is 1.3×10^{-3} M.

So $\dfrac{5.3 \times 10^{-11}}{\left(1.3 \times 10^{-3}\right)^2} = 2.9 \times 10^{-5}$ M

108. Regarding the precipitation of $CaSO_4$ and $PbSO_4$ from a solution 0.010 M in each metal:
    (a) Examine the K_{sp} of the two salts:
        K_{sp} for $CaSO_4 = 4.9 \times 10^{-5}$ and for $PbSO_4 = 2.5 \times 10^{-8}$
        These data indicate that $PbSO_4$ will begin to precipitate first—as it is the less soluble of

        the two salts. This can be found by examining the general Ksp expression for 1:1 salts:
        $K_{sp} = [M^{2+}][SO_4^{2-}]$. With both metal concentrations at 0.010 M, the sulfate ion for

        the less soluble sulfate will be exceeded first.
    (b) When the calcium salt just begins to precipitate, the $[Pb^{2-}] = ?$.

        We know that the metal ion concentration is 0.010 M. So the sulfate ion concentration

        when the more soluble calcium sulfate begins to precipitate is:
        $K_{sp} = [0.010][SO_4^{2-}] = 4.9 \times 10^{-5}$ and $[SO_4^{2-}] = (4.9 \times 10^{-5}/0.010) = 4.9 \times 10^{-3}$ M

        At that point the $[Pb^{2+}]$ would be:
$$K_{sp} = [Pb^{2+}][SO_4^{2-}] = 2.5 \times 10^{-8}$$
        Then $[Pb^{2+}][4.9 \times 10^{-3}] = 2.5 \times 10^{-8}$ and $[Pb^{2+}] = (2.5 \times 10^{-8}/4.9 \times 10^{-3})$ or
        5.1×10^{-6} M.

110. To separate CuS and $Cu(OH)_2$:
    The K_{sp} for CuS is 6×10^{-37} and that for $Cu(OH)_2$ is 2.2×10^{-20}.

    The small size of the K_{sp} for CuS indicates that it is not very soluble. Addition of acid will

    cause the more soluble hydroxide to dissolve, leaving the CuS in the solid state.

Chapter 19
Principles of Reactivity: Entropy and Free Energy

Reviewing Important Concepts

4. Processes product-favored or reactant-favored under standard conditions:
   (a) $Hg\,(\ell) \rightarrow Hg\,(s)$. Since Hg exists at room T as a liquid, we predict reactant-favored.
   (b) $H_2O\,(g) \rightarrow H_2O\,(\ell)$ Water exists mainly as a liquid at room T, and water vapor condenses spontaneously, so product-favored.
   (c) $2\,HgO\,(s) \rightarrow Hg\,(\ell) + O_2\,(g)$ Reactant-favored, since energy is required to carry out this process
   (d) $C\,(s) + O_2\,(g) \rightarrow CO_2\,(g)$ Product-favored since carbon will burn readily to form CO_2.
   (e) $NaCl\,(s) \rightarrow NaCl\,(aq)$ Table salt spontaneously dissolves in water, so product-favored.
   (f) $CaCO_3\,(s) \rightarrow Ca^{2+}\,(aq) + CO_3^{2-}\,(aq)$ Calcium carbonate does not dissolve to any appreciable extent, so reactant-favored.

Practicing Skill
Entropy Comparisons

12. Compound with the higher entropy:
    (a) $CO_2\,(s)$ at $-78°$ vs **CO_2 (g) at 0 °C**: Entropy increases with temperature.
    (b) $H_2O\,(\ell)$ at 25 °C vs **H_2O (ℓ) at 50 °C**: Entropy increases with temperature.
    (c) $Al_2O_3\,(s)$ (pure) vs **Al_2O_3 (s) (ruby)**: Entropy of a solution (even a solid one) is greater than that of a pure substance.
    (d) **1 mol N_2 (g) at 1bar** vs 1 ml N_2 (g) at 10 bar: With the increased P, molecules have greater order.

14. Compound with higher standard entropy:
    (a) $O_2\,(g)$ vs **CH_3OH (g)**: Entropy increases with molecular complexity.
    (b) HF (g) vs HCl (g) vs **HBr (g)**: Entropy increases with molecular size (mass)
    (c) $NH_4Cl\,(s)$ vs **NH_4Cl (aq)**: Entropy of solutions is greater than that of the solid.
    (d) **HNO_3 (g)** vs $HNO_3\,(\ell)$ vs $HNO_3\,(aq)$: Entropy of the gaseous state is very high. The liquid state has relatively lower entropy.

Predicting and Calculating Entropy Changes

16. Entropy changes:

(a) KOH (s) $\rightarrow$ KOH(aq)

$$\Delta S° = 91.6 \frac{J}{K \cdot mol}(1 \text{ mol}) - 78.9 \frac{J}{K \cdot mol}(1 \text{ mol})$$

$$= + 12.7 \frac{J}{K}$$

The increase in entropy reflects the greater disorder of the solution state.

(b) Na (g) $\rightarrow$ Na (s)

$$\Delta S° = 51.21 \frac{J}{K \cdot mol}(1 \text{ mol}) - 153.765 \frac{J}{K \cdot mol}(1 \text{ mol})$$

$$= -102.55 \frac{J}{K}$$

The lower entropy of the solid state is evidenced by the negative sign.

(c) Br_2 (ℓ) $\rightarrow$ Br_2 (g)

$$\Delta S° = 245.42 \frac{J}{K \cdot mol}(1 \text{ mol}) - 152.2 \frac{J}{K \cdot mol}(1 \text{ mol})$$

$$= + 93.2 \frac{J}{K}$$

The increase in entropy is expected with the transition to the disordered state of a gas.

(d) HCl (g) $\rightarrow$ HCl (aq)

$$\Delta S° = 56.5 \frac{J}{K \cdot mol}(1 \text{ mol}) - 186.2 \frac{J}{K \cdot mol}(1 \text{ mol})$$

$$= -129.7 \frac{J}{K}$$

The lowered entropy reflects the greater order of the solution state over the gaseous state.

18. For the reaction: 2 C (graphite) + 3 H_2 (g) $\rightarrow$ C_2H_6 (g)

$$\Delta S° = 1 \cdot S° \, C_2H_6 - [2 \cdot S° \, C \text{ (graphite)} + 3 \cdot S° \, H_2 \text{ (g)}]$$

$$= (1 \text{ mol})(229.2 \frac{J}{K \cdot mol}) -$$

$$[(2 \text{ mol})(5.6 \frac{J}{K \cdot mol}) + (3 \text{ mol})(130.7 \frac{J}{K \cdot mol})]$$

$$= - 174.1 \frac{J}{K}$$

20. Standard Entropy change for compound formation from elements:

(a) HCl (g): Cl_2 (g) + H_2 (g) → 2 HCl (g)

$$\Delta S° = 2 • S° \text{ HCl (g)} - [1 • S° \text{ } Cl_2 \text{ (g)} + 1 • S° \text{ } H_2 \text{ (g)}]$$

$$= (2 \text{ mol})(186.2 \frac{J}{K•mol}) -$$

$$[(1 \text{ mol})(223.08 \frac{J}{K•mol}) + (1 \text{ mol})(130.7 \frac{J}{K•mol})]$$

$$= +18.6 \frac{J}{K}$$

(b) $Ca(OH)_2$ (s): Ca(s) + O_2 (g) + H_2 (g) — $Ca(OH)_2$ (s)

$$\Delta S° = 1 • S° \text{ } Ca(OH)_2 \text{ (s)} - [1 • S° \text{ Ca (s)} + 1 • S° \text{ } O_2 \text{ (g)} + 1 • S° \text{ } H_2 \text{ (g)}]$$

$$= (1 \text{ mol})(83.39 \frac{J}{K•mol}) -$$

$$[(1 \text{ mol})(41.59 \frac{J}{K•mol}) + (1 \text{ mol})(205.07 \frac{J}{K•mol}) + (1 \text{ mol})(130.7 \frac{J}{K•mol})]$$

$$= -293.97 \frac{J}{K} \text{ (or } -294.0 \frac{J}{K} \text{ --given sf restrictions)}$$

22. Standard molar entropy changes for:

(a) 2 Al (s) + 3 Cl_2 (g) → 2 $AlCl_3$ (s)

$$\Delta S° = 2 • S° \text{ } AlCl_3 \text{ (s)} - [2 • S° \text{ Al (s)} + 3 • S° \text{ } Cl_2 \text{ (g)}]$$

$$= (2 \text{ mol})(109.29 \frac{J}{K•mol}) -$$

$$[(2 \text{ mol})(28.3 \frac{J}{K•mol}) + (3 \text{ mol})(223.08 \frac{J}{K•mol})]$$

$$= -507.3 \frac{J}{K}$$

(b) 2 CH_3OH (ℓ) + 3 O_2 (g) → 2 CO_2 (g) + 4 H_2O (g)

$$\Delta S° = [2 • S° \text{ } CO_2 \text{ (g)} + 4 • S° \text{ } H_2O \text{ (g)}] - [2 • S° \text{ } CH_3OH \text{ (ℓ)} + 3 • S° \text{ } O_2 \text{ (g)}]$$

$$= [(2 \text{ mol})(213.74 \frac{J}{K•mol}) + (4 \text{ mol})(188.84 \frac{J}{K•mol})] -$$

$$[(2 \text{ mol})(127.19 \frac{J}{K•mol}) + (3 \text{ mol})(205.07 \frac{J}{K•mol})]$$

$$= +313.25 \frac{J}{K}$$

$\Delta S°$univ and Spontaneity

24. Is the reaction: Si (s) + 2 Cl_2 (g) → $SiCl_4$ (g) spontaneous?

$$\Delta S°_{sys} = 1 • S° \text{ } SiCl_4 \text{ (g)} - [1 • S° \text{ Si (s)} + 2 • S° \text{ } Cl_2 \text{ (g)}]$$

$$= (1 \text{ mol})(330.86 \frac{J}{K•mol}) -$$

$$[(1 \text{ mol})(18.82 \frac{J}{K•mol}) + (2 \text{ mol})(223.08 \frac{J}{K•mol})]$$

$$= -134.12 \frac{J}{K}$$

To calculate $\Delta S°_{surr}$, we calculate $\Delta H°_{sys}$:

$$\Delta H° = 1 \cdot \Delta H° \ SiCl_4 \ (g) - [1 \cdot \Delta H° \ Si \ (s) + 2 \cdot \Delta H° \ Cl_2 \ (g)]$$

$$= (1 \ mol)(-662.75 \frac{kJ}{mol}) - [(1 \ mol)(0 \frac{kJ}{mol}) + (2 \ mol)(0 \frac{kJ}{mol})]$$

$$= -662.75 \frac{kJ}{mol}$$

$$\Delta S°_{surr} = -\Delta H°/T = (662.75 \times 10^3 \ J/mol)/298.15 \ K = 2222.9 \frac{J}{K}$$

$$\Delta S°_{univ} = \Delta S°_{sys} + \Delta S°_{surr} = -134.12 \frac{J}{K} + 2222.9 \frac{J}{K} = 2088.7 \frac{J}{K}$$

26. Is the reaction: $2 \ H_2O(\ell) \rightarrow 2 \ H_2 \ (g) + O_2 \ (g)$ spontaneous?

$$\Delta S°_{sys} = [2 \cdot S° \ H_2 \ (g) + 1 \cdot S° \ O_2 \ (g)] - 2 \cdot S° \ H_2O \ (\ell)$$

$$= [(2 \ mol)(130.7 \frac{J}{K \cdot mol}) + (1 \ mol)(205.07 \frac{J}{K \cdot mol})] - (2 \ mol)(69.95 \frac{J}{K \cdot mol})$$

$$= +326.57 \frac{J}{K} \text{ and for decomposition of 1 mol of water: } +326.57/2 = 163.3 \frac{J}{K}$$

To calculate $\Delta S°_{surr}$, we calculate $\Delta H°_{sys}$:

$$\Delta H°_{sys} = [2 \cdot \Delta H° \ H_2 \ (g) + 1 \cdot \Delta H° \ O_2 \ (g)] - 2 \cdot \Delta H° \ H_2O \ (\ell)$$

$$= [(2 \ mol)(0 \frac{kJ}{mol}) + (1 \ mol)(0 \frac{kJ}{mol})] - (2 \ mol)(-285.83 \frac{kJ}{mol})$$

$$= +571.66 \frac{kJ}{mol} \text{ and for decomposition of 1 mol of water: } +571.66/2 = 285.83 \frac{kJ}{mol}$$

$$\Delta S°_{surr} = -\Delta H°/T = -(285.83 \times 10^3 \ J/mol)/298.15 \ K = -958.68 \frac{J}{K}$$

$$\Delta S°_{univ} = \Delta S°_{sys} + \Delta S°_{surr} = 163.3 \frac{J}{K} + -958.68 \frac{J}{K} = -795.4 \frac{J}{K} \text{ Since this value is}$$

less than zero, the process is not spontaneous.

28. Using Table 19.2 classify each of the reactions:
    (a) $\Delta H°_{system} = -$, $\Delta S°_{system} = -$ Product-favored at lower T
    (b) $\Delta H°_{system} = +$, $\Delta S°_{system} = -$ Not product-favored under any conditions

Effect of Temperature on Reactions

30. For the decomposition of $MgCO_3$ (s)) $\rightarrow$ MgO (s) + CO_2 (g):

    (a) $\Delta S°_{sys} = [1 \cdot S° \ MgO \ (s) + 1 \cdot S° \ CO_2 \ (g)] - 1 \cdot S° \ MgCO_3 \ (s)$

    $$= [(1 \ mol)(26.85 \frac{J}{K \cdot mol}) + (1 \ mol)(213.74 \frac{J}{K \cdot mol})] - (1 \ mol)(65.84 \frac{J}{K \cdot mol})$$

    $$= +174.75 \frac{J}{K}$$

To calculate $\Delta S°_{surr}$ we calculate $\Delta H°_{sys}$:

$$\Delta H°_{sys} = [1 \cdot \Delta H° \text{ MgO (s)} + 1 \cdot \Delta H° \text{ CO}_2 \text{ (g)}] - 1 \cdot \Delta H° \text{ MgCO}_3 \text{ (s)}$$

$$= [(1 \text{ mol})(-601.24\frac{kJ}{mol}) + (1 \text{ mol})(-393.509\frac{kJ}{mol})] - (1 \text{ mol})(-1111.69\frac{kJ}{mol})$$

$$= +116.94\frac{kJ}{mol}$$

$$\Delta S°_{surr} = -\Delta H°/T = -(285.83 \times 10^3 \text{ J/mol})/298.15 \text{ K} = -392.4\frac{J}{K}$$

(b) $\Delta S°_{univ} = \Delta S°_{sys} + \Delta S°_{surr} = +174.75\frac{J}{K} + -392.4\frac{J}{K} = -217.67\frac{J}{K}$ Since this value is less than zero, the process is not spontaneous.

(c) From Table 19.2 we observe that this type of reaction ($\Delta H = +$ and $\Delta S = +$) is spontaneous at higher T (product-favored).

Changes in Free Energy

32. Calculate $\Delta G°_{rxn}$ for :

(a) $2 \text{ Pb (s)} + O_2 \text{ (g)} \rightarrow 2 \text{ PbO (s)}$

$$\Delta H°rxn = (2 \text{ mol})(-219\frac{kJ}{mol}) - [0 + 0] = -438 \text{ kJ}$$

$$\Delta S° = (2 \text{ mol})(66.5\frac{J}{K \cdot mol}) -$$

$$[(2 \text{ mol})(64.81\frac{J}{K \cdot mol}) + (1 \text{ mol})(205.07\frac{J}{K \cdot mol})] = -201.7 \text{ J/K}$$

$$\Delta G°rxn = \Delta H°_f - T\Delta S°$$

$$= -438 \text{ kJ} - (298.15 \text{ K})(-201.7 \text{ J/K})(\frac{1.000 \text{ kJ}}{1000. \text{ J}})$$

$$= -378 \text{ kJ}$$

Reaction is product-favored since $\Delta G < 0$. With the very large negative ΔH, the process is enthalpy driven.

(b) $NH_3 \text{ (g)} + HNO_3 \text{ (aq)} \rightarrow NH_4NO_3 \text{ (aq)}$

$$\Delta H°rxn = (1 \text{ mol})(-339.87\frac{kJ}{mol}) -$$

$$[(1 \text{ mol})(-45.90\frac{kJ}{mol}) + (1 \text{ mol})(-207.36\frac{kJ}{mol})] = -86.61 \text{ kJ}$$

$$\Delta S° = (1 \text{ mol})(259.8\frac{J}{K \cdot mol}) -$$

$$[(1 \text{ mol})(192.77\frac{J}{K \cdot mol}) + (1 \text{ mol})(146.4\frac{J}{K \cdot mol})] = -79.4 \text{ J/K}$$

$$\Delta G°_{rxn} = \Delta H°_{rxn} - T\Delta S°$$

$$= -86.61 \text{ kJ} - (298.15 \text{ K})(-79.4 \text{ J/K})(\frac{1.000 \text{ kJ}}{1000. \text{ J}})$$

$$= -62.95 \text{ kJ}$$

Reaction is product-favored since $\Delta G < 0$. With the very large negative ΔH, the process is enthalpy driven.

34. Calculate the molar free energies of formation for:

(a) CS_2 (g)   The reaction is: C (graphite) + 2 S (s,rhombic) → CS_2 (g)

$$\Delta H°rxn = (1 \text{ mol})(116.7 \frac{\text{kJ}}{\text{mol}}) - [0 + 0] = +116.7 \text{ kJ}$$

$$\Delta S° = (1 \text{ mol})(237.8 \frac{\text{J}}{\text{K}\cdot\text{mol}})$$
$$- [(1 \text{ mol})(5.6 \frac{\text{J}}{\text{K}\cdot\text{mol}}) + (2 \text{ mol})(32.1 \frac{\text{J}}{\text{K}\cdot\text{mol}})] = +168.0 \text{ J/K}$$

$$\Delta G°_f = \Delta H°_f - T\Delta S°$$

$$= (116.7 \text{ kJ}) - (298.15 \text{ K})(168.0 \frac{\text{J}}{\text{K}})(\frac{1.000 \text{ kJ}}{1000 \text{ J}})$$

$$= +66.6 \text{ kJ} \qquad\qquad \text{Appendix value: 66.61 kJ/mol}$$

(b) NaOH (s) The reaction is: Na (s) + $\frac{1}{2}$ O_2(g) + $\frac{1}{2}$ H_2(g) → NaOH (s)

$$\Delta H°_f = (1 \text{ mol})(-425.93 \frac{\text{kJ}}{\text{mol}}) - [0 + 0 + 0] = -425.93 \text{ kJ}$$

$$\Delta S° = (1 \text{ mol})(64.46 \frac{\text{J}}{\text{K}\cdot\text{mol}}) -$$
$$[(1 \text{ mol})(51.21 \frac{\text{J}}{\text{K}\cdot\text{mol}}) + (\frac{1}{2} \text{ mol})(205.07 \frac{\text{J}}{\text{K}\cdot\text{mol}}) + (\frac{1}{2} \text{ mol})(130.7 \frac{\text{J}}{\text{K}\cdot\text{mol}})]$$

$$= -154.6 \text{ J/K}$$

$$\Delta G°_f = \Delta H°_f - T\Delta S°$$

$$= (-425.93 \text{ kJ}) - (298.15 \text{ K})(-154.6 \text{ J/K})(\frac{1.000 \text{ kJ}}{1000. \text{ J}})$$

$$= -379.82 \text{ kJ} \qquad\qquad \text{Appendix value: -379.75 kJ/mol}$$

(c) ICl (g)   The reaction is: $\frac{1}{2}$ I_2 (g) + $\frac{1}{2}$ Cl_2 (g) → ICl (g)

$$\Delta H°_f = (1 \text{ mol})(+17.51 \frac{\text{kJ}}{\text{mol}}) - [0 + 0] = +17.51 \text{ kJ}$$

$$\Delta S° = (1 \text{ mol})(247.56 \frac{J}{K \cdot mol}) -$$

$$[(\frac{1}{2} \text{mol})(116.135 \frac{J}{K \cdot mol}) + (\frac{1}{2} \text{mol})(223.08 \frac{J}{K \cdot mol})]$$

$$= +77.95 \text{ J/K}$$

$$\Delta G°_f = \Delta H°_f - T\Delta S°$$

$$= (+17.51 \text{ kJ}) - (298.15 \text{ K})(+77.95 \text{ J/K})(\frac{1.000 \text{ kJ}}{1000. \text{ J}})$$

$$= -5.72 \text{ kJ} \qquad \qquad \text{Appendix value: } -5.73 \text{ kJ/mol}$$

Free Energy of Formation

36. Calculate $\Delta G°_{rxn}$ for the following equations. Are they product-favored?

(a) $2 \text{ K (s)} + \text{Cl}_2 \text{ (g)} \rightarrow 2 \text{ KCl (s)}$

$$\Delta G°_{rxn} = [2 \cdot \Delta G°_f \text{ KCl(s)}] - [1 \cdot \Delta G°_f \text{ Cl}_2\text{(g)} + 2 \cdot \Delta G°_f \text{ K (s)}]$$

$$= [(2 \text{ mol})(-408.77 \frac{kJ}{mol}] - [(1 \text{ mol})(0 \frac{kJ}{mol}) + (2 \text{ mol})(0 \frac{kJ}{mol})]$$

$$= -817.54 \text{ kJ and per mol of KCl} = -408.77 \text{ kJ/mol}$$

With a $\Delta G < 0$, the reaction is product-favored.

(b) $2 \text{ CuO (s)} \rightarrow 2 \text{ Cu (s)} + \text{O}_2 \text{ (g)}$

$$\Delta G°_{rxn} = [2 \cdot \Delta G°_f \text{ Cu(s)} + \Delta G°_f \text{ O}_2 \text{ (g)}] - [2 \cdot \Delta G°_f \text{ CuO (s)}]$$

$$\Delta G°_{rxn} = [(2 \text{ mol})(0 \frac{kJ}{mol}) + (1 \text{ mol})(0 \frac{kJ}{mol})] - [(2 \text{ mol})(-128.3 \frac{kJ}{mol})$$

$$= +256.6 \text{ kJ and } +128.3 \text{ kJ/mol CuO}$$

With a $\Delta G > 0$, the reaction is not product-favored.

(c) $4 \text{ NH}_3 \text{ (g)} + 7 \text{ O}_2 \text{ (g)} \rightarrow 4 \text{ NO}_2 \text{ (g)} + 6 \text{ H}_2\text{O (g)}$

$$\Delta G°_{rxn} = [4 \cdot \Delta G°_f \text{ NO}_2 \text{ (g)} + 6 \cdot \Delta G°_f \text{ H}_2\text{O (g)}] - [4 \cdot \Delta G°_f \text{ NH}_3 \text{ (g)} + 7 \cdot \Delta G°_f \text{ O}_2 \text{ (g)}]$$

$$\Delta G°_{rxn} = [(4 \text{ mol})(+51.23 \frac{kJ}{mol}) + (6 \text{ mol})(-228.59 \frac{kJ}{mol})] -$$

$$[(4 \text{ mol})(-16.37 \frac{kJ}{mol}) + (7 \text{ mol})(0 \frac{kJ}{mol})] = -1101.14 \text{ kJ}$$

With a $\Delta G < 0$, the reaction is product-favored.

38. Value for $\Delta G°_f$ of $BaCO_3(s)$:

$$\Delta G°_{rxn} = [\Delta G°_f \text{ BaO(s)} + \Delta G°_f \text{ CO}_2\text{(g)}] - [\Delta G°_f \text{ BaCO}_3\text{(s)}]$$

$$+219.7 \text{ kJ} = [(1 \text{ mol})(-520.38 \frac{kJ}{mol}) + (1 \text{ mol})(-394.359 \frac{kJ}{mol})] - \Delta G°_f \text{ BaCO}_3\text{(s)}$$

$$+219.7 \text{ kJ} = -914.74 \text{ kJ} - \Delta G°_f \text{ BaCO}_3\text{(s)}$$

$$-1134.4 \text{ kJ/mol} = +\Delta G°_f \text{ BaCO}_3\text{(s)}$$

Effect of Temperature on ΔG

40. Entropy-favored or Enthalpy-favored reactions?

(a) N_2 (g) + 2 O_2 (g) → 2 NO_2 (g)

$$\Delta H^\circ = (2 \text{ mol})(+33.1 \frac{kJ}{mol}) - [0 + 0] = +66.2 \text{ kJ}$$

$$\Delta S^\circ = (2 \text{ mol})(+240.04 \frac{J}{K \cdot mol}) - $$

$$[(1 mol)(191.56 \frac{J}{K \cdot mol}) + (2 \text{ mol})(+205.07 \frac{J}{K \cdot mol})] = -121.62 \frac{J}{K}$$

$$\Delta G^\circ = (2 \text{ mol})(51.23 \frac{kJ}{mol}) - [(1 \text{ mol})(0 \frac{kJ}{mol}) + (1 \text{ mol})(0 \frac{kJ}{mol})] = 102.46 \text{ kJ}$$

The reaction is **not** entropy OR enthalpy favored. There is **no** T at which ΔG < 0.

(b) 2 C (s) + O_2 (g) → 2 CO (g)

$$\Delta H^\circ = (2 \text{ mol})(-110.525 \frac{kJ}{mol}) - [0 + 0] = -221.05 \text{ kJ}$$

$$\Delta S^\circ = (2 \text{ mol})(+197.674 \frac{J}{K \cdot mol}) - $$

$$[(2 \text{ mol})(+5.6 \frac{J}{K \cdot mol}) + (1 mol)(+205.07 \frac{J}{K \cdot mol})] = +179.1 \frac{J}{K}$$

$$\Delta G^\circ = (2 \text{ mol})(-137.168 \frac{kJ}{mol}) - [(1 \text{ mol})(0 \frac{kJ}{mol}) + (1 \text{ mol})(0 \frac{kJ}{mol})]$$

$$= -274.336 \text{ kJ}$$

This reaction is **both** entropy- and enthalpy-favored *at all temperatures*.

(c) CaO (s) + CO_2 (g) → $CaCO_3$ (s)

$$\Delta H^\circ = (1 \text{ mol})(-1207.6 \frac{kJ}{mol}) - [(1mol)(-635.0 \frac{kJ}{mol}) + (1mol)(-393.509 \frac{kJ}{mol})]$$

$$= -179.1 \text{ kJ}$$

$$\Delta S^\circ = (1 \text{ mol})(+91.7 \frac{J}{K \cdot mol}) - $$

$$[(1mol)(38.2 \frac{J}{K \cdot mol}) + (1mol)(+213.74 \frac{J}{K \cdot mol})] = -160.2 \frac{J}{K}$$

$$\Delta G^\circ = (1 \text{ mol})(-1129.16 \frac{kJ}{mol}) - $$

$$[(1mol)(-603.42 \frac{kJ}{mol}) + (1mol)(-394.359 \frac{kJ}{mol})] = -131.4 \text{ kJ}$$

This reaction is *enthalpy-favored* and will be spontaneous at low temperatures.

(d) 2 NaCl (s) → 2 Na (s) + Cl_2 (g)

$$\Delta H^\circ = [(2 \text{ mol})(0 \frac{kJ}{mol}) + (1mol)(0 \frac{kJ}{mol})] - (2mol)(-411.12 \frac{kJ}{mol})]$$

$$= +822.24 \text{ kJ}$$

$$\Delta S^\circ = [(2\ mol)(+51.21\ \tfrac{J}{K\bullet mol}\ ) + (1mol)(+223.08\ \tfrac{J}{K\bullet mol}\ )] -$$
$$(2mol)(+72.11\ \tfrac{J}{K\bullet mol}\ )] = +181.28\ \tfrac{J}{K}$$

$$\Delta G^\circ = [(2\ mol)(\ 0\tfrac{kJ}{mol}\ ) + (1mol)(\ 0\tfrac{kJ}{mol}\ )] - (2mol)(-384.04\ \tfrac{kJ}{mol}\ )] = +768.08\ kJ$$

This reaction is *entropy-favored* and will be spontaneous at high temperatures.

42. Estimate the temperature to decompose HgS (s) into Hg (ℓ) and S (g):

For this process, we need to find the temperature at which the ΔG° becomes < 0

Calculate ΔH for the process: HgS (s) $\rightarrow$ Hg (ℓ) + S (g)

$$\Delta H^\circ_{rxn} = [\Delta H^\circ_f\ Hg\ (\ell) + \Delta H^\circ_f\ S\ (g)] - \Delta H^\circ_f\ HgS\ (s)]$$
$$\Delta H^\circ_{rxn} = [0 + (1mol)(+278.98\tfrac{kJ}{mol}\ )] - (1\ mol)(-58.2\ \tfrac{kJ}{mol}\ ]$$
$$\Delta H^\circ_{rxn} = +337.2\ kJ$$

$$\Delta S^\circ_{rxn} = [(1mol)(76.02\ \tfrac{J}{K\bullet mol}\ ) + (1mol)(167.83\ \tfrac{J}{K\bullet mol}\ )] -$$
$$(1\ mol)(82.4\ \tfrac{J}{K\bullet mol}\ ] = +161.5\ \tfrac{J}{K}$$

The process becomes spontaneous when ΔG just becomes negative. So calculate the T at which $\Delta G = 0$.

$$\Delta G = \Delta H - T\Delta S\ \text{ and if }\Delta G = 0\text{ then }\Delta H = T\Delta S$$
$$337.2\ kJ(1000\ J/kJ) = T\bullet 161.5\ \tfrac{J}{K}\ \text{ so } T = 2088\ K\ \text{(Wow, that's hot!)}$$

Free Energy and Equilibrium Constants

44. Calculate K_p for the reaction:

$$\tfrac{1}{2}\ N_2(g) + \tfrac{1}{2}\ O_2(g) \rightarrow NO\ (g)\qquad \Delta G^\circ_f = +86.58\ kJ/mol\ NO$$
$$\Delta G^\circ_{rxn} = -RT\ ln\ K_p$$
$$86.58 \times 10^3\ J/mol = -(8.3145\ \tfrac{J}{K\bullet mol}\ )(298.15\ K)\ ln\ K_p$$
$$-34.926 = ln\ K_p$$
$$6.8 \times 10^{-16} = K_p$$

Note that the + value of ΔG°_f results in a value of K_p which is small--reactants are favored.

A negative value would result in a large K_p -- a process in which the products were favored.

46. From the $\Delta G°$ and Kp, determine if the hydrogenation of ethylene is product-favored:
Using $\Delta G°$ data from the Appendix: C_2H_4 (g) + H_2 (g) → C_2H_6 (g)

$$\Delta G° = (1 \text{ mol})(-31.89 \frac{kJ}{mol}) - [(1\text{mol})(68.35 \frac{kJ}{mol}) + (1\text{mol})(0 \frac{kJ}{mol})] = -100.24 \text{ kJ}$$

and since $\Delta G° = -RT \ln K_p$

$$-100.24 \times 10^3 \text{ J/mol} = -(8.3145 \frac{J}{K\bullet mol})(298.15 \text{ K}) \ln K_p$$

$$40.436 = \ln K_p \text{ and } K_p = 3.64 \times 10^{17}$$

The negative value of ΔG means the reaction is product-favored. The large value of ΔG means that Kp is very large.

General Questions on Thermodynamics

48. Calculate the ΔS for (1) C(s) + 2 H_2 (g) → CH_4 (g)

$$\Delta S_1° = (1 \text{ mol})(+186.26 \frac{J}{K\bullet mol}) -$$

$$[(1\text{mol})(+5.6 \frac{J}{K\bullet mol}) + (2\text{mol})(+130.7 \frac{J}{K\bullet mol})] = -80.7 \frac{J}{K}$$

Calculate the ΔS for (2) CH_4(g) + $\frac{1}{2}$ O_2 (g) → CH_3OH(ℓ)

$$\Delta S_2° = (1 \text{ mol})(+127.19 \frac{J}{K\bullet mol}) -$$

$$[(1\text{mol})(+186.26 \frac{J}{K\bullet mol}) + (\frac{1}{2}\text{mol})(+205.07 \frac{J}{K\bullet mol})] = -161.6 \frac{J}{K}$$

Calculate the ΔS for (3) C(s) + 2 H_2 (g) + $\frac{1}{2}$ O_2 (g) → CH_3OH(ℓ)

$$\Delta S_3° = (1 \text{ mol})(+127.19 \frac{J}{K\bullet mol}) -$$

$$[(1\text{mol})(+5.6 \frac{J}{K\bullet mol}) + (2\text{mol})(+130.7 \frac{J}{K\bullet mol}) + (\frac{1}{2}\text{mol})(+205.07 \frac{J}{K\bullet mol})]$$

$$= -242.3 \frac{J}{K}$$

So $\Delta S_1° + \Delta S_2° = (-80.7 \frac{J}{K}) + (-161.6 \frac{J}{K}) = -242.3 \frac{J}{K}$

50. Calculate $\Delta H°_{rxn}$ and $\Delta S°_{rxn}$ for the combustion of ethane:

	C_2H_6 (g)	+	7/2 O_2 (g)	→	2 CO_2(g)	+	3 H_2O (g)
$\Delta H°_f$ (kJ/mol)	-83.85		0		-393.509		-241.83
S° (J/K•mol)	+229.2		+205.07		+213.74		+188.84

$$\Delta H°_{rxn} = [2 \bullet \Delta H°_f \, CO_2(g) + 3 \bullet \Delta H°_f \, H_2O \, (g)] - [1 \bullet \Delta H°_f \, C_2H_6 \, (g) +$$
$$7/2 \bullet \Delta H°_f \, O_2 \, (g)]$$

$$= [(2 \text{ mol})(-393.509 \tfrac{kJ}{mol}) + (3 \text{ mol})(-241.83 \tfrac{kJ}{mol})] -$$

$$[(1 \text{ mol})(-83.85 \text{ kJ/mol}) + 0]$$

$$= -1428.66 \text{ kJ}$$

$$\Delta S^\circ{}_{rxn} = [2 \cdot S^\circ\ CO_2(g) + 3 \cdot S^\circ\ H_2O\ (g)] - [1 \cdot S^\circ\ C_2H_6\ (g) + 7/2 \cdot S^\circ\ O_2\ (g)]$$

$$= [2 \text{ mol})(213.74 \tfrac{J}{K \cdot mol}) + (3 \text{ mol})(188.84 \tfrac{J}{K \cdot mol})] -$$

$$[(1 \text{ mol})(229.2 \tfrac{J}{K \cdot mol}) + (7/2 \text{ mol})(205.07 \tfrac{J}{K \cdot mol})] = +47.1 \text{ J/K}$$

$$\Delta S^\circ{}_{surroundings} \quad = \frac{-\Delta H_{rxn}}{T} \quad \text{(Assuming we're at 298 K)}$$

$$= \frac{1428.66 \text{ kJ}}{298.15 \text{ K}} \cdot \frac{1000 \text{ J}}{1 \text{ kJ}} \quad = \quad 4791.7 \text{ J/K}$$

so $\Delta S^\circ{}_{system} + \Delta S^\circ{}_{surroundings} \quad = \quad +47.1 \text{ J/K} + 4791.7 \text{ J/K}$

$$= \quad 4838.8 \text{ J/K}$$

Since $\Delta H^\circ = -$ and $\Delta S^\circ = +$, the process is product-favored.

This calculation is consistent with our expectations. We know that hydrocarbons burn completely (in the presence of sufficient oxygen) to produce carbon dioxide and water.

52. The formation of Fe_2O_3 (s): $2 \text{ Fe (s)} + 3/2 \ O_2 \text{ (g)} \rightarrow Fe_2O_3 \text{ (s)}$

The $\Delta G^\circ{}_f = -742.2$ kJ/mol (from Appendix L)

The $\Delta G^\circ{}_{rxn}$ when 454 g of Fe_2O_3 is formed:

$$\frac{-742.2 \text{ kJ}}{1 \text{ mol } Fe_2O_3} \cdot \frac{1 \text{ mol } Fe_2O_3}{159.69 \text{ g } Fe_2O_3} \cdot \frac{454 \text{ g } Fe_2O_3}{1} = -2110 \text{ kJ (to 3sf)}$$

54.   (a) Calculate $\Delta G^\circ{}_{rxn}$ for $NH_3 \text{ (g)} + HCl \text{ (g)} \rightarrow NH_4Cl \text{ (s)}$

	NH_3 (g)	HCl (g)	NH_4Cl (s)
S° (J/K•mol)	192.77	186.2	94.85
$\Delta H^\circ{}_f$ (kJ/mol)	-45.90	-92.31	$-314..55$

$$\Delta S^\circ{}_{rxn} = 1 \cdot S^\circ\ NH_4Cl \text{ (s)} - [1 \cdot S^\circ\ NH_3 \text{ (g)} + 1 \cdot S^\circ\ HCl \text{ (g)}]$$

$$= (1 \text{ mol})(94.85 \text{ J/K} \cdot mol) - [(1 \text{ mol})(192.77 \text{ J/K} \cdot mol) +$$

$$(1 \text{ mol})(186.2 \text{ J/K} \cdot mol)] \quad = -284.12 \text{ J/K}$$

$$\Delta H^\circ{}_{rxn} = 1 \cdot \Delta H^\circ{}_f\ NH_4Cl \text{ (s)} - [1 \cdot \Delta H^\circ{}_f\ NH_3 \text{ (g)} + 1 \cdot \Delta H^\circ{}_f\ HCl \text{ (g)}]$$

$$= (1 \text{ mol})(-314.55 \text{ kJ/mol}) -$$

$$[(1 \text{ mol})(-45.90 \text{ kJ/mol}) + (1 \text{ mol})(-92.31 \text{ kJ/mol})]$$

$$= -176.34 \text{ kJ}$$

$$\Delta G^\circ_{rxn} = \Delta H^\circ_f - T \Delta S^\circ_{rxn}$$

$$= -176.34 \text{ kJ} - (298.15 \text{ K})(-284.12 \text{ J/K})(\frac{1.000 \text{ kJ}}{1000 \text{ J}})$$

$$= -176.01 \text{ kJ} + 84.67 \text{ kJ}$$

$$= -91.64 \text{ kJ}$$

$$\Delta S^\circ_{surroundings} = \frac{-\Delta H_{rxn}}{T} \quad \text{(Assuming we're at 298 K)}$$

$$= \frac{176.34 \text{ kJ}}{298.15 \text{ K}} \cdot \frac{1000 \text{ J}}{1 \text{ kJ}} = 591.45 \text{ J/K}$$

so $\Delta S^\circ_{system} + \Delta S^\circ_{surroundings} = -284.12 \text{ J/K} + 591.45 \text{ J/K} = +307.3 \text{ J/K}$

The values for ΔG° for the equation is negative, indicating that is product-favored. The reaction is enthalpy driven ($\Delta H < 0$).

(b) Calculate K_p for the reaction:

$$\Delta G^\circ_{rxn} = -RT \ln K_p$$

$$-91.34 \times 10^3 \text{ J/mol} = -(8.3145 \frac{J}{K \cdot mol})(298.15 \text{ K}) \ln K_p$$

$$36.85 = \ln K_p$$

$$1.00 \times 10^{16} = K_p$$

56. For the following processes:

(a) $NaCl(s) \rightarrow NaCl(aq)$

$$\Delta S_{sys} = S \text{ NaCl (aq)} - S \text{ NaCl (s)}$$

$$= (1 \text{ mol}) \cdot 115.5 \frac{J}{K \cdot mol} - (1 \text{ mol}) \cdot 72.11 \frac{J}{K \cdot mol} = 43.4 \text{ J/K}$$

$$\Delta H_{rxn} = \Delta H_f \text{ NaCl (aq)} - \Delta H_f \text{ NaCl (s)}$$

$$= (1 \text{mol})(-407.27 \frac{kJ}{mol}) - (1 \text{mol})(-411.12 \frac{kJ}{mol}) = +3.85 \text{ kJ}$$

$$\Delta S_{surr} = -\Delta H/T = 3850 \text{ J}/298\text{K} = -12.9 \text{ J/K}$$

$$\Delta S_{univ} = \Delta S_{surr} + \Delta S_{sys} = (-12.9 \text{ J/K}) + (43.4 \text{ J/K}) = +30.5 \text{ J/K}$$

(b) $NaOH(s) \rightarrow NaOH(aq)$

$$\Delta S_{sys} = S \text{ NaOH (aq)} - S \text{ NaOH (s)}$$

$$= (1 \text{ mol}) \cdot 48.1 \frac{J}{K \cdot mol} - (1 \text{ mol}) \cdot 64.46 \frac{J}{K \cdot mol} = -16.4 \text{ J/K}$$

$$\Delta H = \Delta H_f \text{ NaOH (aq)} - \Delta H_f \text{ NaOH (s)}$$

$$= (1 \text{mol})(-469.15 \frac{kJ}{mol}) - (1 \text{mol})(-425.93 \frac{kJ}{mol}) = -43.22 \text{ kJ}$$

$$\Delta S_{surr} = -\Delta H/T = 43220 \text{ J}/298\text{K} = +145.0 \text{ J/K}$$

$$\Delta S_{univ} = \Delta S_{surr} + \Delta S_{sys} = (+145.0 \text{ J/K}) + (-16.4 \text{ J/K}) = +128.6 \text{ J/K}$$

The dissolution of NaOH is an exothermic process, while that of NaCl is endothermic. This creates a much greater ΔS_{univ} for NaOH than for NaCl.

58. Calculate K_p for the formation of methanol from its' elements:

Begin by calculating a ΔG_{rxn}

$\Delta G_{rxn} = \Delta G_{CH3OH(\ell)} - [\Delta G_{C(graphite)} + 1/2 \cdot \Delta G_{O2(g)} + 2 \cdot \Delta G_{H2(g)}]$

$\Delta G_{rxn} = -166.14 \text{ kJ} - (0 \text{ kJ} + 0 \text{ kJ} + 0 \text{ kJ}) = -166.14 \text{ kJ}$

$\Delta G°_{rxn} = -RT \ln K_p$

$-166.1 \times 10^3 \text{ J/mol} = -(8.3145 \frac{J}{K \cdot mol})(298.15 \text{ K}) \ln K_p$

$67.00 = \ln K_p$

$1.3 \times 10^{29} = K_p$

The large value of K_p indicates that this process is product-favored at 298 K. Judging by

the relative numbers of gaseous particles, (without doing a calculation), one can see that ΔS for the reaction is < 0, so higher temperatures would reduce the value of K_p.

Regarding the connection between $\Delta G°$ and K, the more negative the value of $\Delta G°$, the larger the value of K.

60. Calculate the $\Delta S°$ for the vaporization of ethanol at 78.0 °C.

$$\text{The } \Delta S = \frac{\Delta Hvap}{T} = \frac{39.3 \times 10^3 \text{ J}}{351 \text{ K}} = 112 \text{ J/K}$$

62. Estimate the normal boiling point of ethanol:

Ethanol boils when the transition $(\ell) \rightarrow (g)$ is spontaneous. That is signaled by a ΔG that

makes the transition from + values to − values. Calculate the point at which $\Delta G = 0$.

$\Delta S_{sys} = S \text{ CH3CH2OH (g)} - S \text{ CH3CH2OH } (\ell)$

$= (1 \text{ mol}) \cdot 282.70 \frac{J}{K \cdot mol} - (1 \text{ mol}) \cdot 160.7 \frac{J}{K \cdot mol} = 122.0 \text{ J/K}$

$\Delta H = \Delta H_f \text{ CH3CH2OH (g)} - \Delta H_f \text{ CH3CH2OH } (\ell)$

$= (1 \text{mol})(-235.3 \frac{kJ}{mol}) - (1 \text{mol})(-277.0 \frac{kJ}{mol}) = 41.7 \text{ kJ}$

At $\Delta G = 0$, $\Delta H = T \cdot \Delta S$ so 41.7 kJ(1000J/kJ) = T $\cdot$122.0 J/K and solving for T:

T = 341.8 K (or 341.8 − 273 = 68.8 °C) (Compared to literature value of 78 °C)

64. For the decomposition of phosgene:

$$\Delta H^\circ_{rxn} = [\Delta H^\circ_f\ CO\ (g) + \Delta H^\circ_f\ Cl_2\ (g)] - [\Delta H^\circ_f\ COCl_2(g)]$$

$$\Delta H^\circ_{rxn} = [\ (1\ mol.)(-110.525\frac{kJ}{mol})\ +(1mol)(\ 0\frac{kJ}{mol})] - [(1\ mol)(\ -218.8\frac{kJ}{mol})]$$

$$\Delta H^\circ_{rxn} = 108.275\frac{kJ}{mol}$$

$$\Delta S^\circ_{rxn} = [S^\circ\ CO\ (g) + S^\circ\ Cl_2\ (g)] - [S^\circ\ COCl_2(g)]$$

$$\Delta S^\circ_{rxn} = [(1\ mol)(197.674\frac{J}{K\bullet mol}\ ) + (1\ mol)(223.07\frac{J}{K\bullet mol}\ )] -$$

$$[(1\ mol)(\ 283.53\frac{J}{K\bullet mol}\ )] = 137.2\ J/K$$

Using the ΔS data, we can see that raising the temperature will favor the endothermic decomposition of this substance.

66. For the process of melting benzene (at 5.5 °C), the signs of:

(a) ΔH° = positive; Melting is an endothermic process

(b) ΔS° = positive. Liquids have a higher entropy than solids.

(c) ΔG° (at 5.5 °C) = 0 . This is the point at which ΔG changes from + to -

(d) ΔG° (at 0.0 °C) = positive; The process is *nonspontaneous* at this temperature.

(e) ΔG° (at 25.0 °C) = negative The process is *spontaneous*

68. For the reaction of sodium with water:

$$Na\ (s) + H_2O(\ell)\ \rightarrow\ \ NaOH(aq) + \frac{1}{2}\ H_2\ (g)$$

Predict signs for ΔH° and ΔS°:

This one seems easy !! The reaction of sodium with water gives off heat, and the heat frequently ignites the hydrogen gas that is concomitantly evolved. ΔH° = -.
Regarding entropy, the system changes from one with a solid (low entropy) and a liquid (higher entropy) to a solution (*frequently* higher entropy than liquid) and a gas (high entropy). So we would predict that the entropy would increase, i.e. ΔS° = +.
Now for the calculation:

$$\Delta H^\circ_{rxn} = [1 \bullet \Delta H^\circ_f\ NaOH(aq) + \frac{1}{2}\bullet \Delta H^\circ_f\ H_2(g)] - [1 \bullet \Delta H^\circ_f\ Na(s)+1 \bullet \Delta H^\circ_f\ H_2O(\ell)]$$

$$= [(1\ mol)(-469.15\ \frac{kJ}{mol})+ (\frac{1}{2}mol)(0)]\ - [(1\ mol)(0) + (1\ mol)(-285.83\ \frac{kJ}{mol})]$$

$$= -183.32\ kJ$$

$$\Delta S^\circ_{rxn} = [1 \bullet S^\circ \text{ NaOH(aq)} + \frac{1}{2} \bullet S^\circ \text{ H}_2\text{(g)}] - [1 \bullet S^\circ \text{ Na(s)} + 1 \bullet S^\circ \text{ H}_2\text{O}(\ell)]$$

$$= [(1 \text{ mol})(48.1 \frac{J}{K \bullet mol}) + (\frac{1}{2} \text{ mol})(130.7 \frac{J}{K \bullet mol})] -$$

$$[(1 \text{ mol})(51.21 \frac{J}{K \bullet mol}) + (1 \text{ mol})(69.95 \frac{J}{K \bullet mol})]$$

$$= -7.7 \frac{J}{K \bullet mol}$$

As expected, the ΔH°_{rxn} for the reaction is negative! The surprise comes in the calculation for ΔS°_{rxn}. While we anticipate the sign to be positive, we find a slightly negative number—reflecting the order (hence a decrease in entropy) that can occur as solutions occur.

70. For the reaction :

	$C_6H_{12}O_6$ (aq)	→	2 C_2H_5OH (ℓ)	+	2 CO_2 (g)
$S^\circ (\frac{J}{K \bullet mol})$	289		160.7		213.74
$\Delta H^\circ (\frac{kJ}{mol})$	-1260.0		-277.0		-393.509

$$\Delta H^\circ_{rxn} = [2 \bullet \Delta H^\circ_f \text{ C}_2\text{H}_5\text{OH} (\ell) + 2 \bullet \Delta H^\circ_f \text{ CO}_2 \text{ (g)}] - [1 \bullet \Delta H^\circ_f \text{ C}_6\text{H}_{12}\text{O}_6 \text{ (aq)}]$$

$$\Delta H^\circ_{rxn} = [(2 \text{ mol})(-277.0 \frac{kJ}{mol}) + (2 \text{ mol})(-393.509 \frac{kJ}{mol})] - [(1 \text{ mol})(-1260.0 \frac{kJ}{mol})]$$

$$= -81.0 \text{ kJ}$$

$$\Delta S^\circ_{rxn} = [2 \bullet S^\circ \text{ C}_2\text{H}_5\text{OH} (\ell) + 2 \bullet S^\circ \text{ CO}_2 \text{ (g)}] - [1 \bullet S^\circ \text{ C}_6\text{H}_{12}\text{O}_6 \text{ (aq)}]$$

$$\Delta S^\circ_{rxn} = [(2 \text{ mol})(160.7 \frac{J}{K \bullet mol}) + (2 \text{ mol})(213.74 \frac{J}{K \bullet mol})] - [(1 \text{ mol})(289 \frac{J}{K \bullet mol})]$$

$$= 460. \text{ J/K}$$

$$\Delta G^\circ_{rxn} = \Delta H^\circ_f - T \Delta S^\circ_{rxn}$$

$$= -81.0 \text{ kJ} - (298.15 \text{ K})(460. \frac{J}{K})(\frac{1.000 \text{ kJ}}{1000. \text{ J}}) = -218.1 \text{ kJ}$$

The reaction is product-favored.

72. Calculate ΔG° for the process for which $K_p = 0.14$ at 25 °C:

$$\Delta G^\circ_{rxn} = -RT \ln K_p$$

$$\Delta G^\circ_{rxn} = -(8.3145 \frac{J}{K \bullet mol})(298.15 \text{ K}) \ln(0.14)$$

$$\Delta G^\circ_{rxn} = 4,871 \text{ J/mol or } 4.87 \text{ kJ/mol}$$

The table value for this process is: (2mol)(51.23 kJ/mol)- (1 mol)(97.73 kJ/mol) or 4.73 kJ/mol.

74. What is $\Delta G°$ for the equilibrium: butane (g) $\rightleftharpoons$ isobutane (g)   given K = 2.5 at 25 °C

$$\Delta G° = - RT \ln K = - (8.314 \frac{J}{K•mol})(298.15 \text{ K}) \ln (2.5)$$

$$= - 2.3 \times 10^3 \frac{J}{mol} \text{ or } - 2.3 \text{ kJ/mol}$$

76. The key phrase needed to answer the question: "What is the sign......" is "Iodine dissolves readily....". This phrase tell us that $\Delta G°$ is negative.

Enthalpy-driven processes are exothermic. The "neutrality" of the ΔH for this reaction tells us that the process is NOT enthalpy-driven. Since the iodine goes from the solid state to the "solution" state, we anticipate an increase in entropy, and would therefore state that the process is entropy-driven.

78. For the reaction: $2 SO_3 (g) \rightarrow 2 SO_2 (g) + O_2 (g)$

$$\Delta H°_{rxn} = [2 • \Delta H°_f SO_2 (g) + 1 • \Delta H°_f O_2 (g)] - [2 • \Delta H°_f SO_3 (g)]$$

$$= [(2 \text{ mol})(- 296.84 \frac{kJ}{mol}) + 0] - [(2 \text{ mol})(-395.77 \frac{kJ}{mol})] = 197.86 \text{ kJ}$$

$$\Delta S°_{rxn} = [2 • S° SO_2 (g) + 1 • S° O_2 (g)] - [2 • S° SO_3 (g)]$$

$$= [(2 \text{ mol})(248.21 \frac{J}{K•mol}) + (1 \text{ mol})(205.07 \frac{J}{K•mol})] -$$

$$[(2 \text{ mol})(256.77 \frac{J}{K•mol})] = 187.95 \text{ J/K}$$

(a) Is the reaction product-favored at 25 °C ?:

$$\Delta G°_{rxn} = \Delta H°_{rxn} - T \Delta S°_{rxn}$$

$$= 197.86 \text{ kJ} - (298.15 \text{ K})(187.95 \frac{J}{K})(\frac{1.000 \text{ kJ}}{1000 \text{ J}})$$

$$= 141.82 \text{ kJ}$$

The reaction is not product-favored.

(b) The reaction can become product-favored if there is some T at which $\Delta G°_{rxn} < 0$. To see if such a T is feasible, let's set $\Delta G°_{rxn} = 0$ and solve for T! *Remember that the units of energy must be the same, so let's convert units of J (for the entropy term) into units of kJ*

$$\Delta G°_{rxn} = \Delta H°_{rxn} - T \Delta S°_{rxn}$$

$$0 = 197.86 \text{ kJ} - T(0.18795 \frac{kJ}{K})$$

$$T = \frac{197.86 \text{ kJ}}{0.18795 \frac{kJ}{K}} = 1052.7 \text{ K} \text{ or } (1052.7 - 273.1) = 779.6 °C$$

(c) The equilibrium constant for the reaction at $1500\,°C$. Since we know that $\Delta G°_{rxn} = \Delta H°_{rxn} - T\,\Delta S°_{rxn} = -RT \ln K$, we can solve for K if we know $\Delta G°_{rxn}$ at $1500\,°C$

$$\Delta G°_{rxn} = \Delta H°_{rxn} - T\,\Delta S°_{rxn} = -RT \ln K$$

$$= 197.86\ kJ - (1773\ K)(187.95\ J/K)(\frac{1\ kJ}{1000\ J}\ )$$

And solving for $\Delta G°_{rxn}$ gives $-135.4\ kJ$. Substitute into the last part of the equation:

$$-135.4\ kJ = -8.314\ \frac{J}{K\cdot mol} \cdot \frac{1\ kJ}{1000\ J} \cdot 1773\ K\ \cdot \ln K$$

$$K = 9.7 \times 10^3\ \text{or}\ 1 \times 10^4\ (\text{to 1 sf})$$

80. Reaction:

	$H_2S\ (g)$	$+\ 2\ O_2\ (g)\ \rightarrow$	$H_2SO_4\ (\ell)$
$\Delta H°_f$ (kJ/mol)	-20.63	0	-814
$S°$ (J/K $\cdot$ mol)	205.79	205.07	156.9

$$\Delta H°_{rxn} = [(1\ mol)(-814\frac{kJ}{mol}\ )] - [(1\ mol)(-20.63\frac{kJ}{mol}\ ) + 0]$$

$$= -793\ kJ$$

$$\Delta S°_{rxn} = [(1\ mol)(156.9\frac{J}{K\cdot mol}\ )] -$$
$$[(1\ mol)(205.79\frac{J}{K\cdot mol}\ ) + (2\ mol)(205.07\frac{J}{K\cdot mol}\ )]$$

$$= -459.0\ J/K$$

$$\Delta G°_{rxn} = \Delta H°_f - T\,\Delta S°_{rxn}$$

$$= -793\ kJ - (298.15\ K)(-459.0\frac{J}{K}\ )(\frac{1.000\ kJ}{1000\ J}\ )$$

$$= -657\ kJ$$

The reaction is product-favored at $25\,°C$ ($\Delta G° < 0$) and enthalpy-driven ($\Delta H°_{rxn} < 0$)

82. For the decomposition of silver(I) oxide: Is the decomposition product-favored?
It is product-favored **if** $\Delta G°_{rxn}$ is negative. Calculate the $\Delta G°_{rxn}$:

	$2\ Ag_2O\ (s)\ \rightarrow$	$4\ Ag\ (s)$	$+\ O_2\ (g)$
$\Delta H°_f$ (kJ/mol)	-31.1	0	0
$S°$ (J/K $\cdot$ mol)	121.3	42.55	205.07
$\Delta G°$ (kJ /mol)	-11.32	0	0

From these data:

$$\Delta H°_{rxn} = [0 + 0] - [(2mol\ )(-31.1\ kJ/mol)] = +62.2\ kJ$$

$$\Delta S°_{rxn} = [(4 \text{ mol} \bullet 42.55 \tfrac{J}{K \bullet mol}) + (1 \text{ mol} \bullet 205.07 \tfrac{J}{K \bullet mol})] - [2 \text{ mol} \bullet 121.3 \tfrac{J}{K \bullet mol}]$$

$$\Delta S°_{rxn} = +132.7 \text{ J/K}$$

$$\Delta G°_{rxn} = [(4 \text{ mol} \bullet 0 \tfrac{kJ}{mol}) + (1 \text{ mol} \bullet 0 \tfrac{kJ}{mol})] - [2 \text{ mol} \bullet -11.32 \tfrac{kJ}{mol}]$$

$$\Delta G°_{rxn} = +22.64 \text{ kJ (The reaction is not product-favored.)}$$

At what temperature could it become product-favored?

When $\Delta G°_{rxn} = 0$, $\Delta H = T\Delta S$ and $\dfrac{\Delta H}{\Delta S} = T = \dfrac{+62.2 \text{ kJ}}{+132.7 \times 10^3 \tfrac{kJ}{K}} = 469 \text{ K (196 °C)}$

84. Calculate the $\Delta G°$ for the transition of S_8 (rhombic) $\rightarrow$ S_8 (monoclinic)

(a) At 80 °C $\Delta G°_{rxn} = \Delta H° - T \Delta S°$

$\Delta G°_{rxn} = 3.213 \text{ kJ} - (353 \text{ K})(0.0087 \tfrac{kJ}{K})$

$= 0.14 \text{ kJ}$

At 110 °C $\Delta G°_{rxn} = \Delta H° - T \Delta S°$

$\Delta G°_{rxn} = 3.213 \text{ kJ} - (383 \text{ K})(0.0087 \tfrac{kJ}{K})$

$= - 0.12 \text{ kJ}$

The rhombic form of sulfur is the more stable at lower temperature, while the monoclinic form is the more stable at higher temperature. The transition to monoclinic form is product-favored at temperatures above 110 degrees C.

(b) The temperature at which $\Delta G°_{rxn} = 0$:

$\Delta G°_{rxn} = 3.213 \text{ kJ} - (T)(0.0087 \tfrac{kJ}{K})$

$0 = 3.213 \text{ kJ} - (T)(0.0087 \tfrac{kJ}{K})$

$T = \dfrac{3.213 \text{ kJ}}{0.0087 \tfrac{kJ}{K}} = 370 \text{ K} = 96 \text{ °C}$

86. The function of the "Desert refrigerator" is accomplished by the evaporation of water from the cloths on the cloth covering (Swamp coolers--a type of cooler frequently used in dry areas—like the American Southwest--use the same effect.) The evaporation absorbs heat ($\Delta H = +$). As the liquid water evaporates (to form "gaseous" water), entropy increases. ($\Delta S = +$). Since the evaporation is spontaneous at room T, we know that $\Delta G = -$.

The process proceeds toward equilibrium (as do physical and chemical processes) at which time $\Delta G = 0$. So $\Delta G = \Delta H - T\Delta S$. For this condition to occur T must become smaller (the temperature must drop).

88.   (a) Calculate K_p at 25 °C for the reaction:

$$N_2(g) + O_2(g) \rightarrow 2\ NO\ (g)$$

ΔG°_f kJ/mol	0	0	86.58
ΔH°_f kJ/mol	0	0	90.29
ΔS°_f J/K•mol	191.56	205.07	210.76

$$\Delta H^\circ rxn = [(2\ mol)(90.29\ \frac{kJ}{mol})] - [(1\ mol)(0) + (1\ mol)(0)] = 180.58\ kJ$$

$$\Delta S^\circ rxn = [(2\ mol)(210.76\ \frac{J}{K\bullet mol})] -$$
$$[(1\ mol)(191.56\ \frac{J}{K\bullet mol}) + (1\ mol)(205.07\ \frac{J}{K\bullet mol})] = 24.89 J/K$$

$$\Delta G^\circ rxn = [(2\ mol)(86.58\ kJ/mol)] - [0 + 0] = 173.16\ kJ$$

Calculating K_p: $\Delta G^\circ rxn = -RT \ln K_p$

$$173.16 \times 10^3\ J/mol = -(8.3145\ \frac{J}{K\bullet mol})(298.15\ K)\ \ln K_p$$

$$-69.85 = \ln K_p$$

$$4.62 \times 10^{-31} = K_p$$

Note that the + value of ΔG°_f results in a value of K_p which is small--reactants are favored. A negative value would result in a large K_p -- a process in which the products were favored.

(b) $\Delta G^\circ rxn$ at 700 °C: $\Delta G^\circ = \Delta H^\circ - T\Delta S^\circ = 180.58\ kJ - (973\ K)(24.89 J/K)(1kJ/1000\ J)$

$$= 156.4\ kJ$$

The estimate for K_p at 700 °C: $156.4 \times 10^3\ J/mol = -(8.3145\ \frac{J}{K\bullet mol})(973\ K)\ \ln K_p$

$$\ln K_p = -19.33\ and\ K_p = 4 \times 10^{-9}$$

The reaction is still reactant favored, although K is much greater than at 25°C.

(c) With $K_p = 4 \times 10^{-9}$, equilibrium pressures of N_2, O_2 and NO:

$$\frac{P_{NO}^2}{P_{N_2} \bullet P_{O_2}} = 4 \times 10^{-9}$$

If the $P(N_2) = P(O_2) = 1.00$ bar initially, and we allow some amount, x, to form NO, then equilibrium concentrations would be: $P(N_2) = P(O_2) = (1.00-x)$ bar and $P(NO) = x$.

Substituting into the equilibrium expression:

$$\frac{x^2}{(1.00 - x)(1.00 - x)} = 4 \times 10^{-9} \text{ and taking the square root of both sides:}$$

$$\frac{x}{1.00 - x} = 6.32 \times 10^{-5} \quad \text{Solving for } x = 6 \times 10^{-5} \text{ and the equilibrium concentrations are:}$$

$P(N_2) = P(O_2) = 1$ bar (essentially unchanged) $P(NO) = 6 \times 10^{-5}$ bar

90. For the reaction:

	$4 Ag(s)$	$+$	$O_2(g)$	$\rightarrow$	$2 Ag_2O(s)$
$S° (\frac{J}{K \cdot mol})$	42.55		205.07		121.3
$\Delta H° (\frac{kJ}{mol})$	0		0		-31.1
$\Delta G° (kJ/mol)$	0		0		-11.32

(a) Calculate $\Delta H°_{rxn}$, $\Delta S°$, and $\Delta G°_{rxn}$:

$$\Delta H°_{rxn} = [(2 \text{ mol})(-31.1 \frac{kJ}{mol})] - [(4 \text{ mol})(0) + (1 \text{ mol})(0)] = -62.2 \text{ kJ}$$

$$\Delta S° = [(2 \text{ mol})(121.3 \frac{J}{K \cdot mol})] - [(4 \text{ mol})(42.55 \frac{J}{K \cdot mol}) + (1 \text{ mol})(205.07 \frac{J}{K \cdot mol})]$$

$$= -132.7 \text{ J/K}$$

$$\Delta G°_{rxn} = (-62.2 \text{ kJ}) - (298 \text{ K})(-132.67 \text{ J/K})(\frac{1.000 kJ}{1000 J}) = -22.6 \text{ kJ}$$

(b) To calculate P_{O_2} we'll need K_p.

$$\Delta G°_{rxn} = -RT \ln K_p$$

$$-22.6 \times 10^3 \text{ J} = -(8.314 \frac{J}{K \cdot mol})(298.15 \text{ K}) \ln K_p$$

$$+9.11 = \ln K_p \text{ and } K_p = 9.15 \times 10^3$$

$$K_p = \frac{1}{P_{O_2}} \quad \text{(Remember solids are omitted from equilibrium expressions.)}$$

$$\frac{1}{P_{O_2}} = 9.15 \times 10^3 \qquad \text{and } P_{O_2} = 1.1 \times 10^{-4} \text{ bar}$$

(c) T at which $P_{O_2} = 1.00$ bar

If $P_{O_2} = 1.00$ bar, then $K_p = \frac{1}{1.00 \text{ bar}} = 1.00$

Calculate the $\Delta G°$ at which $K_p = 1.00$

$$\Delta G°_{rxn} = -RT\ln K_p$$

The difficulty is that if T is not 25 °C, we need to calculate $\Delta G°$, $(\Delta H - T\Delta S)$, so

$$\Delta H° - T\Delta S° = -RT \ln(1.00)$$

$$\text{or } \Delta H° = -RT \ln(1.00) + T\Delta S°$$

$$\text{or } \frac{\Delta H°}{T} = -R \ln(1.00) + \Delta S°$$

Using $\Delta H°$ and $\Delta S°$ data (part a)

$$\frac{-62.2 \text{ kJ}}{T} = -(8.314 \frac{J}{K \cdot mol})(0) + -132.7 \text{ J/K}$$

$$\text{or } \frac{-62.2 \text{ kJ}}{T} = -132.7 \text{ J/K and } T = \frac{-62.2 \text{ kJ}}{-132.7 \times 10^{-3} \frac{kJ}{K}} = 469 \text{ K or } 196 °C$$

92. If we dissolve a solid (e.g. table salt), the process proceeds spontaneously ($\Delta G° < 0$).

If $\Delta H° = 0$, we can write: $\Delta G° = \Delta H° - T\Delta S°$ and $(-) = 0 - (+)\Delta S°$.

The only mathematical condition for which this equation is true is if $\Delta S° = +$, hence the process is entropy driven.

Chapter 20
Principles of Reactivity: Electron Transfer Reactions

Reviewing Important Concepts

2. For the electrochemical cell involving Mg and Ag:

   (a) Parts of the cell:

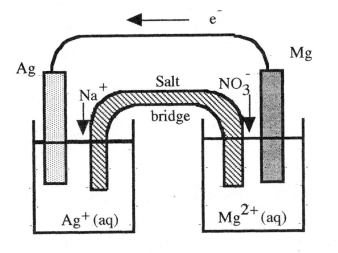

(b) Anode (oxidation):

$$Mg \rightarrow Mg^{2+} + 2e^-$$

Cathode (reduction):

$$2\,Ag^+ + 2\,e^- \rightarrow 2\,Ag$$

Adding gives the net reaction:

$$2\,Ag^+ + Mg \rightarrow Mg^{2+} + 2\,Ag$$

3. Movement of the Na^+ and NO^{3-} ions and electrons in the external circuit (wire) is noted on the diagram in SQ20.2. The salt bridge is necessary because as the oxidation proceeds in the anode compartment., an excess of positive charge builds (as Mg^{2+} are produced). Similarly in the cathode, an excess of negative charge accumulates as(as Ag^+ are reduced to Ag). The ions from the salt bridge counteract these charge buildups.

4. For the lead storage battery:

   Anode consists of Pb plates. Reaction: $Pb(s) + SO_4^{2-}$ (aq) $\rightarrow PbSO_4$ (s) $+ 2e^-$

   Cathode consists of grids of lead filled with PbO_2.

   Reaction: PbO_2 (s) $+ 4\,H_3O^+$ (aq) $+ 2\,H_2SO_4 + 2\,e^- \rightarrow PbSO_4 + 6\,H_2O\,(\ell)$

   The key feature for the rechargeable nature of this cell is the fact that the product, $PbSO_4$, is in contact with the electrodes.

7. For the series of substances given:

   (a) Arrange from easiest to oxidize to hardest to oxidize:

   Table 20.1 lists substances in order of their *ease of reduction* The easiest to reduce (F_2)is at the top left of the table, and the most difficult to reduce(Li^+) is at the bottom left. Their reduced counterparts (right side of the table) have the opposite chemical nature—that is, the Li atom is *easy to oxidize* and the F- is *difficult to oxidize*. Using these principles, the order is: $K > Zn > H_2 > Cu > Cl^-$.

(b) Arrange from easiest to reduce to hardest to reduce:
Using the principles stated in (a), the order is : $Ag^+ > I_2 > H^+ > H_2O > Na^+$

8. Which of the following reactions is product-favored?

(a) $Zn\ (s) + I_2\ (s)\ \rightarrow\ \ \ \ \ \ Zn^{2+}\ (aq) + 2\ I^-\ (aq)$

Compare the relative strengths of oxidizing (or reducing agents). I_2 is a stronger oxidizing agent than Zn^{2+}. Zn is a better reducing agent than I^-. The stronger oxidizing and reducing agents are on the left of the equation, so the reaction is product-favored.

(b) $2\ Cl^-\ (aq) + I_2\ (s) \rightarrow\ Cl_2\ (g)\ + 2\ I^-\ (aq)$

I^- is a stronger reducing agent than Cl^-. Cl_2 is a stronger oxidizing agent than I_2. The stronger oxidizing and reducing agents are on the right of the equation, so the reaction is reactant-favored.

(c) $2\ Na^+\ (aq) + 2\ Cl^-\ (aq) \rightarrow\ 2\ Na\ (s)\ + Cl_2\ (g)$

Common sense will tell you that this reaction is reactant-favored. Imagine having a solution of table salt producing elemental sodium and chlorine spontaneously!
The table confirms our "common sense", showing Na a stronger reducing agent than Cl^-, and Cl_2 a stronger reducing agent than Na^+.

(d) $2\ K\ (s) + 2\ H_2O\ (\ell) \rightarrow\ 2\ K^+\ (aq) + H_2\ (g)\ + 2\ OH^-\ (aq)$

Once again, the knowledge that K is a very reactive metal with respect to water leads to the conclusion that this reaction is product-favored. The table confirms that K is a stronger reducing agent than H_2 and H_2O is a stronger oxidizing agent than K^+.

9. Regarding the signs of $E°$ and $\Delta G°$ for a product-favored oxidation-reduction reaction:

Thermodynamics tells us that for product-favored reactions (of any type), the $\Delta G°$ is less than zero. The relationship $\Delta G° = -nFE°$ tells us that $E°$ must be greater than zero.

11. Relationships relating amp-seconds, coulombs, and F:

The *Faraday* is the charge associated with 1 mol of electrons (96500 coulombs of charge).

The *ampere* is the rate of current flow of 1 coulomb of charge per second.

To electroplate 1.00 mol of Ag from a solution of Ag^+ (aq) with a current of 1.00 amperes (A), we need to know how many moles of electrons we need.

The half-equation we need is:

$Ag^+(aq) + 1e^- \rightarrow Ag(s)$, so 1.00 mol of Ag needs 1.00 mol of electrons. 1.00 mol of electrons has a charge of 96500 C. Our current of 1.00 A produces 1 C/s, so 96,500 C would require 96500 seconds.

Practicing Skills

Balancing Equations for Oxidation-Reduction Reactions

12. Balance the following:

	reactant is	overall process is
(a) $Cr(s) \rightarrow Cr^{3+}(aq) + 3 e^-$	reducing agent	oxidation
(b) $AsH_3(g) \rightarrow As(s) + 3 H^+(aq) + 3 e^-$	reducing agent	oxidation
(c) $VO_3^-(aq) + 6 H^+(aq) + 3 e^- \rightarrow$ $V^{2+}(aq) + 3 H_2O(\ell)$	oxidizing agent	reduction
(d) $2 Ag(s) + 2 OH^-(aq) \rightarrow$ $Ag_2O(s) + H_2O(l) + 2e^-$	reducing agent	oxidation

Note: e^- are used to balance charge; H^+ balances only H atoms; H_2O (or OH^- in base) balances both H and O atoms.

14. Balance the equations (in acidic solutions):

Balancing redox equations in neutral or acidic solutions may be accomplished in several steps. They are:

1. Separating the equation into two equations which represent reduction and oxidation
2. Balancing mass of elements (other than H or O)
3. Balancing mass of O by adding H_2O
4. Balancing mass of H by adding H^+
5. Balancing charge by adding electrons
6. Balancing electron gain (in the reduction half-equation) with electron loss (in the oxidation half-equation)
7. Combining the two half equations

For the parts of this problem, each step will be identified with a number corresponding to the list above. In addition, the physical states of all species will be omitted in all but the final step. While this omission is not generally recommended, it should increase the clarity of the steps involved. In addition when a step leaves a half equation unchanged from the previous step, we have omitted the half equation.

(a) $Ag\ (s) + NO_3^-(aq) \rightarrow NO_2\ (g) + Ag^+\ (aq)$

Oxidation half-equation	Reduction half-equation	
$Ag \rightarrow Ag^+$	$NO_3^- \rightarrow NO_2$	1 & 2
	$NO_3^- \rightarrow NO_2 + H_2O$	3
	$2\ H^+ + NO_3^- \rightarrow NO_2 + H_2O$	4
$Ag \rightarrow Ag^+ + 1e^-$	$2\ H^+ + NO_3^- + 1e^- \rightarrow NO_2 + H_2O$	5 & 6
$2\ H^+(aq) + NO_3^-\ (aq) + Ag\ (s) \rightarrow Ag^+(aq) + NO_2\ (g) + H_2O(\ell)$		7

(b) $MnO_4^-\ (aq) + HSO_3^-\ (aq) \rightarrow Mn^{2+}(aq) + SO_4^{2-}\ (aq)$

Oxidation half-equation	Reduction half-equation	
$HSO_3^- \rightarrow SO_4^{2-}$	$MnO_4^- \rightarrow Mn^{2+}$	1 & 2
$H_2O + HSO_3^- \rightarrow SO_4^{2-}$	$MnO_4^- \rightarrow Mn^{2+} + 4H_2O$	3
$H_2O + HSO_3^- \rightarrow SO_4^{2-} + 3\ H^+$	$8\ H^+ + MnO_4^- \rightarrow Mn^{2+} + 4H_2O$	4
$H_2O + HSO_3^- \rightarrow SO_4^{2-} + 3\ H^+ + 2e^-$	$8\ H^+ + MnO_4^- + 5e^- \rightarrow Mn^{2+} + 4H_2O$	5
$5\ H_2O + 5\ HSO_3^- \rightarrow 5\ SO_4^{2-} + 15\ H^+ + 10e^-$	$16\ H^+ + 2MnO_4^- + 10e^- \rightarrow 2\ Mn^{2+} + 8H_2O$	6
$5\ HSO_3^-\ (aq) + H^+\ (aq) + 2\ MnO_4^-\ (aq) \rightarrow 5\ SO_4^{2-}\ (aq) + 2\ Mn^{2+}\ (aq) + 3\ H_2O\ (\ell)$		7

(c) $Zn\ (s) + NO_3^-(aq) \rightarrow Zn^{2+}\ (aq) + N_2O\ (g)$

Oxidation half-equation	Reduction half-equation	
$Zn \rightarrow Zn^{2+}$	$NO_3^- \rightarrow N_2O$	1
	$2\ NO_3^- \rightarrow N_2O$	2
	$2\ NO_3^- \rightarrow N_2O + 5\ H_2O$	3
	$10\ H^+ + 2\ NO_3^- \rightarrow N_2O + 5\ H_2O$	4
$Zn \rightarrow Zn^{2+} + 2\ e^-$	$10\ H^+ + 2\ NO_3^- + 8\ e^- \rightarrow N_2O + 5\ H_2O$	5
$4\ Zn \rightarrow 4\ Zn^{2+} + 8\ e^-$		6
$10\ H^+\ (aq) + 2\ NO_3^-\ (aq) + 4\ Zn\ (s) \rightarrow 4\ Zn^{2+}(aq) + N_2O\ (g) + 5\ H_2O\ (\ell)$		7

(d) $Cr(s) + NO_3^- (aq) \rightarrow Cr^{3+} (aq) + NO (g)$

Oxidation half-equation	Reduction half-equation	
$Cr \rightarrow Cr^{3+}$	$NO_3^- \rightarrow NO$	1 & 2
	$NO_3^- \rightarrow NO + 2 H_2O$	3
	$4 H^+ + NO_3^- \rightarrow NO + 2 H_2O$	4
$Cr \rightarrow Cr^{3+} + 3 e^-$	$4 H^+ + NO_3^- + 3 e^- \rightarrow NO + 2 H_2O$	5 & 6
$Cr (s) + 4 H^+(aq) + NO_3^- (aq) \rightarrow NO(g) + 2 H_2O(\ell) + Cr^{3+}(aq)$		7

16. Balancing redox equations in basic solutions may be accomplished in several steps. There is only a *slight change* from the "acidic solution".

　1.　Separating the equation into two equations which represent reduction and oxidation

　2.　Balancing mass of elements (other than H or O)

　3.　Balancing mass of O by adding H_2O

　4.　Balancing mass of H by adding H^+

　5.　Balancing charge by adding electrons

　6.　Balancing electron gain (in the reduction half-equation) with electron loss (in the oxidation half-equation)

　7.　Combining the two half equations (removing any redundancies).

　8.　*Add as many OH$^-$ to both sides of the equation as there are H^+ ions, to form water.*

　9.　*Remove any redundancies in H_2O molecules.*

As before, each step will be identified with a number corresponding to the list above, and physical states of all species will be omitted in all but the final step.

(a) $Al (s) + OH^- (aq) \rightarrow Al(OH)_4^- (aq) + H_2 (g)$

Oxidation half-equation	Reduction half-equation	
$Al \rightarrow Al(OH)_4^-$	$OH^- \rightarrow H_2$	1 & 2
$Al + 4 H_2O \rightarrow Al(OH)_4^-$	$OH^- \rightarrow H_2 + H_2O$	3
$Al + 4 H_2O \rightarrow Al(OH)_4^- + 4H^+$	$3 H^+ + OH^- \rightarrow H_2 + H_2O$	4
$Al + 4 H_2O \rightarrow Al(OH)_4^- + 4 H^+ + 3 e^-$	$3 H^+ + OH^- + 2 e^- \rightarrow H_2 + H_2O$	5
$2Al + 8 H_2O \rightarrow 2 Al(OH)_4^- + 8 H^+ + 6 e^-$	$9 H^+ + 3 OH^- + 6 e^- \rightarrow 3 H_2 + 3 H_2O$	6
$2 Al + 5 H_2O + H^+ + 3 OH^- \rightarrow 2 Al(OH)_4^- + 3 H_2$		7 & 8

(combine H$^+$ with 1 OH$^-$ on left side)	
$2Al + 5H_2O + (H_2O) + 2OH^- \rightarrow 2Al(OH)_4^- + 3H_2$	8
$2Al(s) + 6H_2O(\ell) + 2OH^-(aq) \rightarrow 2Al(OH)_4^-(aq) + 3H_2(g)$	9

(b) $CrO_4^{2-}(aq) + SO_3^{2-}(aq) \rightarrow Cr(OH)_3(s) + SO_4^{2-}(aq)$

Oxidation half-equation	Reduction half-equation	
$SO_3^{2-} \rightarrow SO_4^{2-}$	$CrO_4^{2-} \rightarrow Cr(OH)_3$	1 & 2
$H_2O + SO_3^{2-} \rightarrow SO_4^{2-}$	$CrO_4^{2-} \rightarrow Cr(OH)_3 + H_2O$	3
$H_2O + SO_3^{2-} \rightarrow SO_4^{2-} + 2H^+$	$5H^+ + CrO_4^{2-} \rightarrow Cr(OH)_3 + H_2O$	4
$H_2O + SO_3^{2-} \rightarrow SO_4^{2-} + 2H^+ + 2e^-$	$5H^+ + CrO_4^{2-} + 3e^- \rightarrow Cr(OH)_3 + H_2O$	5
$3H_2O + 3SO_3^{2-} \rightarrow 3SO_4^{2-} + 6H^+ + 6e^-$	$10H^+ + 2CrO_4^{2-} + 6e^- \rightarrow 2Cr(OH)_3 + 2H_2O$	6
$H_2O(\ell) + 3SO_3^{2-} + 4H^+ + 2CrO_4^{2-} \rightarrow 2Cr(OH)_3 + 3SO_4^{2-}$		7
$H_2O(\ell) + 3SO_3^{2-} + 4H^+ + 4OH^- + 2CrO_4^{2-} \rightarrow 2Cr(OH)_3 + 3SO_4^{2-} + 4OH^-$		8
$5H_2O(\ell) + 3SO_3^{2-}(aq) + 2CrO_4^{2-}(aq) \rightarrow 2Cr(OH)_3(s) + 3SO_4^{2-}(aq) + 4OH^-(aq)$		9

(c) $Zn(s) + Cu(OH)_2(s) \rightarrow Zn(OH)_4^{2-}(aq) + Cu(s)$

Oxidation half-equation	Reduction half-equation	
$Zn \rightarrow Zn(OH)_4^{2-}$	$Cu(OH)_2 \rightarrow Cu$	1 & 2
$4H_2O + Zn \rightarrow Zn(OH)_4^{2-}$	$Cu(OH)_2 \rightarrow Cu + 2H_2O$	3
$4H_2O + Zn \rightarrow Zn(OH)_4^{2-} + 4H^+$	$2H^+ + Cu(OH)_2 \rightarrow Cu + 2H_2O$	4
$4H_2O + Zn \rightarrow Zn(OH)_4^{2-} + 4H^+ + 2e^-$	$2H^+ + Cu(OH)_2 + 2e^- \rightarrow Cu + 2H_2O$	5 & 6
$2H_2O(\ell) + Zn(s) + Cu(OH)_2(s) \rightarrow Zn(OH)_4^{2-}(aq) + 2H^+(aq) + Cu(s)$		7
$2H_2O(\ell) + Zn(s) + Cu(OH)_2(s) + 2OH^-(aq) \rightarrow Zn(OH)_4^{2-}(aq) + 2H^+(aq) + 2OH^-(aq) + Cu(s)$		8
$Zn(s) + Cu(OH)_2(s) + 2OH^-(aq) \rightarrow Zn(OH)_4^{2-}(aq) + Cu(s)$		9

(d) HS^- (aq) + ClO_3^- (aq) $\rightarrow$ S(s) + Cl^- (aq)

Oxidation half-equation	Reduction half-equation	
$HS^- \rightarrow S$	$ClO_3^- \rightarrow Cl^-$	1 & 2
	$ClO_3^- \rightarrow Cl^- + 3\ H_2O$	3
$HS^- \rightarrow S + H^+$	$6\ H^+ + ClO_3^- \rightarrow Cl^- + 3\ H_2O$	4
$HS^- \rightarrow S + H^+ + 2\ e^-$	$6\ e^- + 6\ H^+ + ClO_3^- \rightarrow Cl^- + 3\ H_2O$	5
$3\ HS^- \rightarrow 3\ S + 3\ H^+ + 6\ e^-$		6
$3\ HS^- + 3\ H^+ + ClO_3^- \rightarrow Cl^- + 3\ H_2O + 3\ S$		7
$3\ HS^- + 3\ H^+ + 3\ OH^- + ClO_3^- \rightarrow Cl^- + 3\ H_2O + 3\ S + 3\ OH^-$		8
$3\ HS^-(aq) + ClO_3^-(aq) \rightarrow Cl^-$ (aq)+ $3\ S$ (s) + $3\ OH^-$(aq)		9

18. For the reaction : $2\ Cr$ (s) + $3\ Fe^{2+}$ (aq) $\rightarrow$ $2\ Cr^{3+}$ (aq) + $3\ Fe$ (s) :

Electrons in the external circuit flow from the <u>Cr</u> electrode to the <u>Fe</u> electrode. Negative ions move in the salt bridge from the <u>iron</u> half-cell to the <u>chromium</u> half-cell. The half-reaction at the anode is <u>$Cr(s) \rightarrow Cr^{3+}(aq) + 3\ e^-$</u> and that at the cathode is <u>$Fe^{2+}(aq) + 2\ e^- \rightarrow Fe(s)$</u>.

Note that the reaction shows that Cr is being oxidized (to Cr^{3+}) by Fe^{2+} which is being reduced (to Fe).The electrons leave the anode (and head to the cathode via the external circuit (wire). With reduction occurring in the cathode half-cell , a net deficit of positive ions accumulates, necessitating the assistance of + ions from the salt bridge. See SQ20.2 and 20.3 for additional questions of this type.

20. Like SQ20.18, we can complete the paragraph in part (c), by deciding on the spontaneous or product-favored reaction between the iron and oxygen half cells. The reduction potentials are:

$$O_2\ (g) + 4\ H^+\ (aq) + 4e^-\ (aq) \rightarrow 2\ H_2O \qquad E° = 1.229V$$
$$Fe^{2+}\ (aq) + 2\ e^- \rightarrow Fe\ (s) \qquad E° = -0.44V$$

(a) Oxidation half-reaction: $2\ Fe$ (s) $\rightarrow 2\ Fe^{2+}$ (aq) + $4\ e^-$
   <u>Reduction half-reaction: $O_2\ (g) + 4\ H^+\ (aq) + 4\ e^-\ (aq) \rightarrow 2\ H_2O$</u>
   Net cell reaction: $2\ Fe$ (s) + O_2 (g) + $4\ H^+$ (aq) $\rightarrow 2\ H_2O$ (ℓ) + $2\ Fe^{2+}$(aq)

How to decide which half-reaction occurs as oxidation? An examination of Table 20.1 (or Appendix M) shows O_2 as a stronger oxidizing agent than Fe^{2+}. Alternatively, Fe is a stronger reducing agent than H_2O. Either of these conclusions points to the direction of reaction which is product-favored.

(b) The anode half-reaction is the oxidation half-reaction: $2\ Fe$ (s) $\rightarrow 2\ Fe^{2+}$ (aq) + $4\ e^-$

At the cathode, the reduction half-reaction: $O_2(g) + 4 H^+ (aq) + 4 e^-(aq) \rightarrow 2 H_2O$

(c) Electrons in the external circuit flow from the <u>Fe</u> electrode to the <u>O_2</u> electrode. Negative ions move in the salt bridge from the <u>oxygen</u> half-cell to the <u>iron</u> half-cell.

Commercial Cells

22. Similarities and differences between dry cells, alkaline batteries, and mercury batteries:

> The three types of cells are non-rechargeable batteries—also called primary batteries. They also share the common anode, Zinc. Alkaline and mercury batteries are in a basic environment, whereas dry cells are in an acidic environment.

Standard Electrochemical Potentials

24. Calculate E° for each of the following, and decided if it is product-favored as written:

$$E°_{cell} = E°_{cathode} - E°_{anode}$$

(a) $2 I^- (aq) + Zn^{2+}(aq) \rightarrow I_2 (g) + Zn (s)$

Cathode reaction: $Zn^{2+}(aq) + 2 e^- \rightarrow Zn (s)$ E° = -0.763 V

Anode reaction: $2 I^- (aq) \rightarrow I_2 (g) + 2 e^-$ <u>E° = +0.535 V</u>

        Cell voltage: E° = -1.298 V(not product-favored)

(b) $Zn^{2+} (aq) + Ni (s) \rightarrow Zn (s) + Ni^{2+} (aq)$

Cathode reaction: $Zn^{2+} (aq) + 2 e^- \rightarrow Zn (s)$ E° = -0.763 V

Anode reaction: $Ni (s) \rightarrow Ni^{2+} (aq) + 2 e^-$ <u>E° = -0.25 V</u>

        Cell voltage: E° = -0.51 V(not product-favored)

(c) $2 Cl^- (aq) + Cu^{2+} (aq) \rightarrow Cu(s) + Cl_2 (g)$

Cathode reaction: $Cu^{2+} (aq) + 2 e^- \rightarrow Cu(s)$ E° = +0.337 V

Anode reaction: $2 Cl^- (aq) \rightarrow Cl_2 (g) + 2 e^-$ <u>E° = +1.360 V</u>

        Cell voltage: E° = -1.023 V(not product-favored)

(d) $Fe^{2+} (aq) + Ag^+ (aq) \rightarrow Fe^{3+} (aq) + Ag (s)$

Cathode reaction: $Ag^+ (aq) + e^- \rightarrow Ag (s)$ E° = +0.80 V

Anode reaction: $Fe^{2+}(aq) \rightarrow Fe^{3+} (aq) + e^-$ <u>E° = +0.771 V</u>

        Cell voltage: E° = 0.029 V (product-favored)

26.   (a) $Sn^{2+} (aq) + 2 Ag (s) \rightarrow Sn (s) + 2 Ag^+ (aq)$

Cathode reaction: $Sn^{2+} (aq) + 2e^- \rightarrow Sn (s)$ E° = -0.14 V

Anode reaction: $2 Ag (s) \rightarrow 2 Ag^+ (aq) + 2e^-$ <u>E° = +0.80 V</u>

        Cell voltage: E° = - 0.94 V (not product-favored)

(b) $2 \text{ Al (s)} + 3 \text{ Sn}^{4+} \text{ (aq)} \rightarrow 3 \text{ Sn}^{2+} \text{ (aq)} + 2 \text{ Al}^{3+} \text{ (aq)}$

Cathode reaction: $3 \text{ Sn}^{4+} \text{ (aq)} + 6e^- \rightarrow 3 \text{ Sn}^{2+} \text{ (aq)}$ $E° = +0.15 \text{ V}$

Anode reaction: $2 \text{ Al (s)} \rightarrow 2 \text{ Al}^{3+} \text{ (aq)} + 6e^-$ $\underline{E° = - 1.66 \text{ V}}$

                                      Cell voltage: $E° = 1.81 \text{ V (product-favored)}$

(c) $\text{ClO}_3^- \text{ (aq)} + 6 \text{ Ce}^{3+} \text{ (aq)} + 6 \text{ H}^+ \text{(aq)} \rightarrow \text{Cl}^- \text{ (aq)} + 6 \text{ Ce}^{4+} \text{ (aq)} + 3 \text{ H}_2\text{O(}\ell\text{)}$

Cathode reaction: $\text{ClO}_3^- \text{ (aq)} + 6 \text{ e}^- + 6 \text{ H}^+ \text{(aq)}$ $E° = +0.62 \text{ V}$

                    $\rightarrow \text{Cl}^-\text{(aq)} + 3 \text{ H}_2\text{O(}\ell\text{)}$

Anode reaction: $6 \text{ Ce}^{3+} \text{ (aq)} \rightarrow 6 \text{ Ce}^{4+} \text{ (aq)} + 6 \text{ e}^-$ $\underline{E° = +1.61 \text{ V}}$

                                        Cell voltage: $E° = -0.99 \text{ V}$

                                                  (not product-favored)

(d) $3 \text{ Cu (s)} + 2 \text{ NO}_3^-\text{(aq)} + 8 \text{ H}^+ \text{(aq)} \rightarrow 3 \text{ Cu}^{2+}\text{(aq)} + 2 \text{ NO (g)} + 4 \text{ H}_2\text{O(}\ell\text{)}$

Cathode reaction: $2 \text{ NO}_3^-\text{(aq)} + 8 \text{ H}^+ \text{(aq)} + 6 \text{ e}^-$ $E° = +0.96 \text{ V}$

                    $\rightarrow 2 \text{ NO (g)} + 4 \text{ H}_2\text{O (}\ell\text{)}$

Anode reaction: $3 \text{ Cu (s)} \rightarrow 3 \text{ Cu}^{2+}\text{(aq)} + 6e^-$ $\underline{E° = +0.337 \text{ V}}$

                                        Cell voltage: $E° = 0.62 \text{ V}$ (product-favored)

28. From the following half-reactions:

    (a) The metal most easily oxidized:

        From the list **Al** is the most easily oxidized metal. Having the most negative reduction potential, Al is the strongest reducing agent of the group, and reducing agents are oxidized as they perform their task.

    (b) Metals on the list capable of reducing Fe^{2+} to Fe

        Zn and **Al** both have more negative reduction potentials than Fe, hence are stronger reducing agents, and can reduce Fe^{2+} to Fe.

    (c) A balanced equation for the reaction of Fe^{2+} with Sn. Is the reaction product-favored?

        $\text{Fe}^{2+} \text{ (aq)} + \text{Sn (s)} \rightarrow \text{Fe (s)} + \text{Sn}^{2+} \text{ (aq)}$; Since Fe is a stronger reducing agent than Sn, this reaction would not have an $E° > 0$, and the reaction would be reactant-favored.

    (d) A balanced equation for the reaction of Zn2+ with Sn. Is the reaction product-favored?

        $\text{Zn}^{2+} \text{ (aq)} + \text{Sn (s)} \rightarrow \text{Zn (s)} + \text{Sn}^{2+} \text{ (aq)}$; Zn is a stronger reducing agent than Sn, this reaction would not have an $E° > 0$, and the reaction would be reactant-favored.

Ranking Oxidizing and Reducing Agents

30. Best element from the group that is the best reducing agent?

     Zn has the most negative standard reduction potential of the group. Recall that the more negative the reduction potential, the stronger a substance is as a reducing agent.

32. Ion from the group that is most easily reduced?

     The specie with the most positive reduction potential is the strongest oxidizing agent of the list, and with that role, becomes the most easily reduced. **Ag^+** fits that role from this list.

34. Regarding the halogens:

     (a) The halogen most easily reduced is the one with the most positive reduction potential **F_2** has the most negative of the halogens.

     (b) MnO_2 has a reduction potential of 1.23 V. Both F_2 and Cl_2 have more positive reduction potentials than MnO_2, and are better oxidizing agents than MnO_2..

Electrochemical Cells Under Nonstandard Conditions

36. The Voltage of a cell that has dissolved species at 0.025 M:

     Calculate the standard voltage of the cell:

     (1) $2 H_2O(\ell) + 2e^- \rightarrow H_2 (g) + 2 OH^- (aq)$ $E° = -0.8277$ V

     (2) $[Zn(OH)_4]^{2-} (aq) + 2 e^- \rightarrow Zn(s) + 4 OH^- (aq)$ $E° = -1.22$ V

     The net equation is given as: $Zn(s) + 2 H_2O (\ell) + 2 OH^- (aq) \rightarrow Zn(OH)_4^- (aq) + H_2 (g)$

     The equilibrium expression (Q) would be: $\dfrac{[Zn(OH)_4^-]P_{H2}}{[OH^-]^2}$

     We'll assume a hydrogen pressure of 1 atm, and we know the concentrations of the other terms: 0.025 M. Note that n (in the Nernst equation) corresponds to 2, since the balanced overall equation indicates that 2 moles of electrons are lost and 2 moles of electrons are gained.

     The cell reaction indicates Zn as a reactant, meaning that reaction (2) runs in the reverse direction, making that process an oxidation, so the E°anode = -1.22 V.

     Calculate $E°_{cell} = E°_{cathode} - E°_{anode} = -0.8277 - (-1.22) = 0.3923$ V

     The Nernst equation $E_{cell} = E°_{cell} - \dfrac{0.0257}{n} \ln \dfrac{[Zn(OH)_4^-]P_{H2}}{[OH^-]^2}$

     $E_{cell} = 0.3923 - \dfrac{0.0257}{2} \ln \dfrac{[0.025] \cdot 1}{[0.025]^2} = 0.3923 - (0.0257 \cdot 3.69/2) = 0.345$ V

38. The voltage of a cell that has Ag in a 0.25 M solution of Ag^+ and Zn electrode in 0.010 M Zn^{2+}:

Calculate the standard voltage of the cell:

(1) Zn^{2+} (aq) + 2e⁻ → Zn (s)   E° = -0.763 V

(2) Ag^+ (aq) + e⁻ → Ag(s)   E° = +0.80 V

The net equation is given as: 2 Ag^+(aq) + Zn (s) → Zn^{2+} (aq) + 2 Ag(s).

The equilibrium expression (Q) would be: $\dfrac{\left[Zn^{2+}\right]}{\left[Ag^+\right]^2}$

The cell will run in the direction that is product favored, so we can calculate

E° cell = E°cathode – E°anode = +0.80 – (-0.763) = 1.563 V

We also note from the balanced equation that 2 moles(n) of electrons are transferred.

Using the Nernst equation $E_{cell} = E°_{cell} - \dfrac{0.0257}{n} \ln\dfrac{\left[Zn^{2+}\right]}{\left[Ag^+\right]^2}$

$E_{cell} = 1.563 - \dfrac{0.0257}{2}\ln\dfrac{[0.010]}{[0.25]^2} = 1.563 - (0.0257/2)\cdot(-1.83) = 1.563 + 0.0235 = 1.58\ V$

40. The voltage of a cell that has Ag in a ? M solution of Ag^+ and Zn electrode in 1.0 M Zn^{2+}:

Calculate the standard voltage of the cell:

(1) Zn^{2+} (aq) + 2e⁻ → Zn (s)   E° = -0.763 V

(2) Ag^+ (aq) + e⁻ → Ag(s)   E° = +0.80 V

The net equation is given as: 2Ag^+(aq) + Zn (s) → Zn^{2+} (aq) + 2 Ag(s).

The equilibrium expression (Q) would be: $\dfrac{\left[Zn^{2+}\right]}{\left[Ag^+\right]^2}$

The cell will run in the direction that is product favored, so we can calculate

E° cell = E°cathode – E°anode = +0.80 – (-0.763) = 1.563 V

We also note from the balanced equation that 2 moles(n) of electrons are transferred.

Using the Nernst equation $E_{cell} = E°_{cell} - \dfrac{0.0257}{n} \ln\dfrac{\left[Zn^{2+}\right]}{\left[Ag^+\right]^2}$

Given the Ecell = 1.48V , we should be able to calculate the value of the "ln term".

Recall that we **know** the concentration of [Zn^{2+}],but **don't know** the [Ag^+].

$1.48 = 1.563 - \dfrac{0.0257}{2}\ln\dfrac{\left[Zn^{2+}\right]}{\left[Ag^+\right]^2}$ or $1.48 = 1.563 - (0.0257/2)\cdot\ln\dfrac{\left[Zn^{2+}\right]}{\left[Ag^+\right]^2}$

311

$$\frac{-(1.48-1.563)2}{0.0257} = \ln\frac{[Zn^{2+}]}{[Ag^+]^2} \quad \text{or } 6.459 = \ln\frac{[Zn^{2+}]}{[Ag^+]^2} \quad ; \text{and} \quad \frac{[Zn^{2+}]}{[Ag^+]^2} = 638.5$$

So $\dfrac{[1.0]}{638.5} = [Ag^+]^2$ and $[Ag^+] = 0.040$ M

Electrochemistry, Thermodynamics, and Equilibrium

42. Calculate $\Delta G°$ and K for the reactions:

(a) $2\,Fe^{3+}(aq) + 2\,I^-(aq) \rightarrow 2\,Fe^{2+}(aq) + 2\,I_2(s)$

Using the potentials: $Fe^{3+}(aq) + e^- \rightarrow 2\,Fe^{2+}$ $E° = 0.771V$ and

$2I^-(aq) \rightarrow 2\,I_2(s) + 2\,e^-$ $E° = 0.535V$, we calculate an $E°_{cell}$.

$E°_{cell} = (0.771V - 0.535V) = 0.236V$

The relationship between $\Delta G°$ and $E°$ is: $\Delta G° = -nFE°$ so.

$\Delta G° = -(2 \text{ mol e})(96{,}500 \text{ C/mol e})(0.236 \text{ V})$ and $1V = 1J/C$ so

$\Delta G° = -(2 \text{ mol e})(96{,}500 \text{ C/mol e})(0.236 \text{ J/C}) = -45548 \text{ J or } -45.5 \text{ kJ}$

$\Delta G° = -RT\ln K$ so $-\Delta G°/RT = \ln K$

$$\frac{45548J}{8.314\,\dfrac{J}{K\bullet mol}\bullet 298K} = \ln K \text{ so } 18.38 = \ln K \text{ and } K = 9 \times 10^7$$

(b) $I_2(aq) + 2Br^-(aq) \rightarrow 2\,I^-(aq) + Br_2(\ell)$

Using the potentials: $Br_2(\ell)\,c \rightarrow 2Br^-(aq)$ $E° = 1.08\text{ V}$ and

$2\,I_2(s) + 2\,e^- \rightarrow 2I^-(aq)$ $E° = 0.535V$

We calculate an $E°_{cell}$, noting that in the cell reaction given, molecular iodine is being

reduced (the cathode) and bromide ion is being oxidized (the anode).

$E°_{cell} = (0.535V - 1.08V) = -0.545V$

The relationship between $\Delta G°$ and $E°$ is: $\Delta G° = -nFE°$ so.

$\Delta G° = -(2 \text{ mol e})(96{,}500 \text{ C/mol e})(-0.545V)$ and $1V = 1J/C$ so

$\Delta G° = -(2 \text{ mol e})(96{,}500 \text{ C/mol e})(-0.545J/C) = +105{,}185 \text{ J or } +110 \text{ kJ (to 2sf)}$

$\Delta G° = -RT\ln K$ so $-\Delta G°/RT = \ln K$

$$\frac{-110{,}000J}{8.314\,\dfrac{J}{K\bullet mol}\bullet 298.15\text{ K}} = \ln K \text{ so } -42.45 = \ln K \text{ and } K = 4 \times 10^{-19}$$

44. Calculate K_{sp} for AgBr using the following reactions:

   (1) $AgBr (s) + 1 e^- \rightarrow Ag(s) + Br^- (aq)$ $E° = 0.0713$ V

   (2) $Ag^+ (aq) + 1e^- \rightarrow Ag(s)$ $E° = 0.7994$ V

Write the Ksp expression for AgBr; $AgBr(s) \rightleftharpoons Ag^+ (aq) + Br^- (aq)$

Note that we can accomplish this as an overall reaction, by reversing equation (2) and

adding that to equation (1). Equation (1) is presently written as a reduction (naturally) and

the *reverse* of Equation (2) would be an oxidation—so the roles for our "cell" are defined.

$E°cell = E°cathode = E°anode = 0.0713$ V $- 0.7994$ V $= -0.7281$V

Using our $\Delta G°$ relationships: $\Delta G° = -nFE° = -RTlnK$ so $lnK = \dfrac{nFE°}{RT}$

So ln K $= \dfrac{1 \text{ mol e} \bullet 96,500 \dfrac{C}{\text{mol e}} \bullet -0.7281\dfrac{J}{C}}{8.314 \dfrac{J}{K \bullet mol} \bullet 298.15 \text{ K}} = -28.36$ and $K = 5 \times 10^{-13}$

46. Calculate the $K_{formation}$ for $AuCl_4^- (aq)$

   (1) $AuCl_4^- (aq) + 3e^- \rightarrow Au(s) + 4 Cl^- (aq)$ $E° = 1.00$ V

   (2) $Au^{3+} (aq) + 3e^- \rightarrow Au (s)$ $E° = 1.50$ V

The formation reaction for the complex is: $Au^{3+} (aq) + 4 Cl^- (aq) \rightleftharpoons AuCl_4^- (aq)$

To achieve this reaction as a net reaction, we need to reverse equation (1) and add it to

equation (2).

The $E°cell$ for that process would be: $E°cathode - E°anode = 1.50 - 1.00 = 0.50$ V

Using our $\Delta G°$ relationships: $\Delta G° = -nFE° = -RTlnK$ so $lnK = \dfrac{nFE°}{RT}$

So ln K $= \dfrac{3 \text{ mol e} \bullet 96,500 \dfrac{C}{\text{mol e}} \bullet 0.50\dfrac{J}{C}}{8.314 \dfrac{J}{K \bullet mol} \bullet 298.15 \text{ K}} = 58.42$ and $K = 2 \times 10^{25}$

48. For the disproportionation of Iron(II):

   (a) The half-reactions that make up the disproportionation reaction:

$Fe^{2+}(aq) + 2 e^- \rightarrow Fe (s)$ $E° = -0.44$ V (cathode)

$\underline{2 Fe^{2+}(aq) \rightarrow 2 Fe^{3+}(aq) + 2 e^-}$ $\underline{E° = 0.771 \text{ V (anode)}}$

$3 Fe^{2+}(aq) \rightarrow 2 Fe^{3+}(aq) + Fe (s)$ $E° = -0.44$ V $- 0.771$ V $= -1.211$V

   (b) Given the voltage calculated, the disproportionation reaction is not product-favored.

(c) $\ln K = \dfrac{2 \text{ mol e} \cdot 96,500\dfrac{C}{\text{mol e}} \cdot -1.211\dfrac{J}{C}}{8.314 \dfrac{J}{K \cdot \text{mol}} \cdot 298.15 \text{ K}} = -94.34$ and $K_{formation} = 1 \times 10^{-41}$

Electrolysis

50 Diagram of a electrolysis apparatus for molten NaCl:

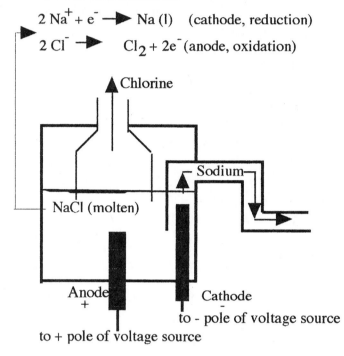

Downs Cell Schematic

$2 \text{ Na}^+ + \text{e}^- \longrightarrow \text{Na (l)}$ (cathode, reduction)

$2 \text{ Cl}^- \longrightarrow \text{Cl}_2 + 2\text{e}^-$ (anode, oxidation)

Chlorine

Sodium

NaCl (molten)

Anode
+

Cathode
-

to - pole of voltage source

to + pole of voltage source

52. For the electrolysis of a solution of KF(aq), what product is expected at the anode: O_2 or F_2

Example 20.11 provides additional help with this concept. The **bottom line** is that the process that occurs in an electrolysis is the one requiring the smaller applied potential.

For an electrolysis: $E°cell = E°cathode - E°anode$. For aqueous KF, those voltages are:

$2 \text{ F}^- \text{ (aq)} \rightarrow \text{F}_2 \text{ (g)} + 2\text{e}^-$ $E° = +2.87$ V

$2 \text{ H}_2\text{O (l)} \rightarrow \text{O}_2 \text{ (g)} + 4 \text{ H}^+ \text{ (aq)} + 4 \text{ e}^-$ $E° = +1.23$ V

At the cathode: $2 \text{ H}_2\text{O(l)} + 2 \text{ e}^- \rightarrow \text{H}_2\text{(g)} + 2 \text{ OH}^-\text{(aq)}$ $E° = -0.83$ V

[K has such a large negative reduction potential, that it *will not be reduced.*]

So the two choices are: $E°cell = E°cathode - E°anode$

For fluorine oxidation: $E°cell = (-0.83 - +2.87) = -3.7$ V

For oxygen oxidation: $E°cell = (-0.83 - +1.23) = -2.06$ V

Oxygen oxidation will require the lower applied potential, and will be produced at the anode.

54. For the electrolysis of KBr (aq):
   (a) The reaction occurring at the cathode: $2\ H_2O(\ell) + 2\ e^- \rightarrow H_2(g) + 2\ OH^-(aq)$
   (b) The reaction occurring at the anode: $2\ Br^-\ (aq) \rightarrow Br_2\ (\ell) + 2e^-$

Counting Electrons

56. Solutions to problems of this sort are best solved by beginning with a factor containing the desired units. Connecting this factor to data provided usually gives a direct path to the answer.

units desired

$\downarrow$

$$\frac{58.69\ g\ Ni}{1\ mol\ Ni} \cdot \frac{1\ mol\ Ni}{2\ mol\ e^-} \cdot \frac{1\ mol\ e^-}{9.65 \times 10^4\ C} \cdot \frac{1\ C}{1\ amp \cdot s}$$

$$\cdot \frac{0.150amps}{1} \cdot \frac{60s}{1\ min} \cdot \frac{12.2\ min}{1} = 0.0334\ g\ Ni$$

The second factor ($\frac{1\ mol\ Ni}{2\ mol\ e^-}$) is arrived at by looking at the reduction half-reaction:

$$Ni^{2+}\ (aq) + 2\ e^- \rightarrow Ni(s)$$

All other factors are either data or common unity factors (e.g. $\frac{60s}{1\ min}$).

58. Follow the pattern from SQ20.56, in starting with a unit that has the units of the desired answer—in this case units of *time*.

$$\frac{1\ amp \cdot s}{1\ C} \cdot \frac{1}{0.66\ amp} \cdot \frac{96500\ C}{1\ mol\ e} \cdot \frac{2\ mol\ e}{1\ mol\ Cu} \cdot \frac{1\ mol\ Cu}{63.546\ g\ Cu} \cdot \frac{0.50\ g\ Cu}{1} = 2300\ s.$$

60. Once again, since the requested answer has units of *hours*, start with a unit that has hours in it.

$$\frac{1\ hr}{3600\ s} \cdot \frac{1\ amp \cdot s}{1\ C} \cdot \frac{1}{1.0\ amp} \cdot \frac{96500\ C}{1\ mol\ e} \cdot \frac{3\ mol\ e}{1\ mol\ Al} \cdot \frac{1\ mol\ Al}{26.98\ g\ Al} \cdot \frac{84\ g\ Al}{1} = 250\ hr$$

General Questions

62. Balanced equations for the following half-reactions:
   (a) $UO_2^+(aq) \rightarrow U^{4+}(aq)$

Reduction half-equation	
$UO_2^+\ (aq) \rightarrow U^{4+}(aq)$	Balance all non H,O

UO_2^+ (aq) $\rightarrow$ U^{4+}(aq) + $2H_2O$	Balance O with H_2O
$4 H^+$(aq) + UO_2^+(aq) $\rightarrow$ U^{4+}(aq) + $2H_2O$ (ℓ)	Balance H with H^+
$4 H^+$(aq) + UO_2^+(aq) + 1 e^- $\rightarrow$ U^{4+}(aq) + 2 H_2O (ℓ)	Balance charge with e^-

(b) ClO_3^- (aq) $\rightarrow$ Cl^-(aq)

Reduction half-equation	
ClO_3^- (aq) $\rightarrow$ Cl^-(aq)	Balance all non H,O
ClO_3^- (aq) $\rightarrow$ Cl^-(aq) + $3H_2O$(ℓ)	Balance O with H_2O
$6 H^+$(aq) + ClO_3^- (aq) $\rightarrow$ Cl^-(aq) + $3H_2O$(ℓ)	Balance H with H^+
$6 H^+$(aq) + ClO_3^- (aq) + 6 e^- $\rightarrow$ Cl^-(aq) + 3 H_2O(ℓ)	Balance charge with e^-

(c) N_2H_4 (aq) $\rightarrow$ N_2 (g)

Oxidation half-equation	
N_2H_4 (aq) $\rightarrow$ N_2 (g)	Balance all non H,O
N_2H_4 (aq) $\rightarrow$ N_2 (g) + 4 H_2O(ℓ)	Balance H with H_2O
$4 OH^-$(aq) + N_2H_4 (aq) $\rightarrow$ N_2 (g) + 4 H_2O(ℓ)	Balance O with OH^-
$4 OH^-$(aq) + N_2H_4 (aq) $\rightarrow$ N_2 (g) + 4 H_2O(ℓ) + 4 e^-	Balance charge with e^-

(d) OCl^-(aq) $\rightarrow$ Cl^-(aq)

Reduction half-equation	
OCl^-(aq) $\rightarrow$ Cl^-(aq)	Balance all non H,O
OCl^-(aq) $\rightarrow$ Cl^-(aq) + OH^- (aq)	Balance O with OH^-
H_2O(ℓ) + OCl^-(aq) $\rightarrow$ Cl^-(aq) + OH^- (aq) + *OH^- (aq)*	Balance H with H_2O rebalance H with OH^-
H_2O(ℓ) + OCl^-(aq) + 2 e^- $\rightarrow$ Cl^-(aq) + 2 OH^-(aq)	Balance charge with e^-

64. Select half-cells that would, when coupled with Zn^{2+}(aq, 1M)/Zn (s) half cell gives voltage of:

For simplicity's sake, I've used Table 20.1 as the source of data for this problem.

(a) 1.1 V; We want $E°$cell to be 1.1v = $E°$cathode $-$ $E°$anode.

Substitute Zn's reduction potential for the cathode and then for the anode and solve for the other quantity.

$E°$cell = $E°$cathode $-$ $E°$anode.

1.1V = -0.763V $-$ $E°$anode. And 1.863V = - $E°$anode, and $E°$anode = -1.863V. An

examination of the Table shows **Al** (with $E°$ = -1.66V) to be one possibility.

Now substitute Zn's reduction potential for the anode:

$E°_{cell} = E°_{cathode} - E°_{anode}$.

$1.1V = E°_{cathode} - (-0.763V)$ rearranging: $1.1 + (-0.763) = E°_{cathode}$;

$0.337 = E°_{cathode}$. **Cu** fits that situation.

(b) 0.5V; We want $E°_{cell}$ to be $1.1v = E°_{cathode} - E°_{anode}$. Following the procedure from

(a) substituting 0.5V as $E°_{cell}$.

$E°_{cell} = E°_{cathode} - E°_{anode}$.

$0.5V = -0.763V - E°_{anode}$. And $1.26V = - E°_{anode}$, and $E°_{anode} = -1.26V$. No half-

cells are close to this value. Repeat using Zn's reduction potential for the anode:

$E°_{cell} = E°_{cathode} - E°_{anode}$.

$0.5V = E°_{cathode} - (-0.763V)$ rearranging: $0.5 + (-0.763) = E°_{cathode}$;

$-0.236 = E°_{cathode}$. **Ni** fits that situation.

66. Examine the reduction potentials for: $\underline{E°}$

	$E°$
Au^+ (aq) + 1 e$^-$ → Au (s)	+1.68
Ag^+ (aq) + 1 e$^-$ → Ag (s)	+0.80
Cu^{2+} (aq) + 2 e$^-$ → Cu (s)	+0.337
Sn^{2+} (aq) + 2e$^-$ → Sn (s)	-0.14
Co^{2+} (aq) + 2 e$^-$ → Co (s)	-0.28
Zn^{2+} (aq) + 2 e$^-$ → Zn (s)	-0.763

To clarify the trends, I have listed them in the descending reduction potential typical of

Reduction Potential Charts, like table 20.1.

(a) Metal ion that is the weakest oxidizing agent? -- Zn^{2+} Strength of oxidizing agents

increase with the *increasing + E°*.

(b) Metal ion that is the strongest oxidizing agent? –Au^+

(c) Metal that is the strongest reducing agent? –Zn (s) (since it's oxidized partner is the

weakest oxidizing agent, the metal is the strongest reducing agent.

(d) Metal that is the weakest reducing agent?—Au (s)

(e) Will Sn(s) reduce Cu^{2+} to Cu (s)? – Yes. Use the "NW to SE rule to see that this will

give a positive $E°_{cell}$.

(f) Will Ag (s) reduce Co^{2+}(aq) to Co (s)? No. See part (e)

(g) Which metal ions from the list can be reduced by Sn(s). Using the logic as in part (e)

Au^+, Ag^+, and Cu^{2+} can be reduced by Sn.

(h) Metals that can be oxidized by Ag^+(aq) –Any metal "below" Ag: Cu, Sn, Co, Zn

68. Examine the following reductions $E°$

   Cu^{2+} (aq) + 2 e^- → Cu (s)   +0.337

   Fe^{2+} (aq) + 2 e^- → Fe (s)   -0.44

   Cr^{3+} (aq) + 3 e^- → Cr (s)   -0.74

   Mg^{2+} (aq) + 2 e^- → Mg (s)   -2.37

   (a) In which of the voltaic cells would the S.H.E. be the cathode?

   $E°_{cell}$ must be positive, and the $E°$ S.H.E.= 0.00V, so for $E°_{cell}$ = 0.00V – ($E°$anode) to

   be positive, the $E°_{anode}$ would have to be *negative*, so the iron, chromium, and

   magnesium half-cells would fit this description.

   (b) Voltaic cell with the highest and lowest voltages?

   For $E°_{cell}$ = 0.00V – ($E°_{anode}$) to have the largest value the $E°_{anode}$ would have to be

   the most negative (Magnesium)

   For $E°_{cell}$ = 0.00V – ($E°_{anode}$) to have the smallest value the $E°_{anode}$ would have to be

   the least negative (Iron)

70.   (a) Cell diagram:

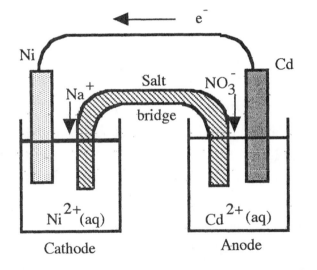

   (b) As shown in part (a), Cd serves as the anode, so it must be oxidized to the 2+ cation.

      Likewise Ni ions would be reduced to elemental Ni.

      The balanced equation is: Ni^{2+}(aq) + Cd(s) → Cd^{2+}(aq) + Ni(s)

   (c) The anode (Cd) serves as the source of electrons to the external circuit, and we label is

      "-". The cathode is labeled "+".

   (d) The $E°_{cell}$ = $E°_{cathode}$ - $E°_{anode}$ = (-0.205) – (-0.403) = + 0.15 V

   (e) As shown on the diagram, electrons flow from Cd to Ni compartments.

   (f) The direction of travel for the sodium and nitrate ions are shown on the diagram.

(g) The K for the reaction: $\Delta G^\circ = -nFE^\circ cell$

$$= -(2 \text{ mol e})(96500 \text{ C/mol e})(+0.15 \text{ J/C}) = 28950 J$$

$\Delta G^\circ = -RT\ln K$ and $\ln K = \dfrac{nFE^\circ}{RT}$

So $\ln K = \dfrac{2 \text{ mol e} \bullet 96,500 \dfrac{C}{\text{mol e}} \bullet 0.15 \dfrac{J}{C}^\circ}{8.314 \dfrac{J}{K \bullet mol} \bullet 298 \text{ K}} = 11.68$ and $K = 1 \times 10^5$

(h) If $[Cd^{2+}] = 0.010M$ and $[Ni^{2+}] = 1.0M$, the value for Ecell = ?

Using the Nernst equation $E_{cell} = E^\circ cell - \dfrac{0.0257}{n} \ln \dfrac{[Cd^{2+}]}{[Ni^{2+}]}$

$E_{cell} = 0.15 - \dfrac{0.0257}{2} \ln \dfrac{[0.010]}{[1.0]} = 0.15 - (0.0257/2) \bullet (-4.605) = 0.15 + 0.0592 = 0.21 \text{ V}$

(i) Lifetime use of battery?

We have 1.0 L of each solution, we should determine the limiting reagent, if there is one. The spontaneous reaction reduces Ni^{2+} and oxidizes Cd..

The Nickel solution contains: $1.0 \text{ L} \bullet 1.0M = 1.0 \text{ mol } Ni^{2+}$ The cadmium electrode weighs 50.0 g, so we have: 50.0 g Cd $\bullet$ (1 mol Cd/112.41 g Cd) = 0.445 mol Cd, so Cd is the limiting reagent. Now we can calculate:

$0.445 \text{ mol Cd} \bullet \dfrac{2 \text{ mol e}}{1 \text{ mol Cd}} \bullet \dfrac{96500 \text{ C}}{1 \text{ mol e}} \bullet \dfrac{1 \text{ amp} \bullet s}{1C} \bullet \dfrac{1}{0.050 \text{ A}} = 1.7 \times 10^6 \text{ s or 480 hr}$

72. For the cell with $E_{cell} = +0.146V$:

(a) What is the reduction potential for the anode reaction?

$E^\circ cell = E^\circ cathode - E^\circ anode$; $+0.146 = (+0.80) - (E_{anode}) = +0.65 \text{ V}$

(b) The K_{sp} for Ag_2SO_4:

Write the equation for the K_{sp} of Ag_2SO_4: $Ag_2SO_4 \rightleftharpoons 2 Ag^+(aq) + SO_4^{2-}(aq)$

Rearrange the two equations to give this as a net equation:

$Ag_2SO_4 + 2 e^- \rightarrow 2 Ag(s) + SO_4^{2-}(aq)$ $E^\circ = 0.65 \text{ V}$

$\underline{2 Ag (s) \rightarrow 2 Ag^+(aq) + 2 e^- \qquad\qquad\qquad E^\circ = 0.80 \text{ V}}$

Note that the reaction for the second reaction was multiplied by 2. Recall that E° are NOT changed as a function of coefficients, so the original reduction potential is retained. Now we can calculate the K_{sp}.

Note that to obtain the K_{sp} expression, we have written the silver reaction as an

oxidation (that is to say, an anode) with the silver sulfate reduction potential known, we can calculate the E_{cell} *for this process* (whether it is positive or not).

The voltage would be –0.146 V *(for the prescribed process)*.

319

$$\Delta G° = - R\,T\,\ln K \text{ and } \ln K = \frac{nFE°}{RT}$$

$$\text{So } \ln K = \frac{2 \text{ mol e} \cdot 96,500 \dfrac{C}{\text{mol e}} \cdot -0.146\dfrac{J}{C}}{8.314 \dfrac{J}{K \cdot mol} \cdot 298.15 \text{ K}} = -11.36 \text{ and } K = 1.2 \times 10^{-5}$$

74. If $E°_{cell} = +2.12V$, the $\Delta G° = ?$

This can be determined easily if we first determine the number of moles of e transferred. Since Zn forms a dipositive cation, and each atom of the chlorine molecule will require 1 electron, n = 2. $\Delta G = -nFE°cell = -(2 \text{ mol e})(96500 \text{ C/mol e})(+2.12 \text{ J/C}) = -409160$ J or –409 kJ.

76. Number of kwh to produce 1.0×10^3 kg of Al:

$$\frac{1.0 \times 10^6 \text{ g Al}}{1} \cdot \frac{1 \text{ mol Al}}{26.98 \text{ gAl}} \cdot \frac{3 \text{ mol e}^-}{1 \text{ mol Al}} \cdot \frac{96500 \text{ C}}{1 \text{ mol e}} \cdot \frac{1 \text{ J}}{1\,V \cdot C} \cdot \frac{5.0 \text{ V}}{1} \cdot \frac{1 \text{ kwh}}{3.6 \times 10^6 \text{ J}} = 1.5 \times 10^4 \text{ kwh}$$

78. To calculate the charge on the Rh^{x+} ion, we need to know two things:

1. How many moles of elemental rhodium are reduced?

2. How many moles of electrons caused that reduction?

Moles of rhodium: $\dfrac{0.038 \text{ g Rh}}{1} \cdot \dfrac{1 \text{ mol Rh}}{102.9 \text{ g Rh}} = 3.7 \times 10^{-4}$ mol Rh

Moles of electrons:

$$0.0100 \text{ amp} \cdot \frac{1C}{1amp \cdot s} \cdot \frac{3600 \text{ s}}{1 \text{ hr}} \cdot \frac{3 \text{ hr}}{1} \cdot \frac{1 \text{ mol e}^-}{96500 \text{ C}} = 1.1 \times 10^{-3} \text{ mol e}^-$$

Recall that our general reduction reactions are written: $M^{+x} + x\,e^- \rightarrow M$

If we know the number of $\dfrac{\text{moles of electrons}}{\text{mol of metal}}$, we know the charge on the cation, hence for

the Rh^{x+} ion we have $\dfrac{1.1 \times 10^{-3} \text{ mole}^-}{3.7 \times 10^{-4} \text{ mol Rh}} = 3.0 \dfrac{\text{mol e}^-}{\text{mol Rh}}$.

The ion is therefore the Rh^{3+} ion!

80. Mass of Zn consumed when 35 amp•hours is used from a dry cell:

$$\frac{65.39 \text{ g Zn}}{1 \text{ mol Zn}} \cdot \frac{1 \text{ mol Zn}}{2 \text{ mol e}} \cdot \frac{1 \text{ mol e}^-}{96500 \text{ C}} \cdot \frac{1 \text{ C}}{1 \text{ amp} \cdot s} \cdot \frac{3600 \text{ s}}{1 \text{ hr}} \cdot \frac{35 \text{ amp} \cdot \text{hr}}{1} = 43 \text{ g Zn}$$

82. Current flowing if 0.052 g Ag are deposited in 450 s:

$$\frac{0.052 \text{ g Ag}}{1} \cdot \frac{1 \text{ mol Ag}}{108 \text{ g Ag}} \cdot \frac{1 \text{ mol e}^-}{1 \text{ mol Ag}} \cdot \frac{9.65 \times 10^4 \text{ C}}{1 \text{ mol e}^-} = 46 \text{ C}$$

and with this charge flowing in 450 s, the current is: $\frac{46 \text{ C}}{450 \text{ s}} = 0.10$ amperes

84. Balanced equations for the following half-reactions:

(States are omitted until the last step, to increase the clarity of the process.)

(a) $HCO_2H \rightarrow HCHO$

$HCO_2H \rightarrow HCHO$	Balance all non H,O
$HCO_2H \rightarrow HCHO + H_2O$	Balance O with H_2O
$2 H^+ + HCO_2H \rightarrow HCHO + H_2O$	Balance H with H^+
$2 H^+(aq) + HCO_2H(aq) + 2 e^- \rightarrow HCHO(aq) + H_2O(\ell)$	Balance charge with e^-

(b) $C_6H_5CO_2H \rightarrow C_6H_5CH_3$

$C_6H_5CO_2H \rightarrow C_6H_5CH_3$	Balance all non H,O
$C_6H_5CO_2H \rightarrow C_6H_5CH_3 + 2 H_2O$	Balance O with H_2O
$C_6H_5CO_2H + 6 H^+ \rightarrow C_6H_5CH_3 + 2 H_2O$	Balance H with H^+
$C_6H_5CO_2H(aq) + 6H^+(aq) + 6 e^- \rightarrow C_6H_5CH_3(\ell) + 2H_2O(\ell)$	Balance charge with e^-

(c) $CH_3CH_2CHO \rightarrow CH_3CH_2CH_2OH$

$CH_3CH_2CHO \rightarrow CH_3CH_2CH_2OH$	Balance all non H,O
$CH_3CH_2CHO \rightarrow CH_3CH_2CH_2OH$	Balance O with H_2O
$CH_3CH_2CHO + 2 H^+ \rightarrow CH_3CH_2CH_2OH$	Balance H with H^+
$CH_3CH_2CHO (aq) + 2 H^+(aq) +2 e^- \rightarrow CH_3CH_2CH_2OH(aq)$	Balance charge with e^-

(d) $CH_3OH \rightarrow CH_4$

$CH_3OH \rightarrow CH_4$	Balance all non H,O
$CH_3OH \rightarrow CH_4 + H_2O$	Balance O with H_2O
$2 H^+ + CH_3OH \rightarrow CH_4 + H_2O$	Balance H with H^+
$2 H^+(aq) + CH_3OH(aq) + 2 e^- \rightarrow CH_{4(g)} + H_2O(\ell)$	Balance charge with e^-

86.   (a) Voltage of a cell consisting of Ag^+/Ag and Fe^{2+}/Fe^{3+}

$$Ag^+(aq) + 1e \rightarrow Ag (s) \qquad E° = 0.80V$$
$$Fe^{3+}(aq) + 1e \rightarrow Fe^{2+}(aq) \qquad E° = 0.771$$

The predicted voltage would be: $E°_{cell} = E°_{cathode} - E°_{anode} = 0.80 - 0.771 = 0.029V$

(b) Net ionic equation: $Ag^+(aq) + Fe^{2+}(aq) \rightarrow Fe^{3+}(aq) + Ag(s)$ (based upon definition of anode and cathode in part (a).

(c) The silver electrode is the cathode and the iron electrode is the anode.

(d) Cell voltage when $[Ag^+] = 0.10M$; $[Fe^{2+}] = [Fe^{3+}] = 1.0$ M:

Using the Nernst equation $\quad E_{cell} = E^{\circ}_{cell} - \dfrac{0.0257}{n} \ln \dfrac{[Fe^{3+}]}{[Fe^{2+}][Ag^+]}$

$E_{cell} = 0.029V - \dfrac{0.0257}{1} \ln \dfrac{[1.0]}{[1.0][0.10]} = 0.029V - (0.0257) \bullet (2.303)$

$= 0.029V - 0.0592 = -0.030V$

The negative voltage indicates the reaction *now occurring* is the *reverse of the one we proposed*: $Fe^{3+}(aq) + Ag(s) \rightarrow Ag^+(aq) + Fe^{2+}(aq)$

88. Easier to reduce water in acid or base?

Using the Nernst equation $\quad E_{cell} = -0.83V - \dfrac{0.0257}{2} \ln \dfrac{[OH^-]^2 \bullet P_{H2}}{1}$

(We'll assume that the pressure of hydrogen gas = 1 bar)

For pH = 7.0, $[OH^-] = 1 \times 10^{-7}$ $\quad E_{cell} = -0.83V - \dfrac{0.0257}{2} \ln \dfrac{[1 \times 10^{-7}]^2 \bullet 1}{1}$

$\qquad\qquad E_{cell} = -0.83V - \dfrac{0.0257}{2}(-32.24) =$

$\qquad\qquad = -0.83V + 0.414 = -0.42$ V

Now for pH=0, $[OH^-] = 1 \times 10^{-14}$ $\quad E_{cell} = -0.83V - \dfrac{0.0257}{2} \ln \dfrac{[1 \times 10^{-14}]^2 \bullet 1}{1}$

$\qquad\qquad = -0.83V - \dfrac{0.0257}{2} \ln (1 \times 10^{-28})$

$\qquad\qquad = -0.83 - \dfrac{0.0257}{2}(-64,47) = -0.0015$ V

At pH = 1, $[OH^-] = 1 \times 10^{-13}$ $\quad E_{cell} = -0.83V - \dfrac{0.0257}{2} \ln (1 \times 10^{-26})$

$\qquad\qquad = -0.83 - \dfrac{0.0257}{2}(-59.87) = -0.06$ V

From these calculations, one can see that it is easier to reduce water in **acid**.

90.   (a) Since A and C reduce H^+ to elemental hydrogen, they are both stronger reducing agents than H_2.

Reducing agent strength: $\quad H_2 < A,C$

(b)　C reduces A, B, and D—hence it is the strongest reducing agent.

　　　　A, B, D $<$ C and from part (a)　above　B, D $<$ H_2 $<$ A $<$ C

(c)　D reduces B^{n+}, so D is a stronger reducing agent than B.

　　　So　B $<$ D $<$ H_2 $<$ A $<$ C

92.　In the electrolysis of 150 g of CH_3SO_2F:

(a) The mass of HF required to electrolyze 150 g of CH_3SO_2F:

$$\frac{150 \text{ g } CH_3SO_2F}{1} \cdot \frac{1 \text{ mol } CH_3SO_2F}{98.10 \text{ g } CH_3SO_2F} \cdot \frac{3 \text{ mol HF}}{1 \text{ mol } CH_3SO_2F} \cdot \frac{20.01 \text{ g HF}}{1 \text{ mol HF}} = 92 \text{ g HF}$$

$$\frac{150 \text{ g } CH_3SO_2F}{1} \cdot \frac{1 \text{ mol } CH_3SO_2F}{98.10 \text{ g } CH_3SO_2F} \cdot \frac{1 \text{ mol } CF_3SO_2F}{1 \text{ mol } CH_3SO_2F} \cdot \frac{152.07 \text{ g } CF_3SO_2F}{1 \text{ mol } CF_3SO_2F}$$
$$= 230 \text{ g } CF_3SO_2F$$

$$\frac{150 \text{ g } CH_3SO_2F}{1} \cdot \frac{1 \text{ mol } CH_3SO_2F}{98.10 \text{ g } CH_3SO_2F} \cdot \frac{3 \text{ mol } H_2}{1 \text{ mol } CH_3SO_2F} \cdot \frac{2.02 \text{ g } H_2}{1 \text{ mol } H_2} = 9.3 \text{ g } H_2$$

(b) H_2 produced at the anode or cathode? Since H is being reduced from +1 to 0, it will be produced at the cathode.

(c) Energy consumed: $\dfrac{1 \text{ kwh}}{3.60 \times 10^6 \text{ J}} \cdot \dfrac{1 \text{ J}}{1 \text{ V} \cdot \text{C}} \cdot \dfrac{8.0 \text{ V}}{1} \cdot \dfrac{250 \text{ C}}{1 \text{ s}} \cdot \dfrac{3600 \text{ s}}{1 \text{ hr}} \cdot \dfrac{24 \text{ hr}}{1} = 48 \text{ kwh}$

94.　Since the reaction depends on the oxidation of elemental hydrogen to water (2 mol e^- per mol H_2), we must determine the amount of H_2 present:

$$n = \frac{(200. \text{ atm})(1.0 \text{ L})}{(0.0821 \frac{\text{L} \cdot \text{atm}}{\text{K} \cdot \text{mol}})(298 \text{ K})} = 8.2 \text{ mol } H_2$$

The amount of time this can produce current:

$$8.2 \text{ mol } H_2 \cdot \frac{2 \text{ mol } e^-}{1 \text{ mol } H_2} \cdot \frac{9.65 \times 10^4 \text{ C}}{1 \text{ mol } e^-} \cdot \frac{1 \text{ A} \cdot \text{s}}{1 \text{ C}} \cdot \frac{1}{1.5 \text{ A}} = 1.1 \times 10^6 \text{ s}$$

(290 hrs)

95.　For the oxidation of glucose:

(a) The quantity of O_2 required to react with glucose (2400 Cal) if $\Delta H° = -2800$ kJ/mol

How much glucose?

$$2400 \text{ kcal} \cdot \frac{4.184 \text{ kJ}}{1 \text{ kcal}} \cdot \frac{1 \text{ mol } C_6H_{12}O_6}{2800 \text{ kJ}} = 3.6 \text{ mol } C_6H_{12}O_6$$

Amount of O_2 that will react:

$$3.6 \text{ mol } C_6H_{12}O_6 \cdot \frac{6 \text{ mol } O_2}{1 \text{ mol } C_6H_{12}O_6} = 22 \text{ mol } O_2$$

(b) Amount of electrons needed:

The conversion of O_2 to H_2O requires 4 mole e^- per mol O_2

$22 \text{ mol } O_2 \cdot 4 \text{ mol } e^- = 88$ mole of electrons

(c) Current flowing per second:

$$\frac{1 \text{ amp} \cdot s}{1 \text{ C}} \cdot \frac{96500 \text{ C}}{1 \text{ mol e}} \cdot \frac{88 \text{ mol e}}{3.6 \text{ mol } C_6H_{12}O_6} \cdot \frac{1 \text{ hr}}{3600 \text{ s}} \cdot \frac{3.6 \text{ mol } C_6H_{12}O_6}{24 \text{ hr}} = 98 \text{ amperes}$$

(d) Given that $E° = 1.0$ V, what is the energy expenditure rate?

$$\frac{98 \text{ amp}}{1} \cdot \frac{1 \text{ C}}{1 \text{ amp} \cdot s} \cdot \frac{1 \text{ J}}{1 \text{ V} \cdot C} \cdot \frac{1 \text{ watt} \cdot s}{1 \text{ J}} = 98 \text{ watts}$$

Chapter 21
The Chemistry of the Main Group Elements

Reviewing Important Concepts

18. The oxides listed in order of increasing basicity:

Metal oxides are basic, nonmetal oxides are acidic. The three listed range from a nonmetal oxide (CO_2) to a metal oxide, SnO_2. So the order is: $CO_2 < SiO_2 < SnO_2$

20. Balanced equations for the following reactions:
    (a) $2 Na (s) + Br_2 (\ell) \rightarrow 2 NaBr (s)$
    (b) $2 Mg (s) + O_2 (g) \rightarrow 2 MgO (s)$
    (c) $2 Al (s) + 3 F_2 (g) \rightarrow 2 AlF_3 (s)$
    (d) $C(s) + O_2 (g) \rightarrow CO_2 (g)$

Practicing Skills

Hydrogen

22. Balanced chemical equation for hydrogen gas reacting with oxygen, chlorine, and nitrogen.
    $2 H_2 (g) + O_2 (g) \rightarrow 2 H_2O (g)$
    $H_2 (g) + Cl_2 (g) \rightarrow 2 HCl (g)$
    $3 H_2 (g) + N_2 (g) \rightarrow 2 NH_3 (g)$

24. (a) $2 SO_2 (g) + 4 H_2O (\ell) + 2 I_2 (s) \rightarrow 2 H_2SO_4 (\ell) + 4 HI (g)$
     (b) $2 H_2SO_4 (\ell) \rightarrow 2 H_2O (\ell) + 2 SO_2 (g) + O_2 (g)$
     (c) $\underline{4 HI (g) \rightarrow 2 H_2 (g) + 2 I_2 (g)}$
     Net: $2 H_2O (\ell) \rightarrow 2 H_2 (g) + O_2 (g)$

26. Thermodynamic data for the reaction of carbon and water to give CO and H_2:

	C(s)	+ H_2O (g)	→ CO (g)	+ H_2 (g)
$\Delta H°$ (kj/mol)	0	-241.83	-110.525	0
$S° \left(\dfrac{J}{K \cdot mol}\right)$	5.6	188.84	197.674	130.7
$\Delta G°$ (kj/mol)	0	-228.59	-137.168	0

$\Delta H°rxn = [\Delta Hf° CO(g) + \Delta Hf° H_2(g)] - [\Delta Hf° C(s) + \Delta Hf° H_2O(\ell)]$

$\Delta H°rxn = [(1 \text{ mol})(-110.525 \text{ kj/mol}) + 0] - [0 + (1 \text{ mol})(-241.83 \text{ kj/mol})] = +131.31 \text{ kJ}$

$$\Delta S°\text{rxn} = [S° \ CO(g) + S° \ H_2(g)] - [S° \ C(s) + S° \ H_2O(\ell)]$$

$$\Delta S°\text{rxn} = [(1 \ \text{mol})(\ 197.674 \frac{J}{K \cdot \text{mol}}) + (1 \ \text{mol})(\ 130.7 \frac{J}{K \cdot \text{mol}})] - [(1 \ \text{mol})(5.6 \frac{J}{K \cdot \text{mol}})$$

$$+ (1 \ \text{mol})(\ 188.84 \frac{J}{K \cdot \text{mol}})] = + 133.9 \ \text{J/K}$$

$$\Delta G°\text{rxn} = [\Delta Gf° \ CO(g) + \Delta Gf° \ H_2(g)] - [\Delta Gf° \ C(s) + \Delta Gf° \ H_2O(\ell)]$$

$$\Delta G°\text{rxn} = [(1 \ \text{mol})(\ -137.168 \ \text{kj/mol}) + 0] - [\ 0 + (1 \ \text{mol})(\ -228.59 \ \text{kj/mol})] = + 91.422 \ \text{kJ}$$

Alkali Metals

28. Equations for the reaction of sodium with the halogens:

$$2 \ Na(s) \ + \ F_2(g) \rightarrow \ 2 \ NaF(s)$$

$$2 \ Na(s) \ + \ Cl_2(g) \rightarrow \ 2 \ NaCl(s)$$

$$2 \ Na(s) \ + \ Br_2(\ell) \rightarrow 2 \ NaBr(s)$$

$$2 \ Na(s) \ + \ I_2(s) \rightarrow \ 2 \ NaI(s)$$

> Physical properties of the alkaline metal halides: (1) ionic solids (2) high melting and boiling points (3) white color (4) water soluble

30. The reaction of lithium, sodium, and potassium with oxygen:

$$4 \ Li(s) \ + \ O_2(g) \rightarrow \ 2 \ Li_2O(s) \qquad\qquad \text{oxide}$$

$$2 \ Na(s) \ + \ O_2(g) \rightarrow \ Na_2O_2(s) \qquad\qquad \text{peroxide}$$

$$K(s) \ + \ O_2(g) \quad \rightarrow \ KO_2(s) \qquad\qquad \text{superoxide}$$

32. In the electrolysis of aqueous NaCl:

(a) The balanced equation for the process:

$$2 \ NaCl(aq) \ + \ 2 \ H_2O(\ell) \rightarrow 2 \ NaOH(aq) \ + \ Cl_2(g) + \ H_2(g)$$

(b) Anticipated masses ratios:

$$\frac{1 \ \text{mol} \ Cl_2}{2 \ \text{mol} \ NaOH} \ = \ \frac{70.9 \ \text{g} \ Cl_2}{80.0 \ \text{g} \ NaOH} = \ 0.88 \ \text{g} \ Cl_2/\text{g} \ NaOH$$

$$\text{Actual:} \ \frac{1.14 \times 10^{10} \ \text{kg} \ Cl_2}{1.19 \times 10^{10} \ \text{kg} \ NaOH} \ = \ 0.96 \ \frac{\text{kg} \ Cl_2}{\text{kg} \ NaOH}$$

The difference in ratios means that alternative methods of producing chlorine are used. One of these is the Kel-Chlor process which uses HCl, NOCl, and O_2 . Other products are NO and H_2O.

Alkaline Earth Elements

34. Balanced equations for the reaction of magnesium with nitrogen and oxygen:

$$3\ Mg(s)\ +\ N_2(g)\ \rightarrow\ Mg_3N_2(s)$$

$$2\ Mg(s)\ +\ O_2(g)\ \rightarrow\ 2\ MgO(s)$$

36. Uses of limestone:

agricultural: to furnish Ca^{2+} to plants and neutralize acidic soils

building: lime (CaO) is used in mortar and absorbs CO_2 to form $CaCO_3$

steel-making: $CaCO_3$ furnishes lime (CaO) in the basic oxygen process. The lime reacts
    with gangue (SiO_2) to form calcium silicate.

The balanced equation for the reaction of limestone with carbon dioxide in water:

$$CaCO_3(s)\ +\ H_2O(\ell)\ +\ CO_2(g)\ \rightleftharpoons\ Ca^{2+}(aq)\ +\ 2\ HCO_3^-(aq)$$

This reaction is important in the formation of "hard water" (not particularly a great
happening for plumbing) and stalagmites and stalactites (aesthetically pleasing in caves).

38. The amount of SO_2 that could be removed by 1200 kg of CaO by the reaction:

$$CaO\ (s)\ +\ SO_2\ (g)\ \rightarrow\ CaSO_3\ (s)$$

$$1.200\times10^6\text{ g CaO}\ \bullet\ \frac{1\text{ mol CaO}}{56.079\text{ g CaO}}\ \bullet\ \frac{1\text{ mol }SO_2}{1\text{ mol CaO}}\ \bullet\ \frac{64.059\text{ g }SO_2}{1\text{ mol }SO_2}\ =\ 1.4\times10^6\text{ g }SO_2$$

Aluminum

40. The equations for the reaction of aluminum with HCl, Cl_2 and O_2:

$$2\ Al(s)\ +\ 6\ HCl(aq)\ \rightarrow\ 2\ Al^{3+}(aq)\ +\ 6\ Cl^-(aq)\ +\ 3\ H_2(g)$$

$$2\ Al(s)\ +\ 3\ Cl_2(g)\ \rightarrow\ 2\ AlCl_3(s)$$

$$4\ Al(s)\ +\ 3\ O_2(g)\ \rightarrow\ 2\ Al_2O_3(s)$$

42. The equation for the reaction of aluminum dissolving in aqueous NaOH:

$$2\ Al\ (s)\ +\ 2\ NaOH\ (aq)\ +\ 6\ H_2O\ (\ell)\ \rightarrow\ 2\ NaAl(OH)_4\ (aq)\ +\ 3\ H_2\ (g)$$

Volume of H_2 (in mL) produced when 13.2 g of Al react:

$$13.2\text{ g Al}\ \bullet\ \frac{1\text{ mol Al}}{26.98\text{ g Al}}\ \bullet\ \frac{3\text{ mol }H_2}{2\text{ mol Al}}\ =\ 0.734\text{ mol }H_2$$

$$V = \frac{(0.734 \text{ mol } H_2)(0.082057 \frac{L \cdot atm}{K \cdot mol})(295.7 \text{ K})}{735 \text{ mm Hg} \cdot \frac{1 \text{ atm}}{760 \text{ mm Hg}}} = 18.4 \text{ L}$$

44. The equation for the reaction of aluminum oxide with sulfuric acid:

$$Al_2O_3(s) + 3 H_2SO_4(aq) \rightarrow Al_2(SO_4)_3 (s) + 3 H_2O(\ell)$$

1.00×10^3 g $Al_2(SO_4)_3 \cdot \dfrac{1 \text{ mol } Al_2(SO_4)_3}{342.1 \text{ g } Al_2(SO_4)_3} \cdot \dfrac{1 \text{ mol } Al_2O_3}{1 \text{ mol } Al_2(SO_4)_3} \cdot$

$\dfrac{102.1 \text{ g } Al_2O_3}{1 \text{ mol } Al_2O_3} \cdot \dfrac{1 \text{ kg}}{1 \times 10^3 \text{ g}} = 0.298 \text{ kg } Al_2O_3$

1.00×10^3 g $Al_2(SO_4)_3 \cdot \dfrac{1 \text{ mol } Al_2(SO_4)_3}{342.1 \text{ g } Al_2(SO_4)_3} \cdot \dfrac{3 \text{ mol } H_2SO_4}{1 \text{ mol } Al_2(SO_4)_3} \cdot$

$\dfrac{98.07 \text{ g } H_2SO_4}{1 \text{ mol } H_2SO_4} \cdot \dfrac{1 \text{ kg}}{1 \times 10^3 \text{ g}} = 0.860 \text{ kg } H_2SO_4$

46. A Lewis electron dot structure for $AlCl_4^-$:

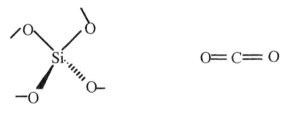

The $AlCl_4^-$ will have a tetrahedral geometry. The aluminum atom would utilize sp^3 hybridization.

Silicon

48. The structure of SiO_2 is tetrahedral with the silicon atom being surrounded by four oxygen atoms. The geometry around C in CO_2 is linear with two groups attached to the C atom.

O=C=O

The energies of four Si - O single bonds is greater than two Si=O double bonds. CO_2, on the other hand, forms discrete C=O double bonds.

The high melting point of SiO_2 is due to the network structure of the SiO_4 tetrahedron, while CO_2 exists as discrete nonpolar molecules—and as a gas at room temperature.

50. In the synthesis of dichlorodimethylsilane:
    (a) The balanced equation: $Si (s) + 2 CH_3Cl (g) \rightarrow (CH_3)_2SiCl_2 (\ell)$

(b) Stoichiometric amount of CH_3Cl to react with 2.65 g silicon:

$$2.65 \text{ g Si} \cdot \frac{1 \text{ mol Si}}{28.09 \text{ g Si}} \cdot \frac{2 \text{ mol } CH_3Cl}{1 \text{ mol Si}} = 0.189 \text{ mol } CH_3Cl$$

$$P = \frac{\left(0.189 \text{ mol } CH_3Cl\right)\left(0.082057 \frac{L \cdot atm}{K \cdot mol}\right)(297.7 \text{ K})}{5.60 \text{ L}} = 0.823 \text{ atm}$$

(c) Mass of $(CH_3)_2SiCl_2$ produced assuming 100 % yield:

$$0.0943 \text{ mol Si} \cdot \frac{1 \text{ mol } (CH_3)_2SiCl_2}{1 \text{ mol Si}} \cdot \frac{129.1 \text{ g } (CH_3)_2SiCl_2}{1 \text{ mol } (CH_3)_2SiCl_2} =$$

$$12.2 \text{ g } (CH_3)_2SiCl_2$$

Nitrogen and Phosphorus

52. The $\Delta G°_f$ data from Appendix L are shown below:

compound:	NO(g)	NO$_2$(g)	N$_2$O(g)	N$_2$O$_4$(g)
$\Delta G°$ ($\frac{kJ}{mol}$)	+86.58	+51.23	+104.20	+97.73

To ask if the oxide is stable with respect to decomposition is to ask about the $\Delta G°$ for the

process: $N_xO_y \rightarrow \frac{x}{2} N_2$ (g) + $\frac{y}{2} O_2$(g)

This reaction is *the reverse* of the ΔG_f for each of these oxides. The ΔG for such a process

would have an *opposite sign* from that data given above. Since that sign would be negative,

the process is product-favored. Hence **all** of the oxides shown above are unstable with

respect to decomposition.

54. Calculate $\Delta H°_{rxn}$ for the reaction: $2 NO(g) + O_2(g) \rightarrow 2 NO_2(g)$

$\Delta H°$ ($\frac{kJ}{mol}$)	+90.29	0	+33.1

$$\Delta H°_{rxn} = [(2 \text{ mol})(+33.1 \frac{kJ}{mol})] - [(2 \text{ mol})(+90.29 \frac{kJ}{mol})] = -114.4 \text{ kJ}$$

The reaction is exothermic.

56.   (a) The reaction of hydrazine with dissolved oxygen:

$$N_2H_4 \text{ (aq)} + O_2 \text{ (g)} \rightarrow N_2 \text{ (g)} + 2 H_2O(\ell)$$

(b) Mass of hydrazine to consume the oxygen in 3.00×10^4 L of water:

$$3.00 \times 10^4 \, L \cdot \frac{3.08 \, cm^3 \, O_2}{0.100 \, L} \cdot \frac{1 \, mol \, O_2}{22400 \, cm^3} \cdot \frac{1 \, mol \, N_2H_4}{1 \, mol \, O_2} \cdot \frac{32.05 g N_2H_4}{1 \, mol \, N_2H_4}$$

$$\uparrow \text{ at STP}$$

$$= 1.32 \times 10^3 \, g \, N_2H_4$$

58. The half equations:

$$N_2H_5^+ \, (aq) \rightarrow N_2 \, (g) + 5 \, H^+ \, (aq) + 4 \, e^-$$

$$IO_3^- \, (aq) \rightarrow I_2 \, (s)$$

are balanced according to the procedure in Chapter 5 to give:

$$5 \, N_2H_5^+ \, (aq) + 4 \, IO_3^- \, (aq) \rightarrow 5 \, N_2 \, (g) + 2 \, I_2 \, (aq) + 12 \, H_2O(\ell) + H^+(aq)$$

$E°$ for the reaction is: $E°$ for the hydrazine equation (oxidation) $= -0.23$ V

$E°$ for the iodate equation (reduction) $= \underline{+ 1.195 \text{ V}}$

$E°_{net}$ $= + 1.43$ V

60. The dot structure for the azide ion:

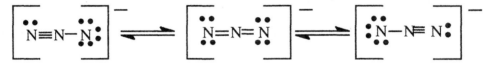

Oxygen and Sulfur

62.   (a) Allowable release of SO_2: (0.30 %)

$$1.80 \times 10^6 \, kg \, H_2SO_4 \cdot \frac{1 \, mol \, H_2SO_4}{98.08 \, g \, H_2SO_4} \cdot \frac{1 \, mol \, SO_2}{1 \, mol \, H_2SO_4} \cdot \frac{64.06 \, g \, SO_2}{1 \, mol \, SO_2}$$

$$\cdot \frac{0.0030 \, kg \, SO_2 \text{ released}}{1.00 \, kg \, SO_2 \text{ produced}} = 3.5 \times 10^3 \, kg \, SO_2$$

(b) Mass of $Ca(OH)_2$ to remove 3.53×10^3 kg SO_2:

$$3.53 \times 10^3 \, kg \, SO_2 \cdot \frac{1 \, mol \, SO_2}{64.06 \, g \, SO_2} \cdot \frac{1 \, mol \, Ca(OH)_2}{1 \, mol \, SO_2} \cdot \frac{74.09 \, g \, Ca(OH)_2}{1 \, mol \, Ca(OH)_2}$$

$$= 4.1 \times 10^3 \, kg \, Ca(OH)_2$$

64. The disulfide ion S_2^{2-} can be pictured as below:

$$\left[\; \overset{\displaystyle \cdot\cdot}{\underset{\displaystyle \cdot\cdot}{\overset{\cdot}{\underset{\cdot}{S}}}} : \overset{\displaystyle \cdot\cdot}{\underset{\displaystyle \cdot\cdot}{\overset{\cdot}{\underset{\cdot}{S}}}} : \; \right]^{2-}$$

Chlorine

66. Calculate the equivalent net cell potential for the oxidation:

$$2\,[Mn^{2+}(aq) \; + \; 4\,H_2O(\ell) \quad \rightarrow \quad MnO_4^-(aq) \; + \; 8\,H^+(aq) \; + \; 5\,e^-] \qquad -1.51 \text{ V}$$
$$2\,BrO_3^-(aq) \; + \; 12\,H^+(aq) \; + \; 10\,e^- \; \rightarrow \quad Br_2(aq) \; + \; 6\,H_2O(\ell)] \qquad +1.44 \text{ V}$$

net $\; 2\,Mn^{2+}(aq) + 2\,H_2O(\ell) \; + \; 2\,BrO_3^-(aq) \rightarrow 2\,MnO_4^-(aq) \hfill -0.07 \text{ V}$
$$+ \; 4\,H^+(aq) \; + \; Br_2 \,(aq)$$

NOTE: The first equation has been reversed from its usual "reduction form", indicating that we want to have this process occur as an oxidation. Recall that the task is to ask if BrO_3^- can oxidize Mn^{2+}! So we arrange the two half-equations to give us a specific net equation, and then use that format to ask, "Will it go this way?" If $E°_{cell}$ is > 0, the answer is **Yes.** If $E°$cell is < 0, the answer is **No.** Given that we calculate $E°_{cell}$ by *subtracting* the $E°_{anode}$ from the $E°_{cathode}$, I have given the reduction potential for the manganese half-reaction a sign that is *opposite* the normal sign. If we add the values, this is the mathematical equivalent of our usual procedure. The negative net potential indicates that this process doesn't favor products with 1.0 M bromate ion.

68. The balanced equation for the reaction of Cl_2 with Br^-

$$2\,Br^- \,(aq) \quad \rightarrow \quad Br_2 \,(\ell) + 2\,e^- \qquad\qquad -1.08 \text{ V}$$
$$Cl_2 \,(g) + 2\,e^- \; \rightarrow \; 2\,Cl^- \,(aq) \qquad\qquad +1.36 \text{ V}$$

net: $\quad Cl_2 \,(g) \; + 2\,Br^- \,(aq) \rightarrow \; 2\,Cl^- \,(aq) + Br_2 \,(\ell) \qquad + 0.28 \text{ V}$

- Note that bromide ions are losing electrons (and donating them to chlorine), causing chlorine to be reduced—so bromide is the reducing agent and chlorine, removing the electrons from bromide ions, is causing the bromide ions to be oxidized—so chlorine is the oxidizing agent.

- Note also that the voltage for the cell we "constructed" is positive, making this process a "product-favored" reaction.

General Questions

70. Describe the elements in the third period:

Atomic No.	Element	(a) Type	(b) Color	(c) State
11	Sodium	metal	grey, shiny	solid
12	Magnesium	metal	grey, shiny	solid
13	Aluminum	metal	grey,shiny	solid
14	Silicon	metalloid	grey,shiny	solid
15	Phosphorus	nonmetal	red, white, black	solid
16	Sulfur	nonmetal	yellow	solid
17	Chlorine	nonmetal	yellow-green	gas
18	Argon	nonmetal	colorless	gas

72. Reactions of Na, Mg, Al, Si, P, S

(a) Balanced equations of the elements with elemental chlorine:

$2\ Na\ (s) + Cl_2\ (g) \rightarrow 2\ NaCl\ (s)$ $Si\ (s) + 2\ Cl_2\ (g) \rightarrow\ SiCl_4\ (\ell)$

$Mg\ (s) + Cl_2\ (g) \rightarrow\ MgCl_2\ (s)$ $P_4\ (s) + 10\ Cl_2\ (g) \rightarrow\ 4\ PCl_5\ (s)$

$2\ Al\ (s) + 3\ Cl_2\ (g) \rightarrow 2\ AlCl_3\ (s)$ $S_8\ (s) + 8\ Cl_2\ (g) \rightarrow 8\ SCl_2(s)$

(b) Bonding in the products:

$NaCl$ and $MgCl_2$ are ionic; all others are covalent. This bonding is attributable—to a first approximation--to the differing electronegativities of the atoms bonded to each other.

(c) Electron dot structures for the products; electron-pair geometry; molecular geometry

Compound	Electron-Pair geometry	Molecular geometry
	Tetrahedral	Tetrahedral
	Trigonal bipyramidal	Trigonal bipyramidal

74. The products expected when molten LiH is electrolyzed:

Molten LiH contains lithium (Li^+) and hydride (H^-) ions.

At the cathode, lithium ions are reduced: $Li^+(\ell) + e^- \rightarrow Li(\ell)$

At the anode, hydride ions are oxidized: $2 H^-(\ell) \rightarrow H_2(g) + 2 e^-$

76. Calculate $\Delta G°$ for the decomposition of the metal carbonates for Mg, Ca, and Ba

$$MCO_3(s) \longrightarrow MO(s) + CO_2(g) \qquad \underline{M}$$

-1028.2		- 568.93	- 394.359	(Mg) $\Delta G° (\frac{kJ}{mol})$	0
- 1129.16		- 603.42	- 394.359	(Ca) $\Delta G°$	0
- 1134.41		- 520.38	- 394.359	(Ba) $\Delta G°$	0

$\Delta G°_{rxn} = \Delta G°_f MO + \Delta G°_f CO_2 - \Delta G°_f MCO_3$

$MgCO_3 = (- 568.93 \frac{kJ}{mol})(1 \text{ mol}) + (-394.359 \frac{kJ}{mol})(1 \text{ mol}) - (-1028.2 \frac{kJ}{mol})(1 \text{ mol})$

$\qquad = 64.9 \text{ kJ}$

$CaCO_3 = (- 603.42 \frac{kJ}{mol})(1 \text{ mol}) + (-394.359 \frac{kJ}{mol})(1 \text{ mol}) - (- 1129.16 \frac{kJ}{mol})(1 \text{ mol})$

$\qquad = 131.38 \text{ kJ}$

$BaCO_3 = (- 520.38 \frac{kJ}{mol})(1 \text{ mol}) + (-394.359 \frac{kJ}{mol})(1 \text{ mol}) - (-1134.41 \frac{kJ}{mol})(1 \text{ mol})$

$\qquad = 219.67 \text{ kJ}$

The relative tendency for decomposition is then $MgCO_3 > CaCO_3 > BaCO_3$.

78.   (a) Since $\Delta G°_{rxn} < 0$ for the reaction to be product-favored, calculate the value for $\Delta G°_f$ MX that will make the $\Delta G°_{rxn}$ zero.

$\Delta G°_{rxn} = \Delta G°_f(MX_n) - n \Delta G°_f(HX)$

for HCl $= \Delta G°_f(MX_n) - n(-95.1 \frac{kJ}{mol})$

so if $n(-95.1 \text{ kJ}) = \Delta G°_f(MX_n)$ then $\Delta G° = 0$

and if $n(-951 \text{ kJ}) > \Delta G°(MX_n)$ then $\Delta G°_{rxn} < 0$.

(b) Examine $\Delta G°_f$ MX values for

metal:	Ba	Pb	Hg	Ti
$\Delta G°_f$ MX:	-810.4	-314.10	-178.6	-737.2
n:	2	2	2	4
n(-95.1):	-190.2	-190.2	-190.2	-380.4

For Barium , Lead, and Titanium, n (-95.1) > $\Delta G°(MX)$; We expect these reactions to be spontaneous.

80.   (a) The N-O bonds are the same length owing to the delocalization of a second pair of electrons between the two N-O bonds. A dot picture of this shows two reasonable structures (resonance hybrids). In essence there is a bond order

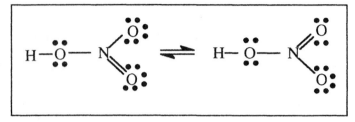

     of 1.5 for these two bonds; compared to a bond order of 1 for the N-OH bond.

     (b) The bond angle for the oxygens involved in the delocalized bond is only slightly larger than anticipated for the trigonal planar geometry (120°). The larger angle reflects the increased electron density (and repulsion) of this bond. The increased bond angle would result in a slightly smaller-than-ideal angle between the two "non-H" oxygens and the O-H bond. The bond angle N-O-H is only slightly less than anticipated for the tetrahedral orientation around the oxygen (two atoms and two lone pairs), a finding consistent with the two lone pairs of electrons on that oxygen.

     (c) The central **N atom has sp^2 hybridization**. The "unhybridized p" orbital on the N can participate in a π - type overlap with the orbitals on the two oxygen atoms, resulting in the pi bond.

<u>82</u>. Reaction scheme:

     Clue

     1.   1.00 g A   +   heat   B   +   gas (P = 209 mm; V = 450 mL; T = 298 K)
            white solid   white solid

     2.   Gas (from 1)   + $Ca(OH)_2$ (aq)   C (s)
                                      white solid

     3.   Aqueous solution of B is basic (turns red litmus paper blue)

     4.   B (aq) + HCl (aq) + heat   D
                                  white solid

     5.   Flame test for B: green flame

     6.   B (aq)   + H_2SO_4 (aq) $\rightarrow$ E
                             white solid

     Clue 5 indicates that **B** is a barium salt.

     Clue 2 suggests that the gas evolved in Clue 1 is CO_2, and that **C** would be $CaCO_3$.

Heating of carbonates liberates CO_2 (g).

Compound **B** is a metal oxide (Clue 3), and probably has the formula BaO.

Clues 3 and 5 suggest the oxide reacts with HCl and H_2SO_4 to form $BaCl_2$ (compound **D**) and $BaSO_4$ (compound **E**) respectively.

Since **B** is most likely BaO, compound **A** must be $BaCO_3$.

One gram of $BaCO_3$ (Molar mass 197) corresponds to 5.06×10^{-3} mol $BaCO_3$.

Compare this amount of substance to the amount of gas liberated when substance **A** is heated. Substitution of data from clue 1 yields:

$$n = \frac{(209 \text{ mm Hg}) (0.450 \text{ L})}{(62.4 \frac{\text{L} \cdot \text{mm Hg}}{\text{K} \cdot \text{mol}}) (298 \text{ K})} = 5.06 \times 10^{-3} \text{ mol gas}$$

This is the quantity of CO_2 anticipated from the thermal decomposition of $BaCO_3$.

The addition of 81.1 mL of 0.125 M HCl corresponds to (0.0811L)(0.125 mol HCl/L)= 0.0101375 mol HCl. This is two times the amount of gas evolved, and corresponds to the reaction: BaO (s) + 2 HCl (aq) $\rightarrow$ $BaCl_2$(aq) + H_2O (ℓ). The stoichiometry matches nicely, since one mol of BaO would require 2 mol of HCl.

(The volume is HCl is incorrect in the first printing of the text.)
Note that the 1.055 g of the solid ($BaCl_2$) (FW 208.23) corresponds to 5.06×10^{-3} mol.

84. Examine the enthalpy change for the reaction:
$$2 N_2 (g) + 5 O_2 (g) + 2 H_2O (\ell) \rightarrow 4 HNO_3 (aq)$$

$\Delta H°_{rxn}$ = [(4 mol)($-207.36 \frac{\text{kJ}}{\text{mol}}$)] -

    [(2 mol)($0 \frac{\text{kJ}}{\text{mol}}$) +(2 mol)($0 \frac{\text{kJ}}{\text{mol}}$) + (2 mol)($-285.83 \frac{\text{kJ}}{\text{mol}}$)]

   = - 257.78 kJ

The reaction is **exothermic** so it is a reasonable "first-guess" that this might be a way to "fix" nitrogen. The only way to be certain is to calculate the $\Delta G°_{rxn}$.

$\Delta G°_{rxn}$ = [(4 mol)($-111.25 \frac{\text{kJ}}{\text{mol}}$)] -

    [(2 mol)($0 \frac{\text{kJ}}{\text{mol}}$) +(2 mol)($0 \frac{\text{kJ}}{\text{mol}}$) + (2 mol)($-237.15 \frac{\text{kJ}}{\text{mol}}$)]

   = 29.2 kJ

The positive value for $\Delta G°_{rxn}$ tells us that the **reaction is not likely at 25 °C,** The

decrease in the number of moles of gas (entropy decrease) is also a factor that works

against this process at 25 ˚C. The favorable $\Delta H°$ does indicate that a lower temperature might be feasible, and below 268 K, the ΔG_{rxn} is favorable, but at this temperature water is a solid—not a good "sign".

86.   (a) Volume of seawater to obtain 1.00 kg Mg:

$$\frac{1.00 \times 10^3 \text{ g Mg}}{1} \cdot \frac{1 \text{ mol Mg}}{24.31 \text{ g Mg}} \cdot \frac{1 \text{ L seawater}}{0.050 \text{ mol Mg}} = 820 \text{ L seawater}$$

Mass of CaO to precipitate the magnesium:

The precipitation equation may be written as two steps:

1. $CaO(s) + H_2O(\ell) \rightarrow Ca(OH)_2(s)$

2. $Ca(OH)_2(s) + Mg^{2+}(aq) \rightarrow Ca^{2+}(aq) + Mg(OH)_2(s)$

The result of these processes is that 1 mole CaO precipitates 1 mole of Magnesium ions.

$$\frac{1.00 \times 10^3 \text{ g Mg}}{1} \cdot \frac{1 \text{ mol Mg}}{24.31 \text{ g Mg}} \cdot \frac{1 \text{ mol CaO}}{1 \text{ mol Mg}} \cdot \frac{56.08 \text{ g CaO}}{1 \text{ mol CaO}} = 2.31 \times 10^3 \text{ g CaO}$$

$$\text{or } 2.31 \text{ kg CaO}$$

(b) $MgCl_2(\ell) \xrightarrow{\text{electricity}} Mg(s) + Cl_2(g)$

Mass of Mg produced at the cathode:

$$1200 \text{ kg MgCl}_2 \cdot \frac{1 \text{ mol MgCl}_2}{95.211 \text{ g MgCl}_2} \cdot \frac{1 \text{ mol Mg}}{1 \text{ mol MgCl}_2} \cdot \frac{24.305 \text{ g Mg}}{1 \text{ mol Mg}} = 310 \text{ kg Mg}$$

Note the absence of a conversion of mass of $MgCl_2$ from kg to grams. Since the answer was requested in units of kg, any conversion to units of grams would have necessitated a conversion back to kg at the end of the calculation. The two conversion factors would cancel each other, and "leaving them out" causes no harm to the integrity of the reasoning (or the answer).

At the anode, chlorine is produced. $[2 \text{ Cl}^- \rightarrow Cl_2 + 2 \text{ e}^-]$
Mass of Cl_2 produced:

$$1200 \text{ kg MgCl}_2 \cdot \frac{1 \text{ mol MgCl}_2}{95.211 \text{ g MgCl}_2} \cdot \frac{1 \text{ mol Cl}_2}{1 \text{ mol MgCl}_2} \cdot \frac{70.906 \text{ g Cl}_2}{1 \text{ mol Cl}_2} = 890 \text{ kg Cl}_2$$

Faradays used in the process:
The reduction of magnesium requires 2 Faradays per mole: $Mg^{2+} + 2e^- \rightarrow Mg$

$$310 \text{ kg Mg} \cdot \frac{1.000 \times 10^3 \text{ g Mg}}{1.0 \text{ kg Mg}} \cdot \frac{1 \text{ mol Mg}}{24.305 \text{ g Mg}} \cdot \frac{2 \text{ F}}{1 \text{ mol Mg}} = 2.5 \times 10^4 \text{ F}$$

The oxidation of chlorine requires 2 Faradays per mole of chlorine: $2 \, Cl^- \rightarrow Cl_2 + 2 \, e^-$

$$890 \text{ kg Cl}_2 \quad \bullet \frac{1.000 \times 10^3 \text{ g Cl}_2}{1.0 \text{ kg Cl}_2} \bullet \frac{1 \text{ mol Cl}_2}{70.906 \text{ g Cl}_2} \bullet \frac{2 \text{ F}}{1 \text{ mol Cl}_2} = 2.5 \times 10^4 \text{ F}$$

The total number of Faradays of electricity used in the process is 2.5×10^4 F.

(c) Joules required per mole of magnesium:

$$\frac{8.4 \text{ kwh}}{1 \text{ lb Mg}} \bullet \frac{3.60 \times 10^6 \text{ J}}{1 \text{ kwh}} \bullet \frac{1 \text{ lb Mg}}{454 \text{ g Mg}} \bullet \frac{24.305 \text{ g Mg}}{1 \text{ mol Mg}} = 1.6 \times 10^6 \frac{\text{J}}{\text{mol Mg}}$$

The reaction, $MgCl_2$ (s) $\rightarrow$ Mg (s) + Cl_2 (g), represents the reverse of the formation of magnesium chloride from its elements (each in their standard states). From Appendix L, the ΔH for the process is +641.62 kJ/mol or 6.4×10^5 J/mol. The difference between this value and the value calculated above may be attributed to the energy required to melt the $MgCl_2$.

88. For the electrolysis of molten sodium chloride:

(a) Mass of sodium produced in one hour:
$$\frac{23 \text{ g Na}}{1 \text{ mol Na}} \bullet \frac{1 \text{ mol Na}}{1 \text{ mole e}^-} \bullet \frac{1 \text{ mole e}^-}{96500 \text{ C}} \bullet \frac{1 \text{ C}}{1 \text{ amp} \bullet \text{s}} \bullet \frac{25000 \text{ amp}}{1} \bullet \frac{3600 \text{ s}}{1 \text{ hr}} \bullet \frac{1 \text{ kg}}{1000 \text{ g}} = 21 \text{ kg/hr}$$

(b) Kilowatt-hours needed to produce 1.00 kg of Na:
$$\frac{1 \text{ kwh}}{3.6 \times 10^6 \text{ J}} \bullet \frac{1 \text{ J}}{1 \text{ V} \bullet \text{C}} \bullet \frac{7.0 \text{ V}}{1} \bullet \frac{96500 \text{ C}}{1 \text{ mol e}^-} \bullet \frac{1 \text{ mol e}^-}{1 \text{ mol Na}} \bullet \frac{1 \text{ mol Na}}{23.0 \text{ g Na}} \bullet \frac{1.00 \times 10^3 \text{ g Na}}{1}$$

$$= 8.2 \text{ kwh}$$

Chapter 22
The Chemistry of the Transition Elements

Practicing Skills

Properties of Transition Elements

16.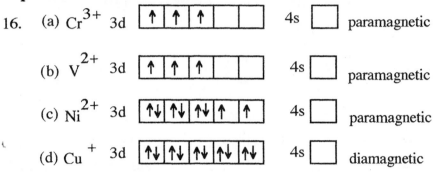

18. Ions from first series transition metals that are isoelectronic with:

    (a) Fe^{3+} has 5 3d electrons so it is isoelectronic with Mn^{2+}

    (b) Zn^{2+} has 10 3d electrons so it is isoelectronic with Cu^+ (see 16(b) above)

    (c) Fe^{2+} has 6 3d electrons so it is isoelectronic with Co^{3+}

    (d) Cr^{3+} has 3 3d electrons so it is isoelectronic with V^{2+} (see 16(b) above)

20. Balance:

    Each is balanced by the procedure shown in CH5 of the Solutions Manual. Refer to that procedure. While each of these is a reduction-oxidation reaction, one can also use the "inspection method" to obtain the balanced equations:

    (a) $Cr_2O_3(s) + 2\,Al(s) \rightarrow Al_2O_3(s) + 2\,Cr(s)$

    (b) $TiCl_4(\ell) + 2\,Mg(s) \rightarrow Ti(s) + 2\,MgCl_2(s)$

    (c) $2\,[Ag(CN)_2]^-(aq) + Zn(s) \rightarrow 2\,Ag(s) + [Zn(CN)_4]^{2-}(aq)$

    (d) $3\,Mn_3O_4(s) + 8\,Al(s) \rightarrow 9\,Mn(s) + 4\,Al_2O_3(s)$

Formulas of Coordination Compounds

22. Classify each of the following as monodentate or multidentate:

(a) CH_3NH_2	monodentate (lone pair on N)
(b) $CH_3C\equiv N$	monodentate (lone pair on N)
(c) N_3^-	monodentate (lone pair on a N atom)
(d) ethylenediamine	multiidentate (lone pairs on terminal N atoms)

(e) Br$^-$ monodentate (lone pair on Br)

(f) phenanthroline   multidentate (lone pairs on N atoms)

24.

	Compound	Metal	Oxidation Number
(a)	$[Mn(NH_3)_6]SO_4$	Mn	+2

Ammonia is a neutral ligand. SO_4^{2-} means that Mn has to have a +2 oxidation no.

(b)	$K_3[Co(CN)_6]$	Co	+3

The 3 K$^+$ ions mean that the complex ion must have a -3 charge. Each CN$^-$ has a –1 charge, so Co must have a +3 charge.

(c)	$[Co(NH_3)_4Cl_2]Cl$	Co	+3

The chloride anion means that the complex ion must have a net +1 charge. While the ammonia ligand is neutral, each Cl has a –1 charge, so Co must have +3.

(d)	$Cr(en)_2Cl_2$	Cr	+2

The ethylenediamine ligand is neutral, so with each Cl having a –1 charge, Cr must be +2.

26. Formula of complex: $[Ni(NH_3)_3(H_2O)en]^{2+}$. The complex has to have a 2+ charge, since each of the three types of ligand are neutral.

Naming Coordination Compounds

28. Formulas for:
- (a) dichlorobis(ethylenediamine)nickel(II) $Ni(en)_2Cl_2$
- (b) potassium tetrachloroplatinate(II) $K_2[PtCl_4]$
- (c) potassium dicyanocuprate(I) $K[Cu(CN)_2]$
- (d) tetraamminediaquairon(II) $[Fe(NH_3)_4(H_2O)_2]^{2+}$

30.

	Formula	Name
(a)	$[Ni(C_2O_4)_2(H_2O)_2]^{2-}$	diaquabis(oxalato)nickelate(II) ion
(b)	$[Co(en)_2Br_2]^+$	dibromobis(ethylenediamine)cobalt(III) ion
(c)	$[Co(en)_2(NH_3)Cl]^{2+}$	amminechlorobis(ethylenediamine)cobalt(III) ion
(d)	$Pt(NH_3)_2(C_2O_4)$	diammineoxalatoplatinum(II)

32. The name or formula for the ions or compounds shown below:
- (a) $[Fe(H_2O)_5(OH)]^{2+}$ Hydroxopentaaquairon(III) ion
- (b) $K_2[Ni(CN)_4]$ Potassium tetracyanonickelate(II)
- (c) $K[Cr(C_2O_4)_2(H_2O)_2]$ Potassium diaquabis(oxalato)chromate(III)

(d) (NH4)2[PtCl4]   Ammonium tetrachloroplatinate(IV)

Isomerism

34. Geometric Isomers of

(a) Fe(NH3)2Cl2

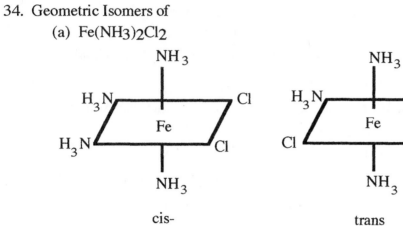

cis-   trans

(b)   Pt(NH3)2(SCN)(Br)

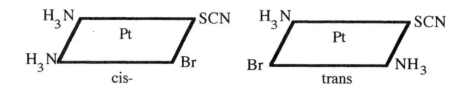

cis-   trans

(c)   Co(NH3)3(NO2)3

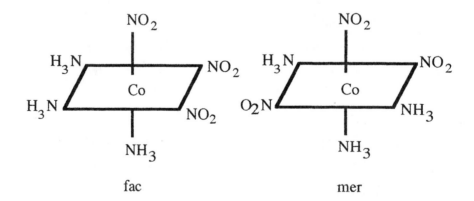

fac   mer

(d) [Co(en)Cl4]⁻

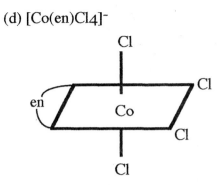

36. Which of the following species has a chiral center?

(a) [Fe(en)₃]²⁺   Yes

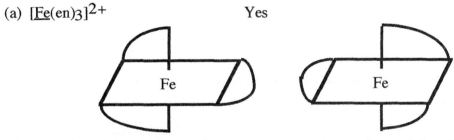

The two mirror images shown above are non superimposable and therefore possess a chiral center.

(b) trans-[Co(en)₂Br₂]⁺   No. The *cis* isomer has a chiral center.

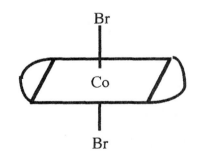

(c) fac-[Co(en)(H₂O)Cl₃]   no

As you can see by the two structures above, a 180° rotation along the Cl-Co-H$_2$O axis, would make these two mirror images superimposable. The *mer* complex would also be superimposable, and therefore possess no chiral center.

(d) Pt(NH$_3$)(H$_2$O)Cl(NO$_2$)   no

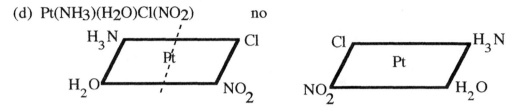

Above are two mirror images of the complex. Rotation of the first complex along the dotted axis by 180° results in the second complex. There are no nonsuperimposable isomers.

Magnetic Properties of Complexes

38. In the name of clarity, the counterion has been omitted. The counterions do not affect the magnetic behavior of the complex ion.

(a) [Mn(CN)$_6$]$^{4-}$

Mn^{2+} has a d^5 configuration

```
low spin
strong field
ligand

___        ___

1↓     1↓      1

paramagnetic
```

(b) [Co(NH$_3$)$_6$]$^{3+}$ Co^{3+} has a d^6 configuration

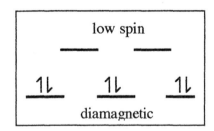

```
low spin

___     ___

1↓    1↓    1↓

diamagnetic
```

(c) [Fe(H$_2$O)$_6$]$^{3+}$

Fe^{3+} has a d^5 configuration (like Mn^{2+}), and has 1 unpaired electron (see diagram in part (a).

(d) $[Cr(en)_3]^{2+}$

Cr^{2+} has the d^4 configuration

low spin

paramagnetic
(2 unpaired)

40. For the following (high spin) tetrahedral complexes, determine the number of unpaired electrons:

(a) $[FeCl_4]^{2-}$ (d^6 ,paramagnetic)	(b)$[CoCl_4]^{2-}$ (d^7, paramagnetic)
(4 unpaired)	(3 unpaired)
(c) $[MnCl_4]^{2-}$ (d^5, paramagnetic)	(d) $[ZnCl_4]^{2-}$ (d^{10}, diamagnetic)
(5 unpaired)	(0 unpaired)

42. For $[Fe(H_2O)_6]^{2+}$

(a) The coordination number of iron is 6. 6 monodentate ligands are attached.

(b) The coordination geometry is octahedral. (Six groups attached to the central metal ion)

(c) The oxidation state of iron is 2+. The charge on the complex is 2+.
Water is a neutral ligand.

(d) Fe^{2+} is a d^6 case. Water is a weak-field ligand (high spin complex) there are 4 unpaired electrons.

(e) The complex would be paramagnetic

44. Aqueous cobalt(III) sulfate is diamagnetic. H_2O provides a large enough Δ_0 to force $[Co(H_2O)_6]^{3+}$ to have no unpaired electrons. An excess of F^- results in the conversion of the hexaaqua complex to the $[CoF_6]^{3-}$ ion. The weaker F^- ligands doesn't separate the d orbital energies as much as H_2O, resulting in four unpaired electrons—a paramagnetic complex.

Spectroscopy of Complexes

46. 500nm light is in the blue region of the spectrum (sidebar on page 950 of your text. The transmitted light, and the color of the solution--is yellow.

General Questions

48. Describe an experiment to determine if:
   Nickel in $K_2[NiCl_4]$ is in a square planar or tetrahedral environment.

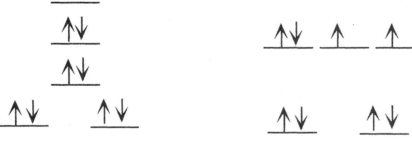

d- orbitals for square-planar complex   d- orbitals for tetrahedral complex

   Square-planar nickel(II) is diamagnetic while tetrahedral nickel(II) is paramagnetic. Measuring the magnetic moment would discriminate between the two.

50. Number of unpaired electrons for high-spin and low-spin complexes of Fe^{2+}:

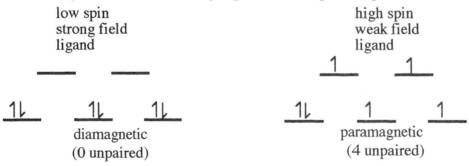

52. Which of the following complexes are square planar?

Of the four complexes, only $[Ni(CN)_4]^{2-}$ is square planar, since such complexes are the geometry assumed by d^8 metal ions. The other complexes are tetrahedral.

54. Geometric isomers of $Pt(NH_3)(CN)Cl_2$:

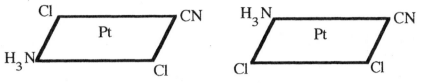

56. Complex absorbs 425-nm light, the blue-violet end of the visible spectrum. So red and green are transmitted , and the complex appears to be yellow.

58. For the high-spin complex $Mn(NH_3)_4Cl_2$ determine:

  (a) The oxidation number of manganese is +2. With a neutral overall complex, and 2 chloride ions (each with a –1 charge) and 4 neutral ammonia ligands, manganese has to be +2.

  (b) With six monodentate ligands, the coordination number for manganese is 6.

  (c) With six monodentate ligands, the coordination geometry for manganese is octahedral.

  (d) Number of unpaired electrons: 5 The ligands attached are weak-field ligands resulting in five unpaired electrons.

  (e) With unpaired electrons, the complex is paramagnetic.

  (f) Cis- and trans- geometric isomers are possible. See SQ21.34a for a similar complex.

60. Structures for cis- and trans- isomers of $CoCl_3 \cdot 4NH_3$

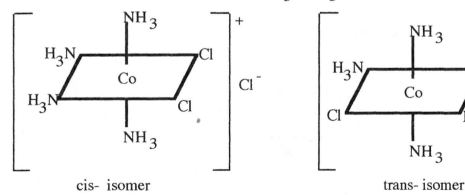

cis- isomer   trans- isomer

cis-tetraamminedichlorocobalt(III) chloride   *trans*-tetraamminedichlorocobalt(III) chloride

62. Formula of a complex containing a Co^{3+} ion, two ethylenediamine molecules, one water molecule, and one chloride ion: $[Co(en)_2(H_2O)Cl]^{2+}$

   The ethylenediamine molecules and the water molecule are neutral, so with a (3+) and (1-) charge, the net charge on the ion is 2+.

<u>64.</u> (a) Structures for the fac- and mer- isomers of $Cr(dien)Cl_3$. In these diagrams the curved lines represent the $H_2N\text{-}CH_2CH_2\text{-}NH\text{-}CH_2CH_2\text{-}NH_2$ ligand--with attachments to the metal ion through the electron pair on the N atoms.

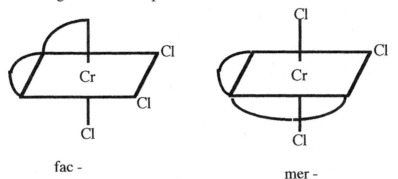

         fac -   mer -

(b) Two different isomers of mer-$Cr(dien)BrCl_2$:

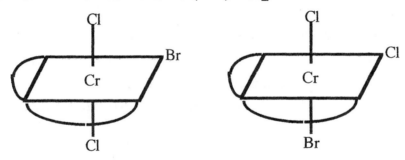

(c) The geometric isomers for isomers of $[Cr(dien)_2]^{3+}$

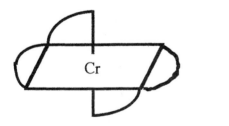 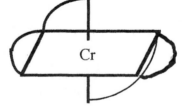 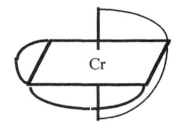

         fac-   mer -

66. Three geometric isomers of $[Co(en)(NH_3)_2(H_2O)_2]^{3+}]$:

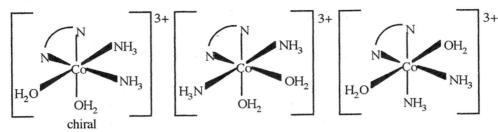

chiral

68. For the two Mn^{2+} complexes:

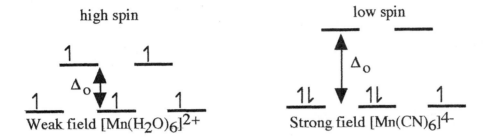

The cyano ligand is a *strong field* ligand, so the difference in energies of the doubly vs triply degenerate d orbitals is greater than with the *weak field* aqua ligand. The differences in these energy levels are referred to as Δ_O, and are shown in the diagram above.

70.   A + $BaCl_2$ →   ppt ($BaSO_4$) ⇒ A = $[Co(NH_3)_5Br]SO_4$

    B + $BaCl_2$ →   no ppt   ⇒ B = $[Co(NH_3)_5SO_4]Br$

Complex A has the sulfate ion in the outer sphere of the transition metal compound. As such, this compound (like many ionic compounds) dissolves in water—and dissociates, liberating sulfate ions. Barium ions react with the sulfate ions to produce the precipitate. Complex B has the sulfate ion as a part of the inner sphere of the compound. This ion is bound tightly to the cobalt ion, and not available to the barium ions—hence no precipitate. The reaction between A and $BaCl_2$:

$[Co(NH_3)_5Br]SO_4(aq) + BaCl_2(aq) \rightarrow BaSO_4 (s) + [Co(NH_3)_5Br]^{2+}(aq) + 2\ Cl^-(aq)$

72. The relative stabilities of the hexaammine complexes with Co^{2+}, Ni^{2+}, Cu^{2+}, and Zn^{2+}:

From Appendix K, the data are:

$Zn^{2+}(aq)\ + 4\ NH_3\ (aq) \rightleftharpoons [Zn(NH_3)_4]^{2+}$ $K_f = 2.9 \times 10^9$

$Cu^{2+}(aq)\ + 4\ NH_3\ (aq) \rightleftharpoons [Cu(NH_3)_4]^{2+}$ $K_f = 6.8 \times 10^{12}$

$Ni^{2+}(aq)\ + 6\ NH_3\ (aq) \rightleftharpoons [Ni(NH_3)_6]^{2+}$ $K_f = 5.6 \times 10^8$

$Co^{2+}(aq)\ + 6\ NH_3\ (aq) \rightleftharpoons [Co(NH_3)_6]^{2+}$ $K_f = 7.7 \times 10^4$

While the general order is followed, the copper tetraammine complex has the largest K_f of

this series, a trend that is generally followed for many transition metal complexes.

74. The five geometric isomers of $Cu(H_2NCH_2CO_2)_2(H_2O)_2$:

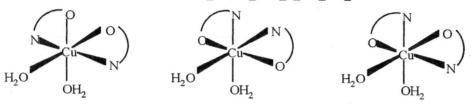

The three isomers above each have a nonsuperimposable mirror image—that is they contain

a chiral center. The two isomers shown below have no chiral center.

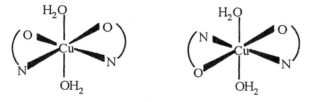

76. Account for differences in K_f for monodentate and bidentate complexes of nickel(II):

For the hexaammine complex:

$$\Delta G = -RT\ln K = \left(-8.314\frac{J}{K\cdot mol}\right)(298.15\ K)\ln(10^8)\left(\frac{1.0\ kJ}{1000\ J}\right) = -45.6\ \frac{kJ}{mol}$$

For the tris(ethylenediammine)complex:

$$\Delta G = -RT\ln K = \left(-8.314\frac{J}{K\cdot mol}\right)(298.15\ K)\ln(10^{18})\left(\frac{1.0\ kJ}{1000\ J}\right) = -102.7\ \frac{kJ}{mol}$$

Note that the *ratios* of G_f values are approximately 2.5:1

The ratio of ΔH for the two complexes is: $\dfrac{-117\ \dfrac{kJ}{mol}}{-109\ \dfrac{kJ}{mol}} = 1.07$

From the value of these two ratios, it is obvious that *entropy* plays a very important role in

the *chelate effect*.

78. For ilmenite ore treatment:

(a) the equation for ilmenite ore reacting with HCl:

The equation can easily be balanced by inspection.

__FeTiO$_3$ (s) + __ HCl(aq) → __FeCl$_2$ (aq) + __TiO$_2$ (s) + __H$_2$O (ℓ)

Note that ilmenite (and Ti in particular) loses one O on forming TiO_2. This O appears in the water molecule. Iron(II) chloride requires 2 Cl species, so we need **2HCl**. We can now write: __$FeTiO_3$ (s) + _2_ HCl(aq) → __$FeCl_2$ (aq) + __TiO_2 (s) + __H_2O (ℓ)

(b) 2~~$FeCl_2$~~ (aq) + 2 ~~H_2O~~(ℓ) + 1/2 O_2 (g) → Fe_2O_3 (s) + 4 ~~HCl~~ (aq)

$\underline{\text{2 FeTiO}_3 \text{ (s)} + 4 \text{ HCl(aq)} \rightarrow 2\text{FeCl}_2\text{(aq)} + 2\text{TiO}_2 \text{ (s)} + 2 \text{ H}_2\text{O (ℓ)}}$

Net: 2 $FeTiO_3$ (s) + 1/2 O_2 (g) → Fe_2O_3 (s) + 2TiO_2 (s)

Note that the *HCl* used as a reactant in the 2nd equation here, *is produced* in the 1st equation shown here.

(c) Mass of iron(III) oxide obtainable from 908 kg of ilmenite:

$$\frac{908 \text{ kg FeTiO}_3}{1} \cdot \frac{1 \text{ mol FeTiO}_3}{151.73 \text{ g FeTiO}_3} \cdot \frac{1 \text{ mol Fe}_2O_3}{2 \text{ molFeTiO}_3} \cdot \frac{159.69 \text{ g Fe}_2O_3}{1 \text{ mol Fe}_2O_3} = 478 \text{ kg Fe}_2O_3$$

Chapter 23:
Nuclear Chemistry

Practicing Skills

Nuclear Reactions

12. Balance the following nuclear equations, supplying the missing particle.
[The missing particle is emboldened.]

(a) $^{54}_{26}Fe + ^{4}_{2}He \rightarrow 2\,^{1}_{0}n + \mathbf{^{56}_{28}Ni}$

(b) $^{27}_{13}Al + ^{4}_{2}He \rightarrow ^{30}_{15}P + \mathbf{^{1}_{0}n}$

(c) $^{32}_{16}S + ^{1}_{0}n \rightarrow ^{1}_{1}H + \mathbf{^{32}_{15}P}$

(d) $^{96}_{42}Mo + ^{2}_{1}H \rightarrow ^{1}_{0}n + \mathbf{^{97}_{43}Tc}$

(e) $^{98}_{42}Mo + ^{1}_{0}n \rightarrow ^{99}_{43}Tc + \mathbf{^{0}_{-1}e}$

(f) $^{18}_{9}F \rightarrow ^{18}_{8}O + \mathbf{^{0}_{1}n}$

14. Balance the following nuclear equations, supplying the missing particle.

[The missing particle is emboldened.]

(a) $^{111}_{47}Ag \rightarrow ^{111}_{48}Cd + \mathbf{^{0}_{-1}e}$

(b) $^{87}_{36}Kr \rightarrow ^{0}_{-1}e + \mathbf{^{87}_{37}Rb}$

(c) $^{231}_{91}Pa \rightarrow ^{227}_{89}Ac + \mathbf{^{4}_{2}He}$

(d) $^{230}_{90}Th \rightarrow ^{4}_{2}He + \mathbf{^{226}_{88}Ra}$

(e) $^{82}_{35}Br \rightarrow ^{82}_{36}Kr + \mathbf{^{0}_{-1}e}$

(f) $\mathbf{^{24}_{11}Na} \rightarrow ^{24}_{12}Mg + ^{0}_{-1}e$

16. $^{235}_{92}U \rightarrow {}^{231}_{90}Th \rightarrow {}^{231}_{91}Pa \rightarrow {}^{227}_{89}Ac \rightarrow {}^{227}_{90}Th \rightarrow {}^{223}_{88}Ra \rightarrow {}^{219}_{86}Rn \rightarrow {}^{215}_{84}Po$

$$+ \qquad + \qquad + \qquad + \qquad + \qquad + \qquad +$$

$$^{4}_{2}He \qquad {}^{0}_{-1}e \qquad {}^{4}_{2}He \qquad {}^{0}_{-1}e \quad {}^{4}_{2}He \quad {}^{4}_{2}He \quad {}^{4}_{2}He$$

and continuing (from Po-215)we have:

$$^{215}_{84}Po \rightarrow {}^{211}_{82}Pb \rightarrow {}^{211}_{83}Bi \rightarrow {}^{211}_{84}Po \rightarrow {}^{207}_{82}Pb$$

$$+ \qquad + \qquad + \qquad +$$

$$^{4}_{2}He \qquad {}^{0}_{-1}e \qquad {}^{0}_{-1}e \qquad {}^{4}_{2}He$$

Nuclear Stability and Nuclear Decay

18. The particle emitted in the following reactions: [The missing particle is emboldened.]

(a) $^{198}_{79}Au \rightarrow {}^{198}_{80}Hg + \mathbf{{}^{0}_{-1}e}$

(b) $^{222}_{86}Rn \rightarrow {}^{218}_{84}Po + \mathbf{{}^{4}_{2}He}$

(c) $^{137}_{55}Cs \rightarrow {}^{137}_{56}Ba + \mathbf{{}^{0}_{-1}e}$

(d) $^{110}_{49}In \rightarrow {}^{110}_{48}Cd + \mathbf{{}^{0}_{1}n}$

20. Predict the probable mode of decay for each of the following:

(a) $^{80}_{35}Br$ (large number of neutrons /proton—beta emission) $^{80}_{35}Br \rightarrow {}^{80}_{36}Kr + {}^{0}_{-1}e$

(b) $^{240}_{98}Cf$ (large isotope- alpha emission) $^{240}_{98}Cf \rightarrow {}^{236}_{96}Cm + {}^{4}_{2}He$

(c) $^{61}_{27}Co$ (mass # > atomic number—beta emission) $^{61}_{27}Co \rightarrow {}^{61}_{28}Ni + {}^{0}_{-1}e$

(d) $^{11}_{6}C$ (more protons than neutrons—positron emission or K-capture)

$$^{11}_{6}C \rightarrow {}^{11}_{5}B + {}^{0}_{-1}e \quad or \quad {}^{11}_{6}C + {}^{0}_{-1}e \rightarrow {}^{11}_{5}B$$

22. Beta particle and positron emission:

(a) Beta particle emission occurs (usually) when the ratio of neutrons/protons is high

Hydrogen-3 has 1 proton and 2 neutrons—**beta particle emission (forms $^{3}_{2}He$)**

Oxygen-16 has 8 protons and 8 neutrons—not expected

Fluorine-20 has 9 protons and 11 neutrons-- **beta particle emission (forms $^{20}_{10}Ne$)**

Nitrogen-13 has 7 protons and 6 neutrons – not expected

351

(b) Position emission occurs when the neutron/proton ratio is too low:

Uranium-238 has 92 protons and 146 neutrons—not expected

Fluorine-19 has 9 protons and 10 neutrons—not expected

Sodium-22 has 11 protons and 11 neutrons—positron emission expected (forms)

Sodium-24 has 11 protons and 13 neutrons—not expected

24. The change in mass (Δm) for ^{10}B is:

$$\Delta m = 10.01294 - [5(1.00783) + 5(1.00867)]$$

$$= 10.01294 - 10.0825 = -0.06956 \text{ g/mol}$$

Binding energy is: $\Delta mc^2 = (6.956 \times 10^{-5} \text{ kg/mol})(3.00 \times 10^8 \text{ m/s})^2 \left(\dfrac{1 \text{ J}}{1 \text{ kg} \cdot \text{m}^2 \cdot \text{s}^{-2}} \right)$

$$= 6.26 \times 10^{12} \text{ J/mol}$$

The **binding energy per nucleon** : $\dfrac{6.26 \times 10^9 \text{ kJ}}{10 \text{ mol nucleons}} = 6.26 \times 10^8 \dfrac{\text{kJ}}{\text{nucleon}}$

The mass change for ^{11}B is:

$$\Delta m = 11.00931 - [5(1.00783) + 6(1.00867)]$$

$$= 11.00931 - 11.09117 = -0.08186 \text{ g/mol}$$

Binding energy is: $\Delta mc^2 = (8.186 \times 10^{-5} \text{ kg/mol})(3.00 \times 10^8 \text{ m/s})^2 \left(\dfrac{1 \text{ J}}{1 \text{ kg} \cdot \text{m}^2 \cdot \text{s}^{-2}} \right)$

$$= 7.367 \times 10^{12} \text{ J/mol}$$

The **binding energy per nucleon**: $\dfrac{7.37 \times 10^9 \text{ kJ}}{11 \text{ mol nucleons}} = 6.70 \times 10^8 \dfrac{\text{kJ}}{\text{nucleon}}$

26. The binding energy per nucleon for calcium-40:

The change in mass (Δm) for ^{40}Ca is:

$$\Delta m = 39.96259 - [20(1.00783) + 20(1.00867)]$$

$$= 39.96259 - 40.3300$$

$$= -0.3674 \text{ g/mol}$$

The energy change is then:

$$\Delta E = (3.674 \times 10^{-4} \text{ kg/mol})(3.00 \times 10^8 \text{ m/s})^2 \left(\dfrac{1 \text{ J}}{1 \text{ kg} \cdot \text{m}^2 \cdot \text{s}^{-2}} \right)$$

$$= 3.307 \times 10^{13} \text{ J/mol or } 3.307 \times 10^{10} \text{ kJ/mol}$$

This energy can be converted into the **binding energy per nucleon:**

$\dfrac{3.307 \times 10^{10} \text{ kJ}}{40 \text{ mol nucleons}} = 8.27 \times 10^8 \dfrac{\text{kJ}}{\text{nucleon}}$

28. Binding energy per nucleon for Oxygen-16

The change in mass (Δ m) for ^{16}O is:

$$\Delta m = 15.99492 - [8(1.00783) + 8(1.00867)]$$
$$= -0.13708 \text{ g/mol}$$

The energy change is then:

$$\Delta E = (1.3708 \times 10^{-4} \text{ kg/mol})(3.00 \times 10^{8} \text{ m/s})^2 \left(\frac{1 \text{ J}}{1 \text{ kg} \cdot \text{m}^2 \cdot \text{s}^{-2}} \right)$$
$$= 1.234 \times 10^{13} \text{ J/mol or } 1.234 \times 10^{10} \text{ kJ/mol}$$

This energy can be converted into the **binding energy per nucleon:**

$$\frac{1.234 \times 10^{10} \text{ kJ/mol}}{16 \text{ mol nucleons}} = 7.71 \times 10^{8} \text{ kJ/nucleon}$$

Rates of Radioactive Decay

30. For ^{64}Cu, t$_{1/2}$ =128 hr

The fraction remaining as ^{64}Cu following n half-lives is equal to $\left(\frac{1}{2}\right)^n$.

Note that 64 hours corresponds to exactly **five** half-lives.

The <u>fraction</u> remaining as ^{64}Cu is $\left(\frac{1}{2}\right)^5$ or $\frac{1}{32}$ or 0.03125.

The mass remaining is:

$$(0.03125)(25.0 \ \mu g) = 0.781 \ \mu g.$$

32.   (a) The equation for β–decay of ^{131}I is:

$$^{131}_{53}I \rightarrow {}^{0}_{-1}e + {}^{131}_{54}Xe$$

(b) The amount of ^{131}I remaining after 40.2 days:

For ^{131}I, t$_{1/2}$ is 8.04 days--so 40.2 days is exactly **five** half-lives:

The fraction of ^{131}I remaining is $\left(\frac{1}{2}\right)^5$ or $\frac{1}{32}$ or 0.03125.

The amount of the original 2.4 μg remaining will be :

$$(0.03125)(2.4 \ \mu g) = 0.075 \ \mu g$$

34. To determine the mass of Gallium-67 left after 13 days, determine the number of half-lives corresponding to 13 days.

$$\frac{13 \text{ days}}{1} \cdot \frac{24 \text{ hrs}}{1 \text{ day}} = 312 \text{ hours}$$

The rate constant is : $k = \dfrac{0.693}{t\frac{1}{2}} = \dfrac{0.693}{78.25 \text{ hr}} = 0.00886 \text{ hrs}^{-1}$

The fraction remaining can be calculated: $\ln(x) = -(0.00886 \text{ hrs}^{-1}) \cdot (312 \text{ hrs})$ and solving for x yields 0.06309 (where x represents the fraction of Ga-67 remaining.

The amount of Gallium-67 remaining is then $(0.06309)(0.015 \text{ mg}) = 9.5 \times 10^{-4} \text{ mg}$

36. For the decomposition of Radon-222:

(a) The balanced equation for the decomposition of Rn-222 with α particle emission.

$$^{222}_{86}\text{Rn} \qquad ^{4}_{2}\text{He} + ^{218}_{84}\text{Po}$$

(b) Time required for the sample to decrease to 20.0 % of its original activity:

Since this decay follows 1st order kinetics, we can calculate a rate constant:

$$k = \dfrac{0.693}{t\frac{1}{2}} = \dfrac{0.693}{3.82 \text{ days}} = 0.181 \text{ days}^{-1}$$

With this rate constant , using the 1st order integrated rate equation, we can calculate the time required:

$$\ln(\dfrac{20.0}{100}) = -(0.181 \text{ days}^{-1}) \cdot t$$

$$\dfrac{-1.609}{-0.181 \text{ days}^{-1}} = t = 8.87 \text{ days}$$

38. The age of the fragment can be determined if:

(a) we calculate a rate constant and,

(b) we use the 1st order integrated rate equation (much as we did in question 32b above)

(a) $k = \dfrac{0.693}{t\frac{1}{2}} = \dfrac{0.693}{5730 \text{ yr}} = 1.21 \times 10^{-4} \text{ yr}^{-1}$

(b) Now we can calculate the time required for the Carbon-14 : Carbon-12 to decay to 72% of that ratio in living organisms.

$$\ln(\dfrac{72}{100}) = -(1.21 \times 10^{-4} \text{ yr}^{-1}) \cdot t$$

$$\dfrac{-0.3285}{-1.21 \times 10^{-4} \text{ yr}^{-1}} = t = 2700 \text{ years (to 2 significant figures)}$$

40.   For the decay of Cobalt-60, $t_{1/2}$ is 5.27 yrs:

(a) Time for Co-60 to decrease to 1/8 of its original activity:

Following the methodology of questions 34,36 and 38, determine the rate constant:

$$k = \frac{0.693}{t\frac{1}{2}} = \frac{0.693}{5.27 \text{ yr}} = 0.131 \text{ yr}^{-1}$$

Substituting into the equation:

$$\ln(\frac{1}{8}) = -0.131 \text{yr}^{-1} \bullet t \quad \text{and}$$

$$\ln(0.125) = -0.131 \text{ yr}^{-1} \bullet t \quad \text{and solving for t} = 15.8 \text{ yrs}$$

A " short-cut" is available here if you notice that 1/8 corresponds to $(\frac{1}{2})^3$. Said another way, one-eighth of the Co-60 will remain after **three half-lives** have passed, so $3 \bullet 5.27$ yrs = 15.8 years !!

(b) Fraction of Co-60 remaining as Co-60 after 1.0 years:

Now we can solve for the fraction on the "left-hand side" of the rate equation:

$$\ln (\text{fraction remaining}) = -k \bullet t = -0.131 \text{ yr}^{-1} \bullet 1.0 \text{ yr}$$

$$\ln (\text{fraction remaining}) = -0.131 \text{ and } e^{-0.131} = \text{fraction remaining}$$

fraction remaining = 0.877 , so **88%** remains after 1.0 years.

42. Graph for P-31 of disintegrations per minute as a function of time for a period of one year

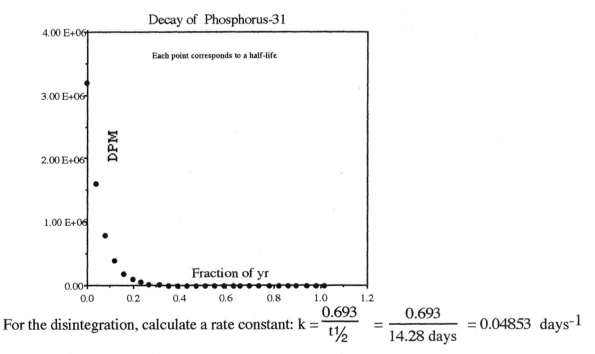

Decay of Phosphorus-31

For the disintegration, calculate a rate constant: $k = \frac{0.693}{t\frac{1}{2}} = \frac{0.693}{14.28 \text{ days}} = 0.04853 \text{ days}^{-1}$

On the graph above is plotted the results of the calculation:

$$\ln\frac{x}{3.2 \times 10^6} = -(0.04853 \text{ days}^{-1})t \text{ I have substituted multiples of 14.28 days, so each}$$

data point corresponds to a half-life.

44. To determine the half-life of polonium-210, we will plot ln (dpm) vs time.

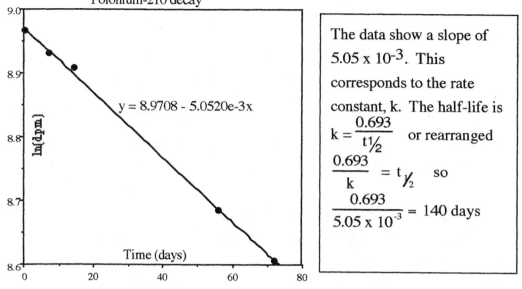

The data show a slope of 5.05×10^{-3}. This corresponds to the rate constant, k. The half-life is $k = \dfrac{0.693}{t_{1/2}}$ or rearranged $\dfrac{0.693}{k} = t_{1/2}$ so $\dfrac{0.693}{5.05 \times 10^{-3}} = 140$ days

Nuclear Reactions

46. For the decay of Americium-241:) $^{239}_{94}Pu + ^{4}_{2}He \rightarrow ^{240}_{95}Am + ^{1}_{1}H + 2\, ^{1}_{0}n$

48. Synthesis of Element 114: $^{242}_{94}Pu + ^{48}_{20}Ca \rightarrow ^{287}_{114}Uuq + 3\, ^{1}_{0}n$

50. Complete the following equations using deuterium bombardment:

    [The missing particle is emboldened.]

   (a) $^{114}_{48}Cd + ^{2}_{1}D \rightarrow \mathbf{^{115}_{48}Cd} + ^{1}_{1}H$ (c) $^{40}_{20}Ca + ^{2}_{1}D \rightarrow ^{38}_{19}K + \mathbf{^{4}_{2}He}$

   (b) $^{6}_{3}Li + ^{2}_{1}D \rightarrow \mathbf{^{7}_{4}Be} + ^{1}_{0}n$ (d) $\mathbf{^{63}_{29}Cu} + ^{2}_{1}D \rightarrow ^{65}_{30}Zn + \gamma$

52. The equation for the bombardment of Boron-10 with a neutron, and the subsequent release of an alpha particle:

$$^{10}_{5}B + ^{1}_{0}n \rightarrow ^{7}_{3}Li + ^{4}_{2}He$$

General Questions

54. The rate constant, $k = \dfrac{0.693}{t_{1/2}} = \dfrac{0.693}{4.8 \times 10^{10}} = 1.44375 \times 10^{-11}\ yr^{-1}$

At some time ,t, we have 1.8×10^{-6} mol Rb. We also have 1.6×10^{-6} mol Sr, which resulted from the decay of an equal amount of Rb. So the initial amount of Rb = $(1.6 + 1.8) \times 10^{-6}$. Substituting into the first-order equation we get:

$$\ln\frac{1.8}{3.4} = -(1.44 \times 10^{-11}\ yr^{-1})t$$

$$\frac{\ln(0.529)}{-1.44 \times 10^{-11} yr^{-1}} = \frac{-0.636}{-1.44 \times 10^{-11} yr^{-1}} = t = 4.4 \times 10^{10}\ yr$$

56. If the ratio of $\frac{Pb\text{-}206}{U\text{-}238}$ is 0.33, what we know is that 1/3 of the uranium has undergone decay. This would imply 2/3 of the uranium remains as U-238. We can then calculate a rate constant and use the 1st order rate equation to calculate the time required for this to occur:

$$k = \frac{0.693}{4.5 \times 10^9\ yr} = 1.54 \times 10^{-10}\ yr^{-1}$$

and $\ln(\frac{67}{100}) = -1.54 \times 10^{-10}\ yr^{-1} \cdot t$

$-0.400 = -1.54 \times 10^{-10}\ yr^{-1} \cdot t$ and $t = \dfrac{-0.400}{-1.54 \times 10^{-10}\ yr^{-1}} = 2.6 \times 10^9\ yr$

58. The decay of Uranium-238 to produce Plutonium-239:

(a) $^{238}_{92}U + ^{1}_{0}n \rightarrow ^{239}_{92}U + \gamma$

(b) $^{239}_{92}U \rightarrow ^{239}_{93}Np + ^{0}_{-1}e$

(c) $^{239}_{93}Np \rightarrow ^{239}_{94}Pu + ^{0}_{-1}e$

(d) $^{239}_{94}Pu + ^{1}_{0}n \rightarrow 2^{1}_{0}n$ + other nuclei + energy

60. The energy liberated by one pound of ^{235}U:

$$\frac{2.1 \times 10^{10}\ kJ}{1\ mol\ ^{235}U} \bullet \frac{1\ mol\ ^{235}U}{235\ g\ U} \bullet \frac{453.6\ g\ U}{1.000\ lb} = \frac{4.1 \times 10^{10}\ kJ}{1.000\ lb\ U}$$

comparing this amount of energy to the energy per ton of coal yields:

$$\frac{4.1 \times 10^{10}\ kJ}{1.000\ lb\ U} \bullet \frac{1\ ton\ coal}{2.6 \times 10^7\ kJ} = 1.6 \times 10^3\ tons\ coal/lb\ U$$

62. This "dilution" problem may be solved using the equation which is useful for solutions:

$$M_c \times V_c = M_d \times V_d$$

where c and d represent the concentrated and diluted states, respectively. We'll use the number of disintegrations/second as our "molarity."

$$2.0 \times 10^6 \text{ dps} \cdot 1.0 \text{ mL} = 1.5 \times 10^4 \text{ dps} \cdot V_d$$

$$130 \text{ mL} = V_d$$

The approximate volume of the circulatory system is 130 mL.

64. If we assume that the catch represents a homogeneous sample of the tagged fish, the problem is rather simple. The percentage of tagged fish in the sample is $\dfrac{27}{5250} = 0.00514$ or 0.51%.

If our 1000 fish represent 0.51% of the fish in the lake, the number is approximately:
$$\dfrac{1000}{0.0051} = 190,000 \text{ fish.}$$

66. For the radioactive decay series of U-238:

(a) Why can the masses be expressed as m = 4n +2 ?

These masses correlate in this fashion since the **principle** mode of decay in the U-238 series is **alpha particle** emission. The mass of an alpha particle is **4** (since it is basically 2 neutrons + 2 protons). Hence the loss of an alpha particle (say from U-238 to Th-234) results in a **mass loss of 4 units**. The series results in a total loss of 8 alpha particles and 6 beta particles. Since the beta particles (electrons) have an insignificant mass (compared to a neutron or a proton), the loss of a beta particle--or for that matter 6 of them--does not significantly affect the masses of the daughter products. These masses correspond to n values: n=(51 -> 59).

(b) Equations corresponding to the decay series for U-235 and Th-232 :

The U-235 series corresponds to the equation **m = 4n+3** with n values: (51 -> 58). The Th-232 series corresponds to the equation **m = 4n**
From an empirical standpoint, the masses of the most massive isotopes in the three series differ by 3 (U-238 $\rightarrow$ U–235) and (U–235 $\rightarrow$ Th-232. So given that the algorithm for the U-238 series is 4n+2, subtracting 3 gave 4n-1 (which also equals 4n+3 — with n reduced by 1; subtracting 3 more gives 4n.

The isotopic masses for these 3 series are summarized in the table below:

n	= 4n+2	= 4n+3	= 4n
51	206		
52	210	207	208
53	214	211	212
54	218	215	216
55	222	219	220
56	226	223	224
57	230	227	228
58	234	231	232
59	238	235	

(c) Identify the series to which each of the following isotopes belong:

Isotope	series
226-Ra:	U-238 (4n + 2)
215-At:	U-235 (4n +3)
228-Th:	Th-232(4n)
210- Bi:	U-238 (4n + 2)

(d) Why is the series "4n+1" missing in the earth's crust?

　　　To occur in the earth's crust, an element must be very stable—that is have a very long half-life or be non-radioactive! From hydrogen to lawrencium, with the exception of two isotopes of hydrogen (protium and tritium), every isotope of every element has a nucleus containing at least one neutron for every proton. With the mass of the neutron and proton being 1, the change in mass number would have to change by a factor of 2—so **4n+1** would not lead to a long-lived isotope and would not be found in the earth's crust.

68. For Protactinium:

　　(a) The series containing Protactinium-231 is the U-235 series. It corresponds to the (4n+3) series—See question 66.

　　(b) A series of reaction to produce Pa-231:

$$^{235}_{92}U \rightarrow {^{231}_{90}}Th + {^4_2}He \text{ followed by the decay } {^{231}_{90}}Th \rightarrow {^0_{-1}}e + {^{231}_{91}}Pa.$$

　　(c) Quantity of ore to provide 1.0 g of Pa-231 assuming 100% yield:

　　　　Stow your calculators! If the ore is 1 part per million, and you want 1.0 g, then you need 1,000,000 g of ore, check?

　　(d) Decay for Pa-231:　　　$^{231}_{91}Pa \rightarrow {^{227}_{89}}Ac + {^4_2}He$